Studies in Applied Philosophy, Epistemology and Rational Ethics

Volume 73

Studies in Applied Philosophy, Epistemology and Rational Ethics (SAPERE) publishes new developments and advances in all the fields of philosophy, epistemology, and ethics, bringing them together with a cluster of scientific disciplines and technological outcomes: ranging from computer science to life sciences, from economics, law, and education to engineering, logic, and mathematics, from medicine to physics, human sciences, and politics. The series aims at [illegible] all the challenging philosophical and ethical themes of contemporary society, making them appropriately applicable to contemporary theoretical and practical problems, impasses, controversies, and conflicts.

Book proposals can be sent to the Series Editor, Prof. Lorenzo Magnani at lmagnani@unipv.it. Please include: A short synopsis of the book or the introduction chapter A Table of Contents The CV of the lead author/editor(s) For more information, please contact the Editor-in-Chief at lmagnani@unipv.it

Our scientific and technological era has offered "new" topics to all areas of philosophy and ethics – for instance concerning scientific rationality, creativity, human and artificial intelligence, social and folk epistemology, ordinary reasoning, cognitive niches and cultural evolution, ecological crisis, ecologically situated rationality, consciousness, freedom and responsibility, human identity and uniqueness, cooperation, altruism, intersubjectivity and empathy, spirituality, violence. The impact of such topics has been mainly undermined by contemporary cultural settings, whereas they should increase the demand of interdisciplinary applied knowledge and fresh and original understanding. In turn, traditional philosophical and ethical themes have been profoundly affected and transformed as well: they should be further examined as embedded and applied within their scientific and technological environments so to update their received and often old-fashioned disciplinary treatment and appeal. Applying philosophy individuates therefore a new research commitment for the 21st century, focused on the main problems of recent methodological, logical, epistemological, and cognitive aspects of modeling activities employed both in intellectual and scientific discovery, and in technological innovation, including the computational tools intertwined with such practices, to understand them in a wide and integrated perspective. **Studies in Applied Philosophy, Epistemology and Rational Ethics** means to demonstrate the contemporary practical relevance of this novel philosophical approach and thus to provide a home for monographs, lecture notes, selected contributions from specialized conferences and workshops as well as selected Ph.D. theses. The series welcomes contributions from philosophers as well as from scientists, engineers, and intellectuals interested in showing how applying philosophy can increase knowledge about our current world. Indexed by SCOPUS, zbMATH, SCImago, DBLP. All books published in the series are submitted for consideration in Web of Science.

Athanassios Raftopoulos

The Origins of Representations: Cognitive and Brain Development

Springer

Athanassios Raftopoulos
Department of Psychology
University of Cyprus
Nicosia, Cyprus

ISSN 2192-6255 ISSN 2192-6263 (electronic)
Studies in Applied Philosophy, Epistemology and Rational Ethics
ISBN 978-3-031-91212-2 ISBN 978-3-031-91213-9 (eBook)
https://doi.org/10.1007/978-3-031-91213-9

This Springer imprint is published by the registered company Springer Nature Switzerland AG
The registered company address is: Gewerbestrasse 11, 6330 Cham, Switzerland

The book is dedicated to Ali, Ettore, Paris, and Natassa.

Preface

The main ideas developed in this book were conceived while I was an invited professor at the University of Bari Aldo Moro during the Fall Semester of the academic year 2022–2023. A year ago, I had received an invitation from a journal to write a paper focused on the origins of human thought. Since, in my mind, the main characteristic of human thought is the use of representations of various kinds, I set out to write a paper concerning the "when and why" representations appeared in the course of the evolution of our species. I read thousands of pages in books and papers, and I wrote a large paper that I submitted to the journal at the end of the fall of 2022. After a few months, I received a rather good review, which, nevertheless, pointed out that I should consider many more books and papers because I had left out some important contributions and discussions on this theme. Graciously enough, the reviewer dedicated a considerable amount of his/her time to provide me with an extended bibliography. In addition, the reviewer suggested that it would be better to write a book instead of a paper, because the issues and arguments involved were too many to be adequately addressed in the confines of a paper.

Back to work, then, and after I had studied the suggested literature to the best of my abilities, I wrote this book developing and expanding the ideas and arguments that were first presented in the short-lived paper. In view of these events, I would like, first, to express my sincere thanks the editor of the journal who gave me the opportunity to start working on this project. I also owe a lot of gratitude to the anonymous reviewer who encouraged me to develop my ideas and suggested a list of books and papers without which this book would have never been completed.

My sincere thanks also goes to my dear friend Luigi Pastore, Associate Professor in the Dipartimento Di Scienze Della Formazione, Psicologia, Comunicazione at the University of Bari, for inviting me to their department to teach. In addition, I would like to thank Sara Dellantonio for her warm hospitality and collaboration on various projects.

At home, I would like to thank my colleagues in the Department of Psychology and the University of Cyprus for granting me the Sabbatical that made possible my tenure in the University of Bari. Finally, and most importantly, I should point out that without my wife's support who took care of everything back at a home for three whole months while I was in Bari, the book would have never been possible.

Nicosia, Cyprus Athanassios Raftopoulos

Competing Interests The author has no competing interests to declare that are relevant to the content of this manuscript.

Contents

Chapter 1
Introduction

Representations are physical entities that re-present something else. They are the cornerstone not only of any activity that requires information that goes beyond what is immediately presented to agents through some environmental input, but also of the perception of, and interaction with, the environment the way humans and other mammals do. Representations are necessary for perception, for perceptual memory, for planning future actions, and for contemplation, reflection, and conjecture. To represent a thing (in a general sense of "thing") means to use in the mind or in an external medium, such as a piece of paper or a piece of stone, a sign that stands for the thing that is represented (the representatum). The power that the sign confers to its user is that it can be used as a proxy of the represented thing in the mind of the user even when it is absent.

Representations come in many varieties, ranging from signs that perceptually resemble their representata in an obvious way for those who have similar perceptual systems (a picture of a cat, for example), to abstract symbols that are related to their representata through a set of elaborate conventions (such as the "V" signal for victory made famous by Sir W. Churchill). Humans are not the only creatures to use representations, as many other species (from whales to apes) employ representations. More to the point, infants use representations but, of course, these are overly simplistic compared to the representations used by adults. One among the reasons for this discrepancy is that the adult brain is much more developed than the infant brain and this, presumably, allows the use of a much more complicated system of representations on the part of the adults. In addition, the human brain is structurally speaking quite different from the brain of the apes, our closest evolutionary relatives, which is why our representational capacities are much more extended and robust.

It is known that in the course of human evolution the human brain underwent a series of important developments, ranging from a gradual increase in size to the appearance of new anatomical areas in the brain, and to the recruiting of older areas to perform new functions. There is also some consensus that to a certain extent phylogeny (that is, the history of the evolution of a species or a group) mirrors

A. Raftopoulos, *The Origins of Representations: Cognitive and Brain Development*, Studies in Applied Philosophy, Epistemology and Rational Ethics 73,
https://doi.org/10.1007/978-3-031-91213-9_1

ontogeny (that is, the developmental history of an organism within its own lifetime). This means that the stages through which a species evolved correspond to the stages through which an organism develops throughout its life span. At a minimum, ontogeny can be used as a source of evidence as to how phylogeny might have developed, although there are cases where ontogeny might mislead phylogenetic considerations, the act of pointing being an example.

A word of caution is needed at this point concerning the relation between phylogeny and ontogeny. As Planer and Sterelny (2021, 54) rightly point out, developmental data about young humans' cognitive capacities cannot be used as evidence concerning the cognitive capacities of early hominins. The main reason is, I think, that modern human infants live and develop in a social environment that is very different from that in which H.E infants were raised. This is putting aside the fact that, as a direct result of social differences and the demands they place for social learning, the infancy in *Homo sapiens* is much more prolonged compared to that of *Homo Erectus* (H.E), which, in turn is more prolonged compared to the infancy periods of chimpanzees. A striking example of such differences is the fact that modern infants are born and develop in a highly rich linguistic and symbolic environment. In such an environment it is much easier to acquire and learn how to use symbols than it was in middle Pleistocene. This makes ontogenetic development different among these two groups of hominins. However, when human infants acquire a certain cognitive capacity very early, say before the age of 3–4 years, before social learning in any of its forms has had time to influence development critically, this entails that this capacity is not that difficult to have, which leads us to suppose that Pleistocene hominins, especially during the middle and late Pleistocene periods, could have them too. In general, when a capacity appears early in a hominin species this generally points to a genetic basis that has gradually evolved in this species' lineage and, thus, if some hominins engaged in activities that require this capacity, it is reasonable to think that they shared this genetic basis. Notice, however, that genetic changes need not be the cause but the consequence of the activities they support. When some members of a group engage in some activities brining in their own style and innovation, if this style is successful and persists it passes on to the next generation through social learning. Individuals that happen to have the genes that facilitate the acquisition of such a behavior are eventually selected by evolution and, thus, the gene becomes dominant.

Moreover, the fact that certain cognitive capacities develop in human infants later than some others allows us to think that is likely that the same pattern occurs in phylogeny. The reason is that if the latest onset of the capacity is due to maturation factors or factors related to the interaction with the environment, it is likely that the same kinds of factors were at place during phylogeny. As we shall see, for example, the *Working memory* (WM) capacity of infants is similar in the beginning to that of chimpanzees but during development it far exceeds that of the chimpanzees. Evidence suggests that the WM capacity of H.E lies between those of humans and chimpanzees, which entails that human infants start lower than H.E and as young human mature and develop they pass from and exceed this capacity. Thus, the ontogenetic development of WM capacity is reflected by the phylogenetic evolution of this same capacity. This is a clear case in which there is evidence that ontogeny in some respects parallels ontogeny, which means that one could use, with due caution, ontogenetic data to make informed extrapolations concerning the development of cognitive capacities during hominin phylogeny.

One can, therefore, plausibly hypothesize that our ancestors from the *Homo erectus* and *Homo habilis* to Modern *Homo sapiens*, and even within the latter family, by having structurally different brains, were able to use representations of different degrees of complexity; the representational systems became increasingly more complex with the evolution of the brain and, in turn, they applied evolutionary pressures for further brain developments. This development did not primarily concern the size of the brain that reached its maximum level 200 Ka, but its structure and organization. In fact, partly because of the development of complex, mainly external, representational systems that accompanied the advent of complex social systems (DeSilva et al., 2021), both the absolute and relative (to the body) brain size undergone a reduction either at the end of the Pleistocene 35 Ka years ago (Ruff et al., 1997), or during the Holocene 10 Ka years ago (Hawks, 2011).

All these presuppose that the earlier *Homo hominins* were using representations, and they were not merely reacting online to environmental inputs the way most animals usually do. I also think it is very plausible that since, starting from Africa, *Homo erectus*[1] were able to colonize the entire world by 1.7 Ma (Finlayson, 2014), as evidenced by fossils found in Georgia, Israel, and, later, 500 Ka in Flores in Java, or even earlier by 2.1 Ma in China (Zhu et al., 2018). To do so, they had to travel great distances and even negotiate vast expanses of oceans. These feats require planning and communication that, in turn, are possible only through the use of representations and, to be precise, through the use of a specific kind of signs, namely symbols, because only symbols can enable some form of language that is necessary for any sort of planning required for this colonization. It is well known that other animal species migrate in long distances as well, crossing on certain occasions the sea. However, these animals did not construct boats or rafts, and did not take water and food with them in order to survive the trip; therefore, no planning was required for their migration. One may go a step further. This sort of planning ahead and the whole set of activities on the part of the group that had to be undertaken before the trip would be greatly facilitated by the development of another sort of tools, namely a communicative tool, in the form of a protolanguage. This is a kind of language that does not have the form of modern languages, but it is much simpler and perhaps lacks formal syntax. That is, it does not allow for the composition of lexical items as hierarchically organized structures; this means that, at best, protolanguages allow the use of small clauses as opposed to sentences (Progovac, 2016). There is, thus an environmental pressure for the development of a protolanguage, since interactions among the members of a community and, especially planning, are greatly enhanced by the existence of a communicative means in the form of a (proto) language.

[1] To simplify things, I do not distinguish between African H.E, Eurasian H.E and Peking Man since I do not aim to discuss whether H.E or some subspecies of H.E is the ancestor of *Homo sapiens*. I also do not address the problem concerning the site of H.E evolution, that is, whether H.E evolved in Asia or in Africa.

Long-distance colonization is not the only activity by Early Pleistocene Hominins, specifically H.E, that points to the beneficial role of a proto-language. As it is likely that H.E communities had, starting 1.5 Ma, established instruction practices for the transmission of knowledge necessary for their knapping techniques (Hiscock, 2014; Planer, 2017), these practices would have been greatly benefited from a protolanguage. In fact, it is difficult to imagine the transmission of such intricate knowledge that includes not only knapping techniques but also a know-how pertaining to quarrying the appropriate stone-material (which occasionally required traveling long-distances) in the absence of some form of language. In this vein, Kolodny and Edelman (2018) argue that there must have been a behavioural context that exerted the selective pressure to drive the sequential adaptations that led to the emergence of language. Since language and the gestures that support it (the earlier the language the more extensive the supporting repertory of gestures) requires both motor skills and verbal communication, these two must be appropriately coupled to produce language. The context in which this coupling occurred had to be one in which each of the systems undergoing coupling was independently necessary or highly beneficial, as well as frequent and recurring over evolutionary time. Such a context was that of teaching tools production techniques, and, as we shall see, H.E did teach tool production techniques, mostly the Acheulian technique that developed 1.7 Ma and undergone considerable elaboration by 1.2 Ma, to the younger members of their groups.

What is more, if, as the evidence seems to suggest, H.E, by 1 Ma, were practicing big game ambush hunting (Kübler et al., 2015), which means that H.E were at the top of the food pyramid (apex predators) (Ben-Dor et al., 2021), planning for the hunting would also be benefited by a protolanguage. In general, all group activities that rely on cooperation require some form of communication, such as gestures, mimicking activities, or language. Language being the more effective form of communication, if the cognitive capacities supporting language (such as WM, attention and its executive functions, a developed theory of mind (TOM)) are present, natural selection will see to it that some form of language emerges. This form of language, in turn, acts as a selective pressure for more advanced cognitive capacities supporting more elaborate forms of language. One might retort at this juncture that other predators, such as lions, practice ambush hunting, but no one would claim that in order to succeed lions need a proto-language, or even a joint intentionality (Tomasello, 2019) that requires some developed form of a theory of mind. Apes may have a theory of mind, but no one has, thus far at least, claim that prides of lions have it too.

So, why should we, based on, say, the practice of ambush hunting attribute these cognitive capacities to H.E? The answer is multifaceted. First, H.E and not lions had a developed PFC that could subserve these capacities. It follows that environmental pressures and the benefit of a proto-language affording communication for planning to hunt work differently in hominins and lions. In other words, hunting in lions would have never led to adaptations leading to proto-language; in H.E it could, and it probably did. In addition, as we shall see, ambush hunting is not the only activity that would benefit for an extended range of cognitive capacities. Emigration, tool production, ambush hunting, all are activities practiced by H.E and success in these activities puts environmental pressures on H.E to develop the cognitive capabilities enabling them, and the neural structures that could support these cognitive capabilities, that would increase the rate of success of these activities.

In addition to this, and much more recently, the cave paintings that appeared about 40 Ka from Australia and Africa to Southern Europe, and perhaps even earlier in Indonesia (50 Ka), that are associated with *Homo sapiens* and which contain both faithful images of animals and even, probably, abstract symbols are another testament to the fact that our ancestors lived in a rich representational world. It should be noted that there is also evidence that *Homo neanderthal* were also painting the walls of caves with the icons of animals or animal parts in Spain—67 Ka (Pike et al., 2012), and, most importantly, they were painting on the walls in caves in France 57 Ka years ago (Marquet et al., 2023) spatially structured, non-figurative marks. More recent evidence pushes possible symbolic activity on the part of Neanderthals back to 75 Ka (Marquet et al., 2023). This is long before *Homo sapiens* arrived in Europe, which undermines the thesis that Neanderthal had considerably diminished cognitive abilities compared to *Homo sapiens*, even though there is no doubt that *Homo sapiens* were much more sophisticated cognitively speaking. I shall return to this point when I discuss the cognitive abilities required for iconic and symbolic representations. As we shall see, such artistic endeavors that may indicate symbolic thinking and some form of iconic representations seem to be extended back in time during the H.E period.

In this book, I attempt to explain the evolution or development of our representational arsenal from its humble beginnings that differ little from what animals do to its modern form that contains highly abstract symbols. To be able to understand the development or evolution of the representational arsenal, one should examine, first, the development of the cognitive capacities that enabled the gradual emergence of representations of increasing complexity and expressive capability. One should also examine both the size-increase and the restructuring of the brain through the evolutions of our species as well as the effects this size-increase and restructuring had on the cognitive capabilities of Hominins. These two factors determine, in synergy with the environment and the social niche, the complexity of the representational systems a species develops and uses. In this book, I address mainly he first facet of the problem, namely, the cognitive development that underlies the emergence and development of our representational systems, even though occasionally I discuss the brain restructuring and the resulting neural "novelties" that enabled the appearance of cognitive innovations.

The most important property of representations is that they refer to other things (be them objects or events). Representations, however, are not the only physical things that are referentially associated with other things. The tracks of an animal on the ground indicate the presence of this animal and the thermostat indicates the temperature in the environment, but, as we shall see, one should refrain from calling them representations even though, through their indicative capacity, they refer to other things (that is, the animal or the temperature respectively). All physical entities that bear referential associations to things in the world are called signs; representations are a particular class of signs. Let me state from the outset that my aim is not to discuss the emergence of conceptual abilities and of concepts, even though representations lead to concepts. My aim is to explain how iconic and symbolic representations were made possible by which cognitive factors in the development of our species.

My aim, furthermore, is to explore the cognitive changes that made possible iconic and symbolic representational systems. I do not intend to examine the way these representational systems, once formed, were further developed, and enriched. In doing so, I deal both with the cognitive changes due to brain genetic changes during the *speciation phase* of the Hominin development, and with cognitive changes during the *tectonic phase* of the Hominin development that were accompanied or caused by epigenetic or cultural factors (Renfrew, 2005). The former concerns the development of language, symbolic thought, self-awareness, autobiographical memory etc., that were accompanied by changes in the genome of our species that led to the increase in brain size and to significant restructuring of the brain. These changes were accompanied, and were influenced, in turn, by cultural factors, so that the cognitive development during this phase could be explained by a synergy of genotype changes and culturally inherited behaviors whose appearances does not require a change in the genome. This period lasted until about 100,000 years ago. Notice that it is not necessary that novel behaviors require some genetic change in order to appear. As West-Eberhard (2003) argues, there is the possibility of adaptive phenotypical plasticity. When a novel behavior discovered by some individual or individuals is successful and due to its success is embraced by the members of a group and gets stabilized, it is transmitted through learning. Genetic variants that enable those who have them to acquire the novel behavior easier and in a more reliable way, will be selected.

Since the modern DNA is the same as that 60,000 years ago and, thus, no genetic change has occurred since then, any cognitive and behavioral innovations, including the famed Human cognitive explosion 40 Ka, cannot be attributed to genetic factors. Any development has to be explained by cultural and environmental factors and their epigenetic effects, that is, how these factors could affect the way genes express themselves. (Unlike genetic changes, epigenetic changes do not change DNA sequences, but they can change the way body and brain read a DNA sequence. Moreover, epigenetic changes can be reversed.) This is the tectonic phase in human development.

In the second chapter, I start by analyzing the notion of sign as a referential association. I concentrate on Peirce's (1997) tripartite system of signs, namely, indices, icons, and symbols, because Peirce was, perhaps, the first to distinguish systematically between the various kinds of signs and discus their interrelations. Peirce thought that indices are referentially more complex than icons in the sense that icons are necessary for fixing the reference of indices. I explain why I disagree with Peirce and argue that icons are a later evolutionary step in that they are representations, while indices are not, and, thus, came first. Then, I distinguish between internal, mental, representations and physical external representations, and I emphasize the distinction between the vehicle of a representation and, when it applies, the representational content of the representation, and the object of the representation. Finally, I explain why indices in Peirce's vocabulary are not representations.

In the third chapter, I shift to modern discussions concerning one of the two main kinds of representations discussed nowadays, namely, symbolic or digital representations that contrast with icons, pictures, or images that are typical examples of the other kind of representations, namely, analog or iconic representations. Words are typical examples of symbolic representations. My aim in this chapter is to explain symbolic representations and discuss their main properties. Bringing to the fore the main properties of symbols will help us understand the profound changes in hominin brain structure and mentality required for symbolic representations to emerge and, mainly through, language shape our brain and behavior. The discussion concerning the main properties of symbols and symbolic systems, namely, the discrete and non-dense character of symbolic representations, the fact that symbolic are assigned referents through conventions, and the fact that symbols compose discursively by means of a set of established syntactic rules, is important for the aims of this book because most of the properties of symbolic representations emanate from these traits of symbols. These properties of symbols, in turn, are necessary for understanding the cognitive capacities required to enable cognizers to have such representations. Having determined these capacities, one may turn to examining when in the evolution of our species and for which reasons these capacities, which constituted cognitive novelties first appeared. If symbols emerged from a background of indices and icons, to explain the cognitive novelty of symbolic cognition one should compare the properties of symbols with those of icons and pinpoint those differences between these two representational forms that require that novel cognitive capacities develop in order for symbolic thought to become possible, that is, the capacities that enable an agent to move from iconic to symbolic representations. A characteristic example is that of "recursiveness", that is, the ability to repeat a construction indefinitely as when one embeds a clause inside a sentence. Advanced symbolic systems are recursive, whereas iconic representations irrespective of their complexity are not recursive. Recursiveness requires that an agent have cognitive abilities that agents who employ only iconic representations need not have; a user of modern recursive languages is endowed with cognitive capacities that a user of a language based on gestures need not have.

To determine the properties of icons, in the fourth chapter, I examine iconic or analog representations highlighting their main properties. I concentrate on their significant difference from symbolic representations concerning both the way icons refer, and the way iconic representations compose to form more complex iconic representation. In this chapter, I also attempt to elucidate the notion of iconic representations because the discussions in the literature tend to be, to say the least, chaotic and perplexing. Researchers usually fail to distinguish between external and mental representations, between representational vehicles and contents, and between individual representations and representational systems, Lyons (2022) being a notable exception.

Having discussed the properties of icons, one can turn to addressing the question of when hominins acquired the cognitive capacities expressing these properties and why. This will allow us to pinpoint the cognitive capacities that agents must possess in order to be able to form iconic representations of their environment. One can also discuss the reasons that *representational* icons succeeded the n*on-representational* indices, that is, what made possible the emergence of representations in the first place.

In Chaps. 3 and 4, I examine the properties of icons and symbols with a view to use their differences to discuss the cognitive innovations required for symbols to emerge from a background of icons. Before that, it is possible that the properties of icons can be used to answer the question of what cognitive factors made possible the emergence of representations from non-representational indices. To achieve both aims, I suggest in Chap. 5 various cognitive changes or innovations realized by specific cognitive mechanisms. It is understood that these mechanisms survived because they played an adaptive role in the evolution of our species. To play such a role, these mechanisms must have enabled functional capacities that were evolutionary speaking advantageous for the survival of our species. This means that not only were these new functional capacities a successful answer to some environmental and cultural pressures, but also that the mechanisms that enabled them were best suited to fulfill these functions compared to other cognitive mechanisms that could have performed these same functions.

Hence, in the fifth chapter, I discuss the cognitive prerequisites that allowed our ancestors to move from indices (which is the first form of non-representational signs used) to representations the first form of which was that of iconic representations. I also briefly address the problem of what necessitated the emergence of representational capacities, that is, why having representations was in evolutionary terms beneficial to a species. Of course, there is no direct empirical evidence about the cognitive capacities of hominins other than modern humans. However, the cognitive capacities of a species determine the actions of the members of the species aimed both at their natural environment but also at their social environment. These actions leave traces, such as tools, iconic or symbolic representations, remains of food that indicate the food sources and, thus, ways in which the food may have been obtained, and other physical traces that paleontology and anthropology can discover and examine. Some of these traces also allow us to draw conclusions concerning the social structures of the group because social structure makes certain activities possible, and the presence of these activities allows us to uncover the enabling social structure.

In other words, the physical traces left behind by the earlier hominis as they interacted with their environment, their life ways, form an empirical basis from which one could draw empirically informed hypotheses concerning the cognitive abilities of these earlier hominins. The reason is that if a set of properties X is necessary for a phenomenon Y to occur and Y occurs, then X also occurs. Therefore, in this chapter, I discuss extensively the empirical evidence concerning the traces left behind by the activity of early hominis, with emphasis on H.E for reasons that will become clear in the discussion. Based on this evidence, I discuss a list of cognitive capacities that the evidence suggests that H.E had. Since, as I explain next, I adopt a gradualist approach to evolutionary phenomena, I think that the cognitive capacities I am examining are more advanced versions of the cognitive capacities of the great apes from which hominis descend, versions that emerged gradually from the baseline that consists in the cognitive capacities of the great apes. Similarly, the cognitive capacities of the early hominins are less elaborated versions of the same cognitive capacities found in modern humans. This gradualist assumption is important because we can establish the cognitive baseline form which hominins started by examining the cognitive capacities of modern great apes and, then attempt to reconstruct the evolutionary steps and the factors that necessitated them, which gradually led to the cognitive capacities of early hominis, most notably, H.E.

In the sixth chapter, I discuss the factors that made necessary, evolutionarily speaking, the emergence of symbols. I start by noting the representational limitations of iconic representational systems, because these limitations paved the way for the development of symbolic thought when the need for this kind of thinking arose. I argue that the main reason why symbolic thought became advantageous to hominins is the need to establish and maintain social relationships that would benefit a group. This was a sufficient environmental pressure to lead to the appearance of a symbolic capacity. I also propose two additional reasons for symbolic thought to develop. These are the need for social learning, especially as it relates to teaching tool-construction techniques, and the need for the group to come up with an effective management of intragroup social tensions and stress that resulted from the increase of the size of the group, and the demand for more cooperative activities.

In the seventh chapter, I address the problem of which cognitive capacities were necessary for symbols to emerge from a background of iconic representations. Firstly, I discuss the relations and differences between symbols and indices. Secondly, I discuss the relations between icons and symbols, and then, I turn to discuss the cognitive factors required for the emergence and usage of symbols. Specifically, I discuss the cognitive factors that enabled the move from iconic representations to symbolic representations. I am searching for a plausible sort of "cognitive revolution", that is a set of processes in the mind, that allowed the move from iconic representations to symbols, and, in this sense, went beyond the cognitive capacities required for the employment of iconic representations. For example, iconic representations are not recursive, whereas advanced symbolic systems are recursive. It follows that agents that use recursive representational systems must have a cognitive capacity that the predecessors who could not use recursive representational systems did not have. Finally, I address the problem of whether H.E possessed some sort of language or protolanguage.

Before I proceed let me say at the outset that this book rests on a set of assumptions that, as such, are not justified in this book. The first is that the book is set in a representationalist framework. That is, throughout the book I assume that organisms of a certain cognitive complexity interact with their environment both in an offline manner, and in the case of perception in an online manner in virtue of representing aspects of the environment. This means that I reject without explaining the reasons for doing so, first, direct theories of reference that state that the symbols we use refer directly to environmental entities without any representational intermediaries. I reject, second, these short-lived, at least I hope, theories situated in the context of dynamic theories of cognition that stipulate that our interactions with the environment do not require the use of representations but are automatically computed on the basis of information that is received from the environment, the way a baseball catcher "knows" how to intercept the ball. Although I fully endorse the dynamic view of cognition, I reject as outlandish the stronger claim concerning the lack of representations.

The second assumption concerns the distinction between gradualist models of cognitive abilities and models of sudden appearance, usually due to some genetic factor. According to gradualism, cognitive abilities evolved slowly, incrementally, and adaptively, following an evolutionary trajectory from the base—line cognitive capacities of the common ancestor of the hominin-chimpanzee line to cognitive capacities of the modern *Homo sapiens*. Thus, an explanation of the cognitive capacities possessed by *Homo sapiens* must be a lineage explanation (Calcott, 2008). This means that, working memory capabilities, for example, gradually increased from the baseline that hominis share with chimpanzees to the working memory capacities of modern humans and did not appear to our species suddenly as a result of the sudden appearance of a gene. When I examine, thus, the evolution of the cognitive capacities required for understanding and using representations in general, and iconic and symbolic representations in particular, I shall attempt to construct an evolutionary line that gradually led from the baseline capacities of apes to those of modern humans, with a special emphasis on the appearance of these capacities in H.E, and I will not attempt to explain these capacities by means of a sudden cognitive revolution. Furthermore, the gradual evolution of cognitive capacities is explained in the context of the interplay among social, economic, and technological evolution within the hominin line. This last remark brings us to the third set of assumptions adopted in the book.

The third set of assumptions underlying the discussion that follows is that of gene-culture coevolution, and that of cognition-culture coevolution. These are different because cognitive change does not require genetic changes, since there can be major changes in the networks in the brain, such as network restructuring or the creation of new networks, that realize and enable these cognitive changes, without any genetic change.

Gene-Culture Co-evolution Theory is the application of *sociobiology*, that is, the general theory of the social organization of species, to humans who transmit culture in a manner leading to quantitative growth across generations. This is a case of *niche construction*, which applies to species in general (such as humans, apes, etc.,) that transform their natural environment to facilitate social interaction and collective behavior (Gintis, 2011). Gene-culture coevolutionary theory is also a branch of theoretical population genetics that models the transmission of genes and cultural traits from one generation to the next exploring their interactions. 'These models have been employed to examine the adaptive advantages of learning and culture, to investigate the forces of cultural change, to partition the variance in complex human behavioral and personality traits, and to address specific cases in human evolution in which there is an interaction between genes and culture.' (Feldman & Laland, 1996).

Cognition-Culture Co-evolution Theory integrates cognitive systems and cultural evolution to examine how developments in one domain entrain developments in the other. Our species has distinctive cognitive differences from all other species. These unusual human characteristics depend on some combination of our cognitive and cultural characteristics, and, so, the evolutionary separation of our lineage from that of the great apes about 8–10 Ma, depends on the co-evolution of those differences in mind and culture, where culture is essentially the ability to learn from each other (Boyd, 2017). Over the past two million years, culture has evolved to enable hominin societies to accumulate local (that is, situated within a specific environmental niche) adaptations that no individual could ever have invented on their own. The changes in hominin cognition and culture as our lineage evolved after the split from great apes were not independent of one another. Consequently, pertaining to humans, there is some fundamental relationship between the evolution of those distinctive features of human cognition and those of human culture (Sterelny, 2020). Language is a characteristic example of gene/culture and cognitive/culture coevolution. It is the result of adaptations based on genetic changes, which may have initially served other non-linguistic purposes as the function of Broca's area indicates. At the same time, language is also shaped by cultural factors and processes (Planer & Sterelny, 2021).

In the book, I assume this conceptual background and try to understand the cognitive capacities that led to the acquisition of representational abilities in terms of the intricate relations between culture, technological achievements, foraging techniques, and other spheres of hominin activity that entrain each other so as to produce through adaptation the cognitive capacities underwriting representation usage.

The last assumption I make in this book is the *Material Engagement Theory*, according to which the way organisms use tools and other artifacts shapes their cognitive mechanisms and rewires their brain either at a genetic or epigenetic scale (Malafouris & Renfrew, 2010; Renfrew, 2005, 2008). This time, however, I discuss MET, and offer two arguments to support it. The first is by showing how the interaction of our visual perceptual systems with the environment in which we live, acting as an information source, has shaped our visual perceptual circuits, and, hence, the cognitive abilities they support. A second case for the imposition of external informational structure on cognition is Planer and Sterelny (2021) very plausible idea that the hierarchical structure inbuilt in the actions involved in the tool-construction technique of H.E somehow shaped the mind in such a way as to allow the emergence of the capacity for hierarchically structured recursive representations, and, eventually, language.

Making these assumptions entails that when I discuss the cognitive innovations that enabled the emergence of representations and the use of various kinds of representations, I take it for granted that an explanation of these changes relies on the coevolution of genes and culture, on the coevolution of cognition and culture, and on the influences of artifacts use on cognition, either at a genetic or epigenetic scale.

References

Ben-Dor, M., Sirtoli, R., & Barkai, R. (2021). The evolution of the human trophic level during the Pleistocene. *American Journal of Physical Anthropology, 175*(72), 27–56.

Boyd, R. (2017). *A different kind of animal: How culture made humans exceptionally adaptable and cooperative*. Princeton University Press.

Calcott, B. (2008). Lineage explanations: Explaining how biological mechanisms change. *British Journal for the Philosophy of Science, 60*, 51–78.

DeSilva, J. M., Traniello, J. F. A., Claxton, A. G., & Fannin, L. D. (2021). When and why did human brains decrease in size? A new change-point analysis and insights from brain evolution in ants. *Frontiers Ecology and Evolution, 9*, 742639. https://doi.org/10.3389/fevo.2021.742639

Feldman, M. W., & Laland, K. N. (1996). Gene-culture coevolutionary theory. *Trends Ecological Evolution, 11*(11), 453–457.

Finlayson, C. (2014). *The improbable primate: How water shaped human evolution*. Oxford University Press.

Gintis H. (2011). Gene-culture coevolution and the nature of human sociality. *Philosophical Transactions Royal Society London B Biological Science, 366*(1566), 878–888. https://doi.org/10.1098/rstb.2010.0310. PMID 21320901; PMCID PMC3048999.

Hawks, J. (2011). Selection for smaller brains in Holocene human evolution

Hiscock P (2014) Learning in lithic landscapes: a reconsideration of the hominid "toolmaking" niche. *Biological Theory*, 9, 27–41.

Kolodny, O., & Edelman, S. (2018). The evolution of the capacity for language: The ecological context and adaptive value of a process of cognitive hijacking. *Philosophical Transactions of the Royal Society London B Biological Science, 373*, 20170052. https://doi.org/10.1098/rstb.2017.0052

Kübler, S., Owenga, P., Reynolds, S. C., Rucina, S. M., & King, G. C. P. (2015). Animal movements in the Kenya Rift and evidence for the earliest ambush hunting by hominins. *Scientific Reports, 5*, Article Number 14011.

Lyons, J. C. (2022). Three grades of iconicity in perception. *Asian Journal of Philosophy, 1*(50). https://doi.org/10.1007/s44204-022-00039-x

Malafouris, L., & Renfrew, C. (2010). The cognitive life of things: Archaeology, material engagement and the extended mind. In L. Malafouris & C. Renfrew (Eds.), *The cognitive life of things: Recasting the boundaries of the mind* (pp. 1–12). McDonald Institute for Archaeological Research, University of Cambridge.

Malafouris, L. (2020). Thinking as "thinging": Psychology with things. *Current Directions in Psychological Science, 29*(1), 3–8.

Marquet, J.-C., Freiesleben, T. H., Thomsen K. J., Calligaro, M., & Macaire, J.-J. (2023). The earliest unambiguous Neanderthal engravings on cave walls: La Roche-Cotard, Loire Valley, France. *PLoS One, 18*(6), e0286568.

Peirce, C. S. (1977). Semiotics and significs. In C. Hardwick (Ed.), Bloomington, Ind: Indiana University Press.

Pike, A. W, Hoffmann, D. L, García-Diez, M., Pettitt, P. B., Alcolea, J., De Balbín, R., González-Sainz, C., de las Heras, C., Lasheras, J. A., Montes, R., & Zilhão, J. (2012). U-series dating of Paleolithic art in 11 caves in Spain. *Science, 336*(6087),1409–1413.

Planer, R. J. (2017). Talking about tools: Did early Pleistocene Hominins have a protolanguage? *Biological Theory, 12*, 211–221.

Planer, R. J., & Sterelny, K. (2021). *From signal to symbol: The evolution of language*. MIT Press.

Progovac, L. (2016). A gradualist scenario for language evolution: Precise linguistic reconstruction of early human (and neandertal) grammars. *Frontiers Psychology, 7*, 1714.

Renfrew, C. (2008). *Prehistory: The making of the human mind*. Phoenix.

Renfrew, C. (2005). Becoming human: The archaeological challenge. *Proceedings of the British Academy, 139*, 217–238.

Ruff, C. B., Trinkaus, E., & Holliday, T. W. (1997). Body mass and encephalization in Pleistocene *Homo. Nature, 387*, 173–176. https://doi.org/10.1038/387173a0

Sterelny, K. (2020). Afterword: Tough questions; hard problems; incremental progress. *Topics in Cognitive Science, 12*(3), 766–783.

Tomasello, M. (2019). *Becoming human*. Harvard University Press.

West-Eberhard, P. (2003). *Developmental plasticity and evolution*. Oxford: Oxford University Press.

Zhu, Z., Dennell, R., Huang, W., Wu, Y., Qiu, S., Yang, S., Rao, Z., Hou, Y., Xie, J., Han, J., & Ouyang, T. (2018). Hominin occupation of the Chinese Loess Plateau since about 2.1 million years ago. *Nature, 559*(7715), 608–612.

Chapter 2
Signs and Representations

I analyze the notion of sign as a referential association. I concentrate on Peirce's tripartite system of signs, namely, indices, icons, and symbols, because Peirce was, perhaps, the first to distinguish systematically between the various kinds of signs and discus their interrelations. Peirce thought that indices are referentially more complex than icons in the sense that icons are necessary for fixing the reference of indices. I explain why I disagree with Peirce and argue that icons are a later evolutionary step in that they are representations, while indices are not, and, thus, came first. Then, I distinguish between internal, mental, representations and physical external representations, and I emphasize the distinction between the vehicle of a representation and, when it applies, the representational content of the representation, and the object of the representation. Finally, I explain why indices, in Peirce's vocabulary, are not representations.

Peirce (1977) was among the first philosophers to discuss systematically the various ways through which one can refer to entities in the world. There are three different categories of such ways that establish referential associations with the world, which Peirce calls "signs", namely indices, icons, and symbols. First, a few words about "signs" and its cogent term "signals".

Signs are physical structures that carry information. This means that when interacting with other physical structures, an explanation of the effects of the said structure on the structures with which it interacts takes recourse to the information carried by the structure, an explanation that goes beyond the purely physical explanation of the interaction between these structures. For example, when uttering a sentence in speech, a series of phonemes uttered by a sender are realised by sound waves that physically characterise the sentence *qua* physical structure. These sound waves interact with the receiver-cells in our ears. To explain the role of the sentence, however, one must factor in the understanding induced in the receiver and this requires that one take recourse to the information, or meaning, carried by the sentence.

A. Raftopoulos, *The Origins of Representations: Cognitive and Brain Development*, Studies in Applied Philosophy, Epistemology and Rational Ethics 73,
https://doi.org/10.1007/978-3-031-91213-9_2

Thus, a "sign" is an object, quality, or event, an entity in general, whose presence or occurrence indicates the probable occurrence of another entity; signs are information-carrying vehicles that refer to, or indicate, some entity. Signs can also consist in a gesture or action used to convey information or instructions. For example, an airplane sign on the road indicates to drivers the location of an airport. When signs function in this manner, they are called "signals." Thus, a "signal" may be a gesture, action, or sound that is used to convey information or instructions. An electrical impulse or radio wave that is used to transmit information is also considered to be a signal. In general, a signal is a means of communication that usually prompts for action (a traffic light changing from red to green signals that it is safe to proceed). To be able to prompt an action, the signal, and the action, or at least the understanding on the part of the receiver of the information the signal conveys, must coincide in time factoring in the reaction time on the part of the receiver. Hence, a signal conveys information from one part in space, the location of the signal, to another part in space, the location of the receiver; signals are information carrying vehicle whose function it is to transmit information over space (Gallistel & King, 2010). Signs, on the other hand, have a more general function in that they need not co-occur in time (although, as we shall see, a kind of signs, namely, indices do in general but not necessarily always) co-occur in space and time with the entity they indicate, which is why many authors refer to them as signals). Signs are repositories of information that awaits interpretation. In what follows I use the term "sign" to denote any information-carrying physical, including mental, entities.

Let us return to Peirce tripartite system of signs. *Indices* are involuntary and non-intentional (i.e., non-representational) objects, and have a form that bears a physical, factual, or causal relationship with what they indicate. The footprint of an animal indicates the animal and is causally related to it. Similarly, smoke is causally and physically related to fire. In addition, a specific cry of velvet monkeys is related to the presence of a specific predator and can be used as an index denoting the presence of a predator. There is a distinction to be drawn between the first two examples and the third one. In the first two cases, the existing factual, physical relationship between the index and the entity it indicates entails that the relationship between these two is not arbitrary, and thus, non-conventional since it is established through some physical process. In addition, it is involuntary.

In the latter case, however, the relation could be construed as arbitrary in the sense that the velvet-monkey cry is not related to the traits of the predator present in a non-trivial physical manner, although the presence of the predator caused the cry, and there is, thus, a causal relation between the index-cry and its referent. As Skyrms (1998, 411) notes, the alarm-calls are not imitations of the calls of the associated predator. It follows that the monkey could have "chosen" some other form of cry to indicate the presence of the predator. In this sense, this signal, the cry, is arbitrary and, thus, conventional because the monkeys could have chosen some other form of cry to alert their conspecifics of the predator. Furthermore, it is voluntarily produced. The distinction is important because in the literature (for this view and discussion see Godfrey-Smith, 2014; Planer & Kalkman, 2021; Skyrms, 2010) there is a long discussion as to whether arbitrariness is a necessary condition for a physical structure being a signal, that is, an information-bearer. If this criterion is accepted, the smoke is not a signal for fire, but the velvet call is a signal. Others reject this view and think that a physical explanation relating the index, and its referent suffices to render the index a signal. Planer (2021) proposes the distinction between conventional and non-conventional indices (or indexes as he calls them) to account for these two different kinds of indices.

Nevertheless, in all cases in which an index occurs, irrespective of whether it is arbitrary or non-arbitrary, an index is a sign that is produced in the spatio-temporal presence of the entity that the index indicates. In other words, indices are triggered by specific environmental states. Moreover, indices are closely related to specific responses, probably learned through associative learning and the ensuing adaptations. This learning is not usually restricted to associating an index with a feature in the environment, as there is a one-to-many correspondence between an index and features in the environment and, thus, the index is underdetermined. Learning an index means, of course, that the animals' attention focuses on some environmental feature, for example, a predator, but this is not learning by associating the signal with the feature but because the animal has learned a perception–action link, which I shall discuss in a few paragraphs and in Chap. 5. (Planer and Sterelny (2021, 14) offer a similar argument from a different perspective.) Once the index is adopted, it is fixed over time and cannot change and so does the reaction with which it is associated. Finally, not only indices are usually used unaccompanied by other signals, but do not even require that other signals exist; they can function in isolation. When we discuss symbolic representations or symbols, we shall see that none of these characteristics of indices applies to symbols and that, in fact, the opposite is true. Symbols, for example, depend on other symbols to have a meaning.

As we saw, Peirce thinks that indices are involuntary signals since they are bearing a causal of physical relation with their referents. If the occurrence of an object is physically related to the appearance of the corresponding index, obviously the existence of the index is not due to some voluntary act but to some physical process. This is correct, but if one allows the possibility that the velvet monkey predator-cries are indices that are caused by the presence of a specific kind of predator, since these cries have resulted on account of some voluntary action on the part of some member of the species sometime in the evolutionary past of the species, it follows that some indices may be voluntary and arbitrary. A word of caution is also needed at this juncture. Peirce's indices should not be confused with "indexicals", such as "I" or "This" in philosophical discussions.

Another important characteristic of indices that results from the fact that they bear a natural, physical or causal relation, to their referents is that indices coincide in space and time with their referents; an index S referring to some object O cannot be detached in space and time from O because the appearance of S requires that O be present. It is easy to see that both conventional and non-conventional indices satisfy this criterion. When smoke appears and is taken as an index for fire, the smoke coincides in space and time with the fire. When the velvet monkey issues a certain cry, a predator is usually currently present, although, as a matter of course, the monkey may be mistaken. This thesis, however, must be taken with a grain of salt. A foot-print on the ground indicates the animal that made it and bears a causal, factual relation to the animal that caused it, and thus, it is an index, but the animal that left the print may be long gone, and, thus, the object that the sign indicates is detached in space and time. Things get more complicated by the fact that, as we shall see, there are iconic and symbolic representations that are not detached from their context, namely the perceptual representations of early vision and indexical symbols, respectively.

Icons are non-arbitrary representational signs that resemble (are similar to) what they stand for or represent. They physically evoke through their resemblance what they refer to. In other words, icons physically resemble their referents. They are voluntary and naturally produced in the sense that they are not invented, although, as we shall see, both of these claims are debated. In icons there is a non-arbitrary, natural linkage between the form and the meaning, that is, between the iconic sign and its referent. Thus, syntax parallels semantics and, in fact, it is the referent that determines the form of the sign since the latter must resemble in some way the former. Unlike indices, icons are detached from their referents in space and time; they have what sometimes is called "displaced reference" which allows signs to the ability to *refer* to entities and events that are removed in space or time from the immediate communicative context. Let me say, in passing, that it is usually thought that only *Homo sapiens* have this capacity, but recent studies (Lameira & Call, 2018; Lyn et al., 2014) suggest that apes have it too. This is important because if the claim is correct, this means that the capacity for using signs that have displaced reference belongs to the baseline cognitive capacities from which Hominis evolved and, thus, that 'displaced reference in language is likely to have originally piggybacked on akin behaviors in an ancestral hominid' (Lameira & Call, 1). This, in turn, is significant because the capacity to use signs with displaced reference is a perquisite for having some form of language (or, proto-language) as a communicative tool.

I have argued that icons exemplify detached reference, that is, they can refer to items with which they are not concurrent in space and time. This may be taken to entail that when a seemingly iconic signal, or an iconic representation, is inextricably tied to a certain spatio-temporal context, then this signal is not an icon but an index. Think of a sign showing an airplane in the main entrance to an airport. This is an iconic sign in that it indicates its meaning through some abstract resemblance to the function of an airport, which is to serve airplanes. This icon, however, is tied to the space and time of its occurrence because it means "airport here and now." The same icon could have used in the highway to indicate the direction to the airport, but in this case the reference id displaced. In the third chapter, I explain why this view is wrong by claiming that there are iconic, perceptual representations that are constitutively contextual. In any case, external iconic representations, at least, are detached representations in the sense explicated above, and, thus, allow for what Morin et al. (2020) call "asynchronous communication".

We saw in the introduction that Peirce thought that indices are referentially more complex than icons since icons are necessary for fixing the reference of indices. Peirce thinks that indices require icons and, in this sense, icons are the most basic kind of signals, the idea being that to establish reference, indices require icons. For a signal to be recognized as such and such, that is, as indicating such and such an object, the organism perceiving the signal must be able to perceive it as having some resemblance to signals that it has encountered in the past. This means that the organism must have the capacity to compare the current icon of the perceived signal with the icon of a signal that it had encountered in the past. This, given Peirce's construal of iconic signals, entails that the organism must be able to perceive the sign as an icon first in order to be able to establish its referent, that is, the object that the index indicates in the world.

This view is wrong (see also Everett, 2017; Planer & Sterelny, 2021). First, indices are not representational and, thus, are cognitively less demanding, which entails that they appeared earlier in the evolution that gradually led to Hominins. Animals that do not have the cognitive capacities to understand and use icons can nevertheless extract information and use animal signals that function as indices. Given the linear, gradual evolutionary trajectory, it seems plausible that early hominis also used first indices and then icons. For example, the first uses of ochre as pigment dating back to 400 Ka, were probably used to indicate some characteristic of its bearer, say, its clan membership or some other trait and were not meant to be icons. For this reasons it is in general assumed in the literature that there is a historical, phylogenetic, progression from indices to icons and, thereby, to symbols. The reason is that indices are not representational, while icons and symbols are representational, and thus, require a more sophisticated cognitive apparatus. All animals use indices but, most of them (apes being most probably the exception) lacking representational capacities cannot use icons or symbols. Hominins used first indices, icons, and then moved on to symbols. There is, thus, a historical progression from indices to icons and, thereby, to symbols; animals use indices, and hominins used first indices, then icons (even in written language the first forms were iconic as in Egyptian hieroglyphics and Sumerian cuneiforms), and then they moved on to symbols (the words, for example, of languages).

Second, Peirce's reasoning why icons precede indices is itself flawed. Peirce thought that a signal can be recognized as indicating such and such an object, if the organism perceiving the signal is able to perceive it as having some resemblance to signals that it has encountered in the past. In other words, the organism must compare experiences and determine whether they are relatively and saliently similar. This means that the organism must have the capacity to compare the current icon of the perceived signal with the icon of a signal that it had encountered in the past. However, this is false. The animal probably has either learned by association what an index indicates, or has inherited the index-reaction, which is a perception–action, link and need not do any comparison of icons. As we shall see in Chap. 5, breaking such perception–action links is necessary for the capacity to represent to emerge. Moreover, the comparison of signals, current and past, means that an index to function must be associated and, thus, presupposes, some other index, but one of the most prominent traits of indices is that they can be used in the absence of any other signals.

Symbols finally are representational signs that refer to something arbitrarily. They are inventions and represent something by means of some culturally defined convention; it is culture that transforms things into symbols and meanings (Everett, 2017,

60). In symbolic signs, the form is independent of the referent since their relation is arbitrary and they need not resemble what they refer to, although some signs may resemble some aspect of their referent (as the word *meaw* in Asiatic languages means cat). Being conventional, symbols are also arbitrary constructions. Finally, symbols are patio-temporally independent of their referents; their meaning does not depend on spatio-temporal coincidence with their referents. This phenomenon in which a sign can refer to entities that are removed in space and time from the immediate communicative or perceptual situation is known as "displaced reference." Having "displaced reference", symbols store information about, and can refer to, entities (be it objects or events) that are spatially and temporally displaced from the symbols. This is a significant difference from indices, a difference on whose significance I elaborate at a later chapter, because it seems that appealing to an arbitrary or conventional relation to the thing signified is not sufficient to distinguish symbols from indices. The reason is that it is likely that there are arbitrary or conventional indices as well, such as the prey-cries of the velvet monkeys.

Gärdenfors (1995) uses the term "cued representations" for those representations that are tied to the here and now and opposes them to the "detached representations" that are not thus constrained. In view of our discussion, it is natural to conclude that indices belong to the former category (leaving aside the problem of whether one could call them representations), while icons and symbols belong to the latter, and, indeed, this is currently the prevalent view. A word of caution is needed here to guard us from jumping to the conclusion that all icons are detached from the context in which they occur. The reason is that, as we have seen, there are iconic representations, namely the perceptual representations of early vision (Raftopoulos, 2019, 2022) that are constitutively tied to the context of their occurrence, and the same holds for iconic signs that specify some place and are, thus, tie to a specific spatio-temporal context. This caution should be extended to symbols as well because there are symbols, such as the indexicals "I", "Here", "This" etc., whose reference can be fixed only within a context.[1] Concerning symbols, and in particular graphic signs that constitute graphic codes, Morin et al. (2020) argue that one of the uses of these codes is "asynchronous communication", by which they mean that communication can occur even when sender and receiver of the code are displaced in space and time. This differs from face-to-face communication where the sender and receiver share, at least, the physical context in which their encounter takes place.

---

[1] In philosophy, indexicals have a referent and a semantic content, the latter usually being the grammatical meaning of the indexical, which is used in a context to determine the referent of the demonstrative. However, the grammatical meaning of the indexical is external to the semantic content of the indexical, which is restricted to its referent. Kaplan (1989), for example, the meaning of "I" determines that when persons utter "I" they use the term to denote the persons speaking, that is, themselves. Thus, when S utters X, the context, that is, the fact that it is S who utters "I", fixes the referent of "I" as the person uttering this term. Note that this view is controversial. Recanati (1997), for example, talks of the psychological mode of presentation involved in "I". Raftopoulos (2009) argues that contra Kaplan, perceptual demonstratives have a non-conceptual mode of presentation that is used to fix their referents. Be that as it may, I shall not pursue this problem any further because it has no bearing to the aim of this book.

Finally, Sterelny (2003, 4) uses the term "decoupled representations" to denote those representations that track features of the environment, and which do not drive specific behavioral responses. At a first glance, indices are coupled representations since an animal that sees the footprint of a predator or smells a predator (these are typical indices) has been conditioned to react by fleeing. On the other hand, upon seeing an icon or a symbol, agents need not engage in any behavioral response, or if they do, the icon or the symbol does not necessarily prompt a specific genetically pre-specified behavioral response, which means that icons and symbols are decoupled representations. I qualified this thought with the "at a first glance" because of a scenario in which a hunter sees the footprint of a predator, say, a tiger. The hunter may decide to proceed to find the tiger or may decide not to do so. This means that the index in this scenario does not drive a specified response of any sort, which means that it is problematic to consider indices in general as coupled representations. The footprint is still an index and yet, it does not prompt a specific behavioral response. Therefore, the term "decoupled representations" does not designate only icons and symbols, but also indices.

Peirce, as we shall see next, relates all signals to the capacity of receivers to read or interpret them, the key terms being that of the "interpretant". In modern discussions of signals, however, the emphasis is put more on the distinctive traits of the signals that allow their categorization in different kinds (say, symbols vs. icons) that on the interpretive processes. Among other things, this allows us to talk about signals abstractly, separating them from a particular interpretative process in the mind of a particular receiver. So, for example, the "V" sign is held to be a signal for victory independently of any receivers noting the fact that "V" is the first letter of the word "victory."

Peirce in discussing semiosis, introduces the notion of the "interpretant" and claims that semiosis involves a sign, its object, and its interpretant.

> A *representamen* is a subject of a triadic relation to a second, called its *object*, for a third, called its *interpretant*, this triadic relation being such that the *representamen* determines its interpretant to stand in the same triadic relation to the same object for some interpretant. 1.541 (1903)

The interpretant is what makes possible to use a sign so as to interpret it, that is, to understand what it refers to. Representations go hand in hand with a process that uses the representation because without such a process the mere presence of a sign, of an object to which the sign purports to refer, and of set of rules that fixes the reference, merely creates the potential for representation. This potentiality is actualized only when some process in an agent uses this triadic scheme for some purpose. Peirce's "interpretant" is such a process. The interpretant, thus, is what allows taking a *representamen* (Peirce's term for a sign or signifier) for the sign of an object, that is, what allows us to understand the sign: 'The interpretant constitutes the only possible mediation between a sign and its possible individual objects (4.127)'.

One could argue that, in cognitive terms, an interpretant is whatever process enables one to infer the reference from some sign in a context, that is, an interpretant is the mediator that brings a sign and its signified together. This holds true both in cases in which the sign refers directly to something and in cases in which meaning interferes and fixes the referent. Thus, one could say that for Peirce a physical entity S is a sign for some other entity O when S causes a receiver R of S to "think" of O. I put "to think" in quotation marks because there need not be any thinking as a cognitive activity involved in establishing O as the referent of S. This reference could be established in a purely visual or otherwise perceptual manner (that is, without thinking or concepts for that matter being involved), as in the case of velvet monkeys hearing the alarming sound indicating the presence of a specific kind of predator. In this case, a monkey does not think that a predator is approaching, as that would require that the monkey possess the concepts predator among other things, which it does not. The monkey has been conditioned to engage in avoidance behavior when it hears this specific call.

Peirce thought that interpretants come in different degrees of complexity and they can also be of categorically different kinds too. Furthermore, he did not confine interpretants only to what goes on consciously in the head. The process that interprets a sign and allows one to take it as a sign for some other object may be an agent (or, to be precise, the idea that the sign produces in the mind of the agent), as when one reads a book, or a set of processes at the sub-personal level like those that work in perception to construct the percept or to react unconsciously to some signal, or an external "tool".

The role of interpretant in the process of representation entails that when one apply the terms index, icon, or symbol to particular things, calling a particular picture an icon, a thermometer an indicator, or a word in some language a symbol, one usually means that this is how they were designed to be interpreted, or are highly likely to be interpreted. As we shall see, even a strong resemblance does not make one thing an icon of another. Only when there is the appropriate interpretant in work so that the features of the sign bring the signified object into mind of a receiver on account of this resemblance is the relationship iconic. At a minimum, the receiver of sign S that resembles some object O must be able to perceive S and detect the similarity with O in order for S to cause the receiver to think of O. Thus, similarity by itself does not cause iconicity, nor does iconicity consist in the physical relationship of similarity (more about that when we discuss icons in detail), since almost anything could be considered an icon of anything else, depending on the vagueness of the similarity considered. It is the kind of a process, as determined by the interpretant, in which the capacity to detect or recognize similarity and, thus, an, icon is based.

The same point can be made for each of the other two modes of signs, namely, indices and symbols. Neither physical or nomological connection, nor involvement in some conventional activity dictates that an object is an indexical or a symbolic sign, respectively. Only when do these objects invoke another through an interpretant are we justified in calling their relationship indexical or symbolic. That a sign is amenable to an iconic or indexical interpretation, in no way does this diminish this sign's capacity of being interpreted symbolically as well. These modes of reference are not mutually exclusive alternatives. Although at any one time only one of these modes may be prominent, the same sign can be an icon, an index, or a symbol depending on the interpretive process.

> [R]eference is not intrinsic to a word, sound, gesture, or hieroglyph; it is created by the nature of some response to it. Reference derives from the process of generating some cognitive action, an interpretive response; and differences in interpretive responses not only can determine different references for the same sign, but can determine reference in different ways. We can refer to such interpretive responses as interpretants. (Deacon, 1997, 63)

These considerations concerning the role of interpretant suggest that for Peirce a physical entity S is a sign referring to some other entity O on account of the fact that when a n agent P receives S, S causes P to think of O. If S causes P to think of O because S is causally or physically related to O, then S is an index. If S causes P to think of O because S physically resembles O, then S is an icon. Finally, if S causes P to think of O because S bears a semantic, meaning relation to O that is established through some convention, then S is a symbol. This account employs the term "think" but, following Deacon, one can generalize this view and say, instead of S causing P to think O, that S generates an interpretive response on the part of P that targets O. This is more general because the interpretive response need not be an act of "thinking" that involves concepts. This is required if one wish to endorse the view that non-conceptual organisms, including human infants, are still capable to interpreting signs. That thought is not required to interpret signals applies to symbols as well, despite the fact that a symbol refers through some semantic relation to its signified entity. This semantic relation, as Skyrms (2010) notes, may result from basic reinforcement learning or some process of natural selection; processes that do not require any cognitive sophistication.

The interpretant, whether it be an agent or a process, utilizes some aspects of the form of the sign, that is, some syntactic aspects of it but not others, to work out its interpretation. To give a couple of examples, an agent that reads a sentence relies on the orientation of the letters to understand its meaning, that is, the proposition it expresses, but the size of the letters is irrelevant. Similarly, to understand a drawing, the configuration of the objects in it and their forms are necessary to interpret the drawing, but the material out of which the canvas is made is irrelevant to the meaning of the drawing.

Pierce introduced three kinds of signs. Not all signs are representations. Pierce, for example, thought that indices, unlike icons and symbols, are non-intentional physical entities, that is, non-representational entities. The annual rings of a tree may indicate the age of a tree but do not represent the age of the tree; they are "indicators" (Dretske, 2000). To be a representation means to be able to represent something for someone who interprets the sign. In philosophy, a physical state is a representation if it has a representational content, that is, if it represents the world as being a certain way. In this case the state has a correctness condition; that is, the world can make the state come out true or false. Peacocke (1992) calls the correctness condition 'the minimal account of content possession.' Bermudez (1995, 344–347) argues that a physical state should be semantically evaluable is a necessary but not a sufficient condition for the sign to have representational content. Bermudez (ibid., 348–349) adds some further interrelated conditions that the content of a mental state must meet in order to be representational. The first criterion is that a physical state is a representational sign and, thus, has a representational content, if its content could in principle figure in explanations of intentional behavior—to wit, in the explanation of the connection between sensory input and behavioral output that cannot be explained by invoking some lawful regularity (in other words, an invariant connection), as when the behavior is the result of some dedicated transducers. Second, representational content should allow cognitive integration, in the sense that it should be able to interact with other representational states. Third, representational content must have structure. This means that its constituent elements could be constituent parts of other contents, and the content could be decomposed into its elements that could be recombined with the elements of other states having representational content.

The further requirements proposed by Bermudez's severely restrict the set of possible representations but allow many states of the perceptual system (and perhaps sub conscious cognitive states) to be representations (see Raftopoulos (2009) for a discussion). These representations may be parts of motor schemes that are automatically activated (as in the human dorsal visual system) and are, thus tied, to these motor schemes. In this book, I am interested in a more restricted class of representations, namely those that can be decoupled from any motor-action schemes and be used in the absence of these schemes. The reasons for this will come to the fore in our discussions in this book.

The representations that are discussed in this book presuppose that agents could have them only if they have self-awareness, even in a very primitive form, because only then could representations be factors in intentional behavior that requires some control over the representations. Also, because only then the representations could have contents parts of which could be combined with the contents of other representations to form new representations. This leaves open the problem of whether—or which—mammals have representations. Self-consciousness requires some form of theory of mind, and one may argue that only agents who have some form of a theory of mind are capable of having representations, even though many creatures have neural states that present the world as being some or other way, that is, even though many creatures have consciousness. One may wish to call these neural states that present the world as being some way, and which are, thus, semantically evaluable, representations. This is a significant problem for philosophy, but it is outside the scope of the book to discuss it. In the book, I adopt Bermudez's criteria because I wish to distinguish between states that are involved in motor control and are coupled with actions (which, I repeat, one might call "representations"), from those states that I call "representations" that are involved in intentional behavior, that is a behavior that goes beyond the involuntary activation of some motor skill. For the philosophy informed reader, the thesis adopted in the book is that the representations I am interested in are carried by agents who have both a "thing" and a "fact" awareness (Dretske, 1993, 2006) of their surroundings, or, equivalently, a "phenomenal" and a "cognitive access" awareness (Block, 2007, 2014) of their surroundings.

According to the third requirement above, for a sign to be a representation it must have a representational content that is structured from components that can be recombined to form other representations. As we shall see, the content of the representation is determined by the contents of the atomic constituents (primitive symbols, in the case of symbolic representations, or suitable iconic parts in the case of iconic representations) and the way these are combined, that is, the compositionality of the system at issue. It is a thorny issue how the atomic constituents acquire their content, that is, their meaning and reference. There are several contenders in philosophy, the most prominent being that the meaning of the atomic symbols is determined by their functional, computational role in the system of representations, or by a certain history of selection as in teleosemantics or biosemantics (Millikan, 1989; Papineau, 1987), or by means of an asymmetrical causal relation with the environment ala Fodor. Fortunately, an answer to this question does not have a bearing on the topic discussed in this book and, thus, can be put aside.

An index that indicates through some physical or causal connection that another animal is present refers directly to this animal. Hence, it has no representational content; it only affords the perceiving animal with potentialities of interaction with the indicated animal (Gibson's (1979) affordances). These, however, are arguably not representational contents (Raftopoulos, 2021). Even if indices had representational contents, for example, if they presented an animal-predator in the perceiver's mind owing to the close physical connection of the print to the animal-predator that caused it and to the role this indication plays in the cognitive economy of the perceiving animal (in which case, the mental state of the perceiving animal is about the indicated animal because it is the biological function of this state to enable the perceiving animal to avoid the predator), it is highly doubtful that the perceiving animal could decompose the representation of the indicated animal into parts and recombine them with other representational parts to form a new representation. There is no need for that; in evolutionary terms, the footprint has a certain function, namely, to indicate to the perceiving animal potentialities of interaction. Nothing more need to be added to the sign, and, thus, no representational structure of the sort discussed in the third criterion is required.

Finally, recall that indices have a physical form that bears a physical, factual, or causal relationship with what they indicate or refer to. The print on the ground indicates to the animal that sees it the presence of another animal that left the print through some lawful regularity or invariant connection of the print to the animal that caused it, and through some lawful connection between the perceived print and the relevant cognitive activity in the mind of the perceiving animal. That is, the animal cannot fail to perceive the print as indicating the animal that caused it, given that it perceives the sign in the first place. This means that indices fail the minimal criterion for a sign to be a representation.

It will be recalled that representations are ways of referring to entities in the world. Representations involve two sets of elements. The signifiers (e.g., signs, such as words, pictures, etc.) that do the referring, and the signified things, that is, the entities to which the signs refer. These two sets are associated by a physical or causal relationship (indices), or by a relationship of physical resemblance (icons), or by a conceptual relationship (symbols), These relationships establish the semantics that map individual elements in the representing world onto those in the represented world. In linguistic semantics there are, at least, two sorts of semantic or meaning relationships, the transparent and the opaque relationship. The relationships that link signifiers and signified by virtue of their similarity are "transparent" to the extent that they require no additional cognitive (as opposed to sensory) resources to "see" the one by means of experiencing the other. Those that link signifiers and signified by virtue of some arbitrary coding or mapping are "opaque" because they require knowledge of the code. The transparent kind of signifier is often referred to as an icon, and the opaque variety is often referred to as a symbol.

Representations being physical entities that represent something have both a referent (the thing to which they refer) and a meaning (recall that in this book a representationalist view of mental entities is assumed). Frege (1982) introduced, albeit using a somewhat different terminology, this distinction. For example, the picture of a cat refers to a cat, and its meaning is whatever information concerning cats the picture brings into the viewer's mind. Sometimes, this is called the "cognitive significance" or the "mode of presentation" of the sign for the viewer who interprets the picture. The meaning of the sign is also referred to as the "content" of the sign or representation. Signs, *qua* physical entities, in their capacity to represent have a physical form and a content. The former is called the representational vehicle, and the latter the representational content of the representation. The properties of a vehicle are the physical properties of the medium in which the representation is expressed; this is the level of the syntax of the representations. The content of the representation, that is, its meaning or the way the representatum is presented to the agent constitutes the semantic level of the representation. One need to pay attention to this distinction because the signified object should be distinguished from the meaning or significance of the sign; the signified thing is external to the sign, while the meaning of the representation is internal to, or part of the definition of, the sign itself. Thus, one must be careful to distinguish between the thing to which a sign refers (the object of intention, in philosophical parlance) and the thing as it participates in the meaning of the sign (the intentional object), although, as philosophy again shows, on many occasions the thing itself, or, to be precise, some encounter in the past with the thing itself, may be part of the meaning or content (Kripke, 1980; Putnam, 1983) of a concept or symbol. In this sense, the modes of presentation of an object in the representation are object involving (see Raftopoulos (2009) for a discussion of the literature). I return to this problem after I have introduced another important distinction between external and internal representations.

I distinguish mental, internal representations from external, physical representations expressed in a publicly accessible medium. This qualification is needed because even though mental representations are also realized in a physical medium, namely, the brain whose states are accessible through brain scanning techniques, these states are accessible only to scientists and, moreover, their representational contents are not publicly accessible. In contradistinction, external representations are thoroughly publicly accessible since, first, their form is publicly accessible, and second, by lacking representational content and referring directly to their representata their content is publicly accessible as well. Note that one could still talk of the content of external representations, but the reader should bear in mind that this is not the representational content of the representation, but the thing represented.

Mental representations have brain states as representational vehicles and they represent some entity in the world through their representational content. The reader should recall that in the introduction I stated that I assume that some representational account of mental representations is correct versus say, Russellian or possible worlds accounts of representations. There is, thus, a threefold distinction between representational vehicles, representational contents, and the represented domain. When I refer to the content of a mental representation, I mean the representational content of the representation and not the represented entity. Truth conditions or the degree of accuracy of a mental representation is a matter of fit of the representational content of the representation to the represented world. A visual perceptual brain state correctly represents a visual scene if its representational content presents the scene accurately.

External representations, on the other hand, are characterized by a representational vehicle and the domain they purport to represent; they have no representational content. In external representations, there is a representational vehicle, and the "content" of the representational vehicle is the represented domain. Truth conditions or the degree of accuracy of an external representation is a matter of fit of the representational vehicle to the represented domain. A regular Hg thermometer shows the correct temperature of item X if the state in which it is after the measurement, that is, a determinate height of Hg, closely corresponds, that is, it is within an acceptable margin of error, to the actual temperature of X.

Let us approach this from a different perspective. There is a distinction to be made between external representations, that is, representations that exist out there in the world outside our minds (even if they constitutively form part of our mental representations—I leave aside any discussion about narrow and wide contents), and mental representations. One has access to physical representational vehicles, one perceives them, whereas we do not have direct access to our mental representations, although though neuroimaging one could acquire an access to the brain representational vehicles realizing these representations. To understand and use an external representation one must have access to its vehicle, while to understand a mental representation no such access is required. In external representations, on the one hand, there is the representational vehicle that belongs to the representing world, the represented domain that belongs to the represented world, and the vehicle represents the represented domain. The content of the representation is the object of the representation. In mental representations, on the other hand, in addition to the representational vehicle and the represented domain, a third entity intervenes, namely the representational content of the representation that differs from the represented domain in that the representational content belongs to the representing world.

There is a further reason for which even though for external representations the represented domain may be taken to fill the role of the content of the representation, for mental representations such as the belief that "a is F" this cannot be true. The content of a mental state is supposed to be "in the head" no matter how right externalism (in the philosophy of mind) about contents may be, and thus, even if it is true that the mind extends to the world in the sense that the world, or our conceptual frameworks and linguistic practices, contribute to the individuation of the contents of mental states (that is, that the semantic content of a mental state depends on constitutive or causal relations with external entities), the fact remains that a mental state is, at least partly, in the head, while the represented domain is in its entirety out there in the world. So, we have to construe of representational contents as mental contents, which, as such, involve modes of presentation (*mops*), akin to Frege's modes of presentation or senses in thought. (These contents cannot be Russellian propositions because according to this interpretation representational states do not have representational content but refer directly to some part of the represented domain, and they cannot be sets of possible worlds either).

There is a further complication affecting the distinction between mental and external representations; this concerns iconic representations such as pictures, photographs, drawings, etc. These are external representations in that their vehicles are publicly accessible, but unlike other typical external representations, such as thermometers, they do not refer directly to the represented domain because they show what they represent through their representational content. Pictures in general show what they depict by eliciting in their viewers mental representational states whose representational contents present the represented domain as being in some way. Thus, they utilize the representational capacities of a viewer's visual system in order to be understood, which is different from using visual perception to read of a measurement in, say, a thermometer. Ayers (2019, 77–78) argues that pictorial representations convey information about what is represented 'by exploiting the content of the experience of seeing it, that is, how the object or scene looks from a particular point of view'.

Seeing a picture, one perceptually represent three things: (i) the two-dimensional surface of the picture: (ii) the three-dimensional scene visually represented by the picture surface; and (iii) the depicted scene. As we saw when we discuss the distinction between the object of a representation and the representational object, (ii) and (iii) may be different. Borrowing an example from Nanay (2022), when one sees a caricature of Mick Jagger in which his lips are most prominent, the representational content (the thick lips) differs from the actual lips of the person who is the object of the representation. The threefold distinction among representational vehicle, representational content, and represented entity applies to these sorts of external representations and for this reason I examine them along mental representations.

Thus, in these types of iconic representations there is a distinction between the object of representation and the representational content; the content is not the object of the representation. The idea that in pictorial representations, whether it be paintings, drawings, or photographs, there is something more than the representational carrier or vehicle and the scene it depicts or represents, which is required for understanding what the image depicts and how it depicts it, is expressed by Haugeland (1998, 203, ft. 1).

> This use of the term 'content' is not altogether standard. Most contemporary authors (and I, in other essays in this volume) mean by the "content" of a representation something distinct from the object it represents, and which determines that object... Here, however, I mean by 'content' that which the representation represents—the "object" itself—but as it is represented to be (whether it is that way or not). Thus, it is a possible object—which may in fact be actual, or similar to something actual, or neither.

Haugeland uses "content" to mean the way the object of the representation is represented as being in the representation; this is an "intentional object" or "possible" object as Haugeland names it. If the representation is accurate, the intentional object is actualized as the physical object that the representation purports to represent.

Hyman (2006) talks of a picture's "internal subject" that differs from its external subject, which is the object that the picture depicts. The internal subject of a picture is the object of a picture as it is depicted in the picture, a clear reminder of the distinction between the object of a representation and the way this object is represented, that is, the intentional object. Hopkins (1995, 1998) discusses the "seeing-in content" of an image, which differs from what the image depicts; the seeing-in content in a caricature of Obama's does not have Obama's outline shape and, yet it depicts Obama. One of the motives behind this view is that an image can represent the same external subject (the object of representation) in quire diverse ways, that is under different aspects (what in mental states we would call modes of presentation). These aspects are the contents of the image (what I call the representational contents of the image *qua* representation), which means that the same image-subject can be represented by different image-objects, that is by different representational contents; images can present their subjects under different aspects (Kulvicki, 2014, 37).[2]

For these reasons, I think that in mental representations and in external iconic, imagistic representations such as pictures, paintings, and photographs, the distinction between syntax and semantics corresponds to the distinction between representational vehicles (neural or mental) and representational contents, in contrast to external

---

[2] An interesting question concerns the distinction between external iconic representations (such as a Hg analog thermometer) whose content is some represented item, and those external imagistic iconic representations where there is a distinction between the representational content, that is, the item as it is represented in the representation, and the object (or subject) of the representation, that is, the represented item itself. What is, thus, the relevant difference between a mercury thermometer and a painting? One might be tempted to say that the thermometer has a derivative intentionality, whereas a painting, even though it needs an observer to interpret it, still, it somehow has its own intentionality in the sense that the creator of the painting put, as it were, the depicted visual scene in the painting, which, thus, acquired its intentionality independent of any prospective viewer. To put it in different words, that the painter created an image of the visual scene that presents it from some perspective matters in how the visual scene will be viewed by its viewers. This immediately puts in play the distinction between the visual scene and the way it is represented. The designers of the thermometer, on the other hand, did not put temperatures in the thermometer, nor did they choose to represent temperatures from some perspective which could have been different (once the representing means, i.e., mercury, had been chosen).

I suspect that underlying this line of thinking is the fact that paintings tend to resemble in some respects their depicted scenes, while other non-imagistic, iconic external representations map onto their domains in the appropriate way that renders them iconic (or analog) representations, but they do not resemble their domains. Specifically, pictures and other visual images represent things by capturing visual appearances, and this is what ties them to visual perceptual representations. Pylyshyn (2003) remarks that mental images and pictures alike are episodes (mental in the former case, physical or external in the latter case) in which one simulates seeing something, which means that we can picture or mentally imagine only the sorts of scenes that we could see. This means that when we see a picture of a scene, we employ the same visual mechanisms that we would have applied had we visually encountered the scene itself. In Chap. 4, I argue that there are two main conceptions of iconic representations; one takes a representation to be iconic if it physically resembles what it represents, while the other takes a representation to be iconic if it maps in an appropriate way onto its representatum.

representations, which as such do not involve representational contents, in which the syntax/semantics distinction corresponds to the distinction between representational vehicles and the represented item (in this case, the object of the representation and the representational object coincide).

This claim needs some further explaining because one might object that the "meaning" of, say, a perceptual state is the perceptual scene represented in so far as it is this scene that renders the content of the perceptual state correct or incorrect, the same way it is a worldly state of affairs that renders the content of a sentence. i.e., its proposition, true of false. It is, thus, a worldly state of affairs that determines the semantics either of a perceptual state or of a sentence. Notice, however, that even in the linguistic case there is a distinction between the worldly fact that determines the truth value of a sentential-like vehicle and the proposition expressed by this vehicle, and the same distinction is found in the perceptual case in the form of the distinction between perceptual content and visual scene, which explains why one could be in a visual perceptual state with content as of a certain scene S even though S fails to obtain. Put differently, there is one fact, the perceptual content that a perceiver is having, and there is another distinct fact, namely the visual scene itself, assuming that the scene obtains, and it is the latter fact that determines the truth value of the former fact. This is a description of the distinction between the object of representation and the representational object that I mentioned above. This distinction affects the way one construes the semantics of mental representations.

To the extent that the semantics of mental representations pertain to the representational content of the representational vehicle, semantics do not concern, directly at least, truth evaluative facts, that is, truth value assignments to the constituents of the representation and the representation itself. The facts that determine the truth values of representations differ from the facts that determine the meanings of these representations. This can also be concluded by noticing that these two may differ, as is obvious in cases of visual illusions or mistaken beliefs, in which cases the representational contents that play a role in the cognitive life of the organism differ from the states of affairs they purport to (re)present. In visual illusions, when one sees (for some reason or other) a red apple as green, the cognitive role of the perceptual state is determined by the greenness of the apple and not by real redness that determines the truth value of the representation, which explains why an organism might choose not to eat a green fruit as it considers it not ripe enough, even though the apple is really red and, thus, ripe. In addition, since the apple is really red and not green, the representation of it as green is incorrect, hence its characterization as a visual illusion. Thus, the cognitive significance or meaning of mental representations depends on their representational content and not on the conditions of their truth evaluability; semantics in mental representations is tied to the representation's representational content.

The distinction between the object of representation and representational or intentional object[3] is not, of course, a novel distinction. In modern philosophy, it has its

[3] Nanay (2022, 2) seems to reiterate this distinction. Nanay discusses pictures, and the representational content of a picture in examining the sensory individuals involved in a picture. 'Whenever

roots in the Husserlian (Husserl, 1898–1925/2006) tradition that has been taken up recently by Wiesing (2009). Wiesing (2009, 33–36) distinguishes between the "image-carrier" (what it is usually called "the representational vehicle", although Wiesing does not accept this term), the "image-object", which is the representation that stands for something else (which explains why Wiesing does not use the term representational vehicle for the image-carrier), and the "image-subject" which is the thing the "image-object" representation or stands for, which is for Wiesing the content of the representation. Wiesing (2009, 35) affirms that the image-carrier displays the image-object or representation, and since the image-objects stands or represents the image-subject, the image-carrier is related to the content or image-subject.

Kulvicki (2014, 19) contrasts this view with that of Wollheim's (1993), according to whom the image-carrier, that is, the surface of a painting, is the representation, and what we see in the painting is the content, that is, what is represented by the painting. Wollheim seems to endorse the view that the representational content of a pictorial representation like a painting is what one sees in the painting, that is, the scene represented. Wiesing, on the other hand, introduces an intermediary between the painting (the image-carrier) and the depicted scene (the image-subject) in the form of the image-object or representation, which means that when viewers see a painting, they see what the painting displays, that is, the image-object, and through this they see in the painting the depicted scene (the image-subject). When someone sees a painting, they are aware of both the image-carrier and the image-object, but there is a marked difference between these two. The former has a real physical presence, while the latter does not; it has an artificial presence but 'it is exclusively visible and is not subject to the laws of physics' (Wiesing, 2009, 35) and in this sense, even though the image-object is experienced as having visible qualities in space, it is not experienced as being in the same space as the one the viewer occupies. Philosophers try to capture this by saying that the image-object has an intentional existence, or it is a "possible object", as Haugeland puts it. In other words, it is, as I have designated it, an intentional entity, a representational content.

In my vocabulary, in which I adopt the standard designation of the representational vehicle as the representation, Wiesing distinguishes among representational vehicle (his image-carrier), the content of the representation or representational object (his image-object), and what the representation is about or the object of the representation

we have a perceptual representation, this representation attributes properties to something. This is the sensory individual, which may be different from what our perceptual reports are about.' The distinction between the sensory individual and the item that our reports concerns points to the distinction between the representational object and the object of the representation (the item the reports are about). Later on, however, we read 'When we look at a picture, we perceptually represent the depicted sensory individual. When we see an apple in a picture, we perceptually attribute properties (of, say, being red and being round) to the depicted ordinary object: the apple.' This may seem to be at odds with the previous statemen but this is not necessarily so. Nany may be alluding to the fact that even though we perceptually represent an object and this representational object is the content of our perceptual state, we attribute whatever properties we perceive to external object because visual perception presents this object as being a certain way (which is another way to express the transparency of experience. The fact that perceiver do so does not entail that the ordinary object that they perceive is the content of their representational visual states.

(the image-subject). Notice, however, that Wiesing does not designate the image-object as the representational content, as I do, but reserves the term "content" for the image-subject, which in my terms is the object of the representation. The representational vehicle and the representational object (in my account), and the image-carrier and the image-object (in Wiesing's account) both belong to the representing world, while the object of representation (in my account) and the image-subject (Wiesing) belongs to the represented world.

To recapitulate one of the main points of the discussion in this chapter, the distinction between the vehicle of the representation and the content of representations is especially important. Rarely in the literature on representations do the authors specify what they mean by "representation". A representation is a sign or symbol that refers to some entity in the world. As such, it is a physical or mental sign that refers to something else; in other words, it is a referential vehicle. In this sense, a representation is a vehicle. At the same time, the referantum of a representation is, generally speaking, the content of the representation. Beck (2019, 223) complains, justifiably, that some authors use "representation' to mean the content, and others use it to mean the vehicle of a representation. Things are complicated by the fact that although there is indiscriminate talk of the content of the representation, rarely is there a distinction made between the content of external, physical, representations, and the content of mental representations. This is unfortunate because there is a fundamental difference between these two sorts of contents. Physical representations refer directly to their represented domains and there is nothing that intervenes between the representation and the represented domain.

The signs on a thermometer, for example, indicate temperatures in the environment. Of course, to read off the representation, the human mind must intervene, but this is external to the representation. In mental representations (in which class I include external imagistic representations for the reasons explained above), in contradistinction, there is an intermediate between the representing symbol and the representatum, which is either the way the symbol shows the referantum (for pictorial and perceptual representations), or the way the referantum is described by the representation (for cognitive, sententially structured representations). In other words, between the representing symbol and the represented domain intervenes the mode of presentation of the latter in the former. This is called the representational content of the representation. In the case of a perceptual state in which one sees a colored object, say, a red cube, there is the brain state that realizes the representation, the mental state in which the perceiver is in as a result of having this brain state, the way the red cube is presented in the mental representation, and the red cube itself. The way the red cube is presented in the mental state is the representational content of the state *qua* representational vehicle. Thus, external representations do not have a representational content, even though some refer to the represented domain of external representations as the "content" of the representation, while mental representations do have a representational content.

References

Ayers, M. (2019). *Knowing and seeing: Groundwork for a new empiricism*. Oxford University Press.

Beck, J. (2019). Perception is analog: The argument from Weber's law. *Journal of Philosophy, 116*(6), 319–349.

Bermudez, J. L. (1995). Nonconceptual content: from perceptual experience to subpersonal computational states. *Mind and Language, 10*(4), 333–369.

Block, N. (2007). Consciousness, accessibility, and the mesh between psychology and neuroscience. *Brain and Behavioral Sciences, 30*, 481–548.

Block, N. (2014). Seeing-as in the light of vision science. *Philosophy and Phenomenological Research, 89*(3), 560–572.

Deacon, T. W. (1997). *The symbolic species: the co-evolution of language and the brain*. W. W. Norton & Company.

Dretske, F. (1993). Conscious experience. *Mind, 102* (406), 263–283 [Reprinted in Noë & Thompson. (2002). *Vision and mind*. MIT Press].

Dretske, F. (2000). *Perception, knowledge, and belief: Selected essays*. Cambridge University Press.

Dretske, F. (2006). Perception without awareness. In T. Gendler & J. Hawthorne (Eds.), *Perceptual experience* (pp. 147–180). Oxford University Press.

Everett, D. L. (2017). *How language began*. Liverlight.

Gallistel, C. R., & King, A. P. (2010). *Memory and the computational brain: Why cognitive science will transform neuroscience* (Vol. 6). Wiley.

Gärdenfors, P. (1995). Cued and detached representations in animal cognition. *Behavioral Processes, 35*(1–3), 263–273.

Gibson, J. J. (1979). *The ecological approach to visual perception*. Houghton Mifflin.

Godfrey-Smith, P. (2014). Signs and symbolic behavior. *Biological Theory, 9*, 78–88.

Haugeland, J. (1998). Representational genera. In J. Haugeland (Ed.), *Having thought* (pp. 171–207). Harvard University Press.

Hopkins, R. (1995). Explaining depiction. *The Philosophical Review, 104*(3), 425–455.

Hopkins, R. (1998). *Picture, image, and experience*. Cambridge University Press.

Husserl, E. (1898–1925/2006). *Phantasy, image consciousness, and memory (collected works (1898–1925)*. J. Brough, translation. Dordrecht: Springer.

Hyman, J. (2006). *The objective eye*. Chicago University Press.

Kaplan, D. (1989). Demonstratives. In J. Almog, J. Perry, & H. Wettstein (Eds.), *Themes from Kaplan*. Oxford University Press.

Kripke, S. A. (1980). *Naming and necessity*. Harvard University Press.

Kulvicki, J. V. (2014). *Images*. Routledge.

Lameira, A. R., & Call, J. (2018). Time-space-displaced responses in the orangutang vocal system. *Science Advances, 4*, eaau3401, 1–7.

Lyn, H., Russell, J. L., Leavens, D. A., Bard, K. A., Boysen, S. T., Schaeffer, J. A., & Hopkins, W. D. (2014). Apes communicate about absent and displaced objects: Methodology matters. *Animal Cognition, 17*(1), 1–18. https://doi.org/10.1007/s10071-013-0640-0

Millikan, R. (1989). Biosemantics. *Journal of Philosophy, 86*, 281–297.

Morin, O., Kelly, P., & Winters, J. (2020). Writing, graphic codes, and asynchronous communication. *Topics in Cognitive Science, 12*(3), 727–743.

Nanay, B. (2022). What do we see in pictures? The sensory individuals of picture perception. *Philosophical Studies*. https://doi.org/10.1007/s11098-022-01864-9

Papineau, D. (1987). *Reality and representation*. Blackwell.

A study of concepts. Oxford University Press.

Peirce, C. S. (1977, 1903). Logic as semiotics: The theory of signs. In J. Buchler (Ed.), *The philosophical writings of Peirce* (pp. 98–119) (1955). Dover.

Peirce, C. S. (1977). *Semiotics and significs* (C. Hardwick, Ed.). Indiana University Press.

Planer, R. J. (2021). What is symbolic cognition. *Topoi, 40*, 233–244.

Planer, R. J., & Kalkman, D. (2021). Arbitrary signals and cognitive complexity. *British Journal for the Philosophy of Science, 72*(2). https://doi.org/10.1093/bjps/axz018.

Planer, R. J., & Sterelny, K. (2021). *From signal to symbol: The evolution of language*. MIT Press.

Putnam, H. (1983). Reference and truth. In *Realism and reason: Philosophical papers, volume* 3. Cambridge University Press.

Pylyshyn, Z. (2003). *Seeing and visualizing: It's not what you think*. MIT Press.

Raftopoulos, A. (2019). *Cognitive Penetrability and the epistemic role of perception*. Palgrave, McMillan.

Raftopoulos, A. (2009). *Perception and cognition: How do psychology and neural sciences inform philosophy*. MIT University Press, a Bradford Book.

Raftopoulos, A. (2021). Précis of "cognitive penetrability and the epistemic role of perception." *Rivista Internazionale di Filosofia e Psicologia, 11*(3), 371–380.

Recanati, F. (1997). *Direct reference: from language to thought*. Blackwell

Skyrms, B. (1998). Salience and symmetry-breaking in the evolution of convention. *Law and Philosophy, 17*, 411–418.

Skyrms, B. (2010). *Signals: Evolution, learning, and information*. Oxford University Press.

Sterelny, K. (2003). *Thought in a hostile world. The evolution of human cognition*. Blackwell Publishing Ltd.

Wiesing, L. (2009). *Artificial presence* (N. F. Schott, Trans.). Stanford University Press.

Wollheim, R. (1993). *The mind and its depths*. Harvard University Press.

Chapter 3
Symbolic or Digital Representations

I discuss one of the two main kinds of representations, namely, symbolic or digital representations (words are typical examples of symbolic representations) that contrast with icons, pictures, or images that are examples of analog or iconic representations. My aim is to explain symbolic representations and discuss their main properties. Bringing to the fore the main properties of symbols will help us understand the profound changes in hominin brain structure and mentality required for symbolic representations to emerge and, mainly through, language shape our brain and behavior. The discussion concerning the main properties of symbols and symbolic systems, namely, their discrete and non-dense character, the fact that symbols refer through conventions, and the fact that symbols compose discursively by means of a set of established rules, is important for the aims of this book because these properties are necessary for understanding the cognitive capacities required to enable cognizers to have such representations. One, then, may turn to examining when in the evolution of our species and for which reasons these capacities first appeared. If symbols emerged from a background of indices and icons, to explain the cognitive novelty of symbolic cognition one should compare the properties of symbols with those of icons and pinpoint those differences between these two representational forms that require that novel cognitive capacities develop in order for symbolic thought to become possible.

Before I proceed, I must clarify a terminological problem. Peirce, as we saw, distinguished among indices, icons, and symbols. In the modern literature, however, one finds a distinction between iconic symbolic representation and digital symbolic representations. The reason is that both iconic and symbolic, in the traditional sense, representations consist of symbols. Both iconic and digital representations involve symbols since every representation is a symbol of some sort. A picture of a cat, which is a pictorial, iconic representation, is a symbol standing for, or referring to, a cat, and "Cat" is a digital symbol for cats. For this reason, Blachowicz (1997), Goodman

A. Raftopoulos, *The Origins of Representations: Cognitive and Brain Development*, Studies in Applied Philosophy, Epistemology and Rational Ethics 73,
https://doi.org/10.1007/978-3-031-91213-9_3

(1984), Haugeland (1987, 1998), Jackendoff (1987) and others, draw a distinction between digital or linguistic-like symbolic representations, and iconic or non-linguistic symbolic representations to substitute for the symbolic/iconic distinction. Goodman (1984, 16), to give an example, distinguishes between *digital linguistic* symbol systems and *analog nonlinguistic* symbol systems, noting that the difference between these two types of symbolic systems lies in the type of processing mechanisms characterizing the two systems. Even though in the past I have accepted this revised vocabulary, in this book I retain the standard distinction between "icons" and "symbols", the reason being that outside philosophical circles and in all other disciplines the standard icon/symbol distinction is being employed. The exception is when I discus Goodman's seminal work on representations; in that context I have to adopt Goodman's vocabulary. The reader should bear in mind that Goodman's digital symbolic representations are Peirce's symbols, while iconic symbolic representations are Peirce's icons. In general, whenever the term "digital symbol" occurs, the reader could substitute it with "symbol".

According to Goodman (1976), a symbolic representational system is digital (in the terminology adopted here, a system of signs sign is symbolical) if it contains discrete symbols, that is, signs that refer through a convention. A digital thermometer that represents temperatures and consists in a series of numerical values shown on a screen is a typical case of a digital representational system because it contains discrete symbols (the numerical values, *qua* rational numbers, are discrete symbols), and these symbols refer to temperature by convention. What characterizes a digital symbolic system is that it refers by means of symbols functioning in a syntactically and semantically non-dense, that is, discrete or differentiated, way. A digital symbolic notation is discrete or differentiated, according to Goodman (1976, 148–152), if it is semantically and syntactically disjoint, as opposed to semantically and syntactically dense. A representational system is differentiated if 'for every two characters K and K′ and a mark m that does not actually belong to both, determination either that m does not belong to K or does not belong to K′ is theoretically possible' (Goodman, 1976, 135–136). Goodman's qualification that a symbolic representation is digital if it contains discrete symbols that refer by convention is justified because if containing discrete symbols was sufficient to qualify a representation as digital, pictures should be deemed digital because they, too, contain discrete symbols (the parts of the picture) and this would be for the majority of researchers counter-intuitive. However, the discrete symbolic representations of which a picture consists, that is, the parts of the picture, refer not by convention but because they physically resemble their representata and this differentiates them from, say, digital thermometers whose parts, the numerical values, represent by convention. As we shall see, things get more complicated because there are representations that nowadays are considered to be iconic and whose parts refer by convention. Think of a sketch that portrays the configuration of a tank column by means of line segments arrayed in space in a way that mimics the actual spatial arrangement of the tanks. The line segments do not represent tanks by physically resembling them in some way, but do so conventionally. I discuss cases like this in the next section and the next chapter.

Jackendoff (1987, 181–182) holds that digital representations, which he calls algebraic representations, represent their objects by means of arbitrary symbols that have no bearing to the physical appearance of the objects, the way the linguistic symbol "Cat" bears no bearing whatsoever to the animal it represents. In algebraic representations, spatial relations are formal relations between pairs of symbols and, thus, have a formal character that does not distinguish them from other representations that happen to have the same formal character; for example, the algebraic representation x *is to the left of* y is parallel in formal character to the representation *x possesses y*. Finally, in algebraic representations, 'one can only list some finite number of stipulated distances ("Point A is x units from point B") or compute new ones by trigonometric principles such as the Pythagorean Theorem.'

Haugeland (1998, 78) claims that a digital representational system is differentiated in the sense that for any candidate token, m, and for at least one of any pair of types, K and K′, in the disjoint set of types that purports to be digital, it is theoretically possible to determine that the candidate token m is not a token of that type. This means that for any m, m is never equivocal between two distinct types since one of the types can be positively ruled out. Haugeland (1998, 78) explicitly defines as digital any system or devise that meets the following conditions: (i) it consists in a set of types; (ii) there is a set of feasible procedures for writing and reading tokens of those types; (iii) a set of suitable conditions that determine how tokens of the types can be operated upon is specified, such that, under those conditions, the procedures for writing and reading of the types are positive and reliable. By a positive procedure, Haugeland (1998, 77) means a procedure that can succeed absolutely and without qualification, in the sense that there is no approximation involved. To use Goodman's account, a token *m* either is not K or is not K′; there is no approximation involved. A procedure is reliable if under certain conditions, it can be counted on to succeed virtually any time, that is, it can be counted on to either produce or read, that is, identify, correctly any token of any type of the digital system.

What often goes unnoticed in the literature is that the definition of digitalness in terms of non-density or differentiation can be applied only to representational systems and not to individual representations even when the latter consist of parts (that is, have an internal structure). The individual arithmetic values of a digital thermometer that represent specific temperatures are neither dense nor non-dense, since the definition of density requires that there exist at least two different representational values/types, which means that density can be defined only with respect to representational systems that consists in these types *qua* parts. At the same time, each numerical value is a discrete entity since it is type-differentiated, that is syntactically differentiated, from the other numerical values. This definition, thus, gives a clear explanation why the so-called digital thermometers are correctly characterized as digital. As we shall see, the term "dense" refers to *linear orders*, which are sets endowed with the relation "less than" that satisfies some axioms (transitivity, etc.) A dense linear order is one in which between any two elements there's another one. Thus, the definition of digitalness in terms of density can be applied only to sets of representations, that is, to representational systems. Let us see why this is so besides the mathematical restriction presented above.

Individual representations can consist of parts; pictures consist of parts, and so do sentences. How does non-density or differentiation apply to them? Let us start with representations that are standardly deemed to be non-digital but iconic, such as pictures. The application of the abovementioned definition of digitalness in terms of non-density should entail that pictures are not digital symbols because they do not satisfy the definition. One can approach this from two perspectives. The first is to take the picture of a visual scene and think of the various changes that have occurred in it in the past, or may occur in it in the future, and consider the set of the pictures that represent all these different versions of the visual scene, or if one thinks that each change creates a new visual scene, consider the set of the pictures that could result from various changes in a certain visual scene. Let us say that the visual scene consists in a house and a tree and suppose that the house one month ago was red1 but now is has been repainted red2 (a different shade of red), and think of the two pictures that represent the two versions of the visual scene or the two visual scenes. This is a representational system since it is a determinable entity which by itself represents nothing specific; only a specific picture is a determinate value of the determinable that represents a particular actualized visual scene from the set of the possible scenes that the whole system represents.

Is this set of individual representations dense or non-dense, and, hence, digital or not? If to answer this question one applies the definitions of density discussed above, one has to determine whether between two different representations of the visual scenes, that is, between two different pictures, there exists a third one that is an intermediate picture of an intermediate visual scene. If the house in our example was painted red* that is a shade between r1 and r2, could there be a picture of the red* painted house? The answer is yes, which means that the set of the pictures we are discussing forms a dense set and, therefore, such a representational system is not digital. The same considerations also apply to the representational parts of which a picture consists. Take, for example, the specific color of the house, say red1. It does not make sense to ask whether this representation is dense or not, but it makes sense to ask whether the representational system that consists in the different values that redness could take is dense or non-dense; the answer, is that this system is dense. From this perspective, to determine density one has to move from pictures to a set of pictures, that is, from individual representations to representational systems and apply the definition of density and the result is that the set of pictures is dense and, thus, it is not a digital system. This means that the definition of digitalness in terms of non-density gives the correct answers that the traditional accounts of iconicity and digitalness would expect but only with respect to representational systems.

However, the question can also be, and this is the other perspective from which one can approach the problem, whether the definition of digitalness in terms of density could be applied to a picture as an individual representation. If someone opts for this, they have to determine whether the parts of which the picture consists form a dense set. This would be the case if between any two parts of the picture there is always a third part. But what is a part of a picture? A picture *qua* vehicle consists of pixels and obviously these form a non-dense set, which means that the picture should be deemed to be a digital representation. Considering a picture as the representational content of the vehicle, that is, considering what the picture shows, say a visual scene, one notes that this content is not dense as well, since a visual scene contains individuals who form a non-dense set and, thus, it is not the case that between any two individuals there is always a third. It follows that the picture is a digital representation. There is almost no one who would argue that a picture is a digital representation, although, as we shall see, it can be digitalized. It follows that the definition of digitalness in terms of density can be applied only to representational systems and not to individual representations. Note that individual symbolic representations may consist of parts too; sentences, for example, consist of words, which, in turn, consist of letters that can be thought of as the atomic constituents of the words. Individual representations, thus, can be complex symbols.

The preceding discussion concerning the discrete and non-dense character of symbolic representations, as well as the fact that symbolic are assigned referents through conventions is important for the aims of this book because, as we shall see next, most of the properties of symbolic representations emanate from these two traits of symbols. These properties of symbols, in turn, are necessary for understanding the cognitive capacities required on the part of cognizers to enable them to have such representations. Having determined these capacities, one may turn to examining when in the evolution of our species and for which reasons these capacities, which constituted cognitive novelties first appeared.

The use of discrete or differentiated symbols to form a complex digital representation imposes several properties on digital symbolic representations or symbols *simpliciter*. First, since symbolic systems consist of discrete symbols, symbolic systems are necessarily not dense but discontinuous (although we shall see that this is not mathematically speaking correct); symbolic systems are differentiated. This means that between any two symbols K and K' in a symbolic system there is not an intermediate symbol K*. If K and K' represent entities O and O', and between these two entities there exists a third entity O* that is represented by K*, this entity cannot be represented in the system because the system does not include the symbol K*. Take digital thermometers. Since digital representations are differentiated and not dense, any replacement of an analog by a digital system involves some sort of deletion. 'For example, to use a digital thermometer with readings in tenths of a degree is to recognize no temperature as lying between 90 and 90.1 degrees' (Goodman, 1978, 15).

Second, since symbols refer only through some convention, any composition of such symbols to form a complex symbol refers through conventions and does not bear any other relation to its referendum. Consider the symbol/concept 'CAT'. 'CAT''s structure is that of a simple concatenation of the less complex symbols "C", "A", and "T" that themselves refer by convention to letters in the alphabet, and, so, no part of 'CAT' refers to cat body parts or to their features, and there is no natural correspondence between the constituents of the representing symbolic form and the constituent parts of its content, that is, and the constituent parts of a cat. 'The word dog, in contrast (to iconic representations, *my comment*) contains no information about ears or any other part of a dog.' (Carey, 2009, 135).

Third, the discreteness of symbolic representations is important for their function as concepts and, thus, for the foundation of thought, which presupposes concepts: ‘the notion of an atomic discrete system is the basis … of all systems of thought … for which a notation is available.’ (Pylyshyn, 1984, 51). This is so because concepts are in a sense “on-off” switches, the “on” meaning that a token belongs to a certain type (concept), and the “off” meaning that the token does not belong to that type. So, for any token m of some type it is theoretically possible to determine whether m belongs to the type L or does not belong to L; Digital symbolic representations allow this because they are differentiated in the manner defined by Goodman. Thus, ‘differentiated linguistic symbols (composed of differentiated characters) are like conceptual switches.’ (Blachowicz, 1997, 65).

Fourth, any successful implementation of a discrete, symbolic representational scheme (notice that I moved from discussing individual representations to discussing representational schemes) requires that each token syntactic state of the implementing representational medium, that is, each token representational vehicle, be reliably typed as an instance of one and only one representational type. This is because discrete symbolic representational systems do not allow for meaningful states ‘between’ or ‘within’ those identified in the scheme (Recall Goodman’s ‘differentiation criterion’ for digital systems). Consider the set of natural numbers represented by a written notational representational system. A written notation, say the numeral ‘3’, is a token syntactic state of the representing notational system. This token is typed as an instance of one and only one representational type, to wit, 3 that is instantiated by various tokens ‘3’. It cannot belong to other types (say 3.01, or 2.99) because between types 3 and 4, and types 2 and 3 of this notational system no other type is permitted as a meaningful representation. Contrast this with a system representing magnitudes. Assume K to be the result of a measurement of a given length with the use of a typical ruler that falls in the vicinity of token 3 on the ruler. Given the inherent error in the measurement, K could indicate that the length is 3 cm long, or that it is 2.9, or 3.1 cm long. This is not possible in a digital measurement of the length that measures lengths in terms of natural numbers, where any token measurement belongs only to one type.

Initially, discussions about symbolic representations, following Goodman’s pioneering work, focused mainly on discreteness and reference by convention as the defining properties of symbols. However, the last twenty years or so there has been a shift of focus to the compositionality of the class of representations that are symbolic, with the main line of thought being that a representational system is symbolic if its representations compose by means of the combination of their parts, which, as we shall see next, conforms to the principle of ‘semantic constituency’, and the composition is discursive. As Planer (2021, 242) states, symbolic systems are combinatorial and discursive. One of the reasons of this refocusing is that there may be iconic representations that are non-dense and, thus, density cannot be used to differentiate between digital and iconic representations. The composition is governed by a set of rules that constitute a syntax. Complex symbols are, thus, syntactically structured. A clear example of this is the various natural languages and symbolic logic. In both cases, complex symbols are formed by putting together individual symbols following the rules of the grammar of the language, or the rules of symbolic logic. This sort of compositionality is discursive since a complex symbol thus formed may be used as an input symbol for the formation of a more complex symbol.

Haugeland (1998, 174) defines digital, or logical as he calls them, representations those that have a discursive compositional semantics in the sense that complex digital structures have a recursively specifiable syntax and determinate atomic constituents. This means that the semantic significance of the whole is determined by its syntax and the semantic significances of its constituents; this type of semantics is called concatenative or compositional semantics. The semantic significances of the constituents are fixed arbitrarily in the representational scheme, as are the rules of the syntax (they depend on the language at hand).

According to Fodor (2007), who echoes Haugeland (1998), digital, that is, symbolic, representational systems are discursive, logical, or, equivalently, propositional representations, that are compositional in a logical sense. A representation is *syntactically compositional* if and only if its syntactic analysis is exhaustively determined by the grammar of the relevant language together with the syntactic analyses of its lexical primitives. In other words, a representation is syntactically compositional if its formation follows the rules of some formal logic. A representation is *semantically compositional* if and only if its semantic interpretation is exhaustively determined by its syntax together with the semantic interpretations of its lexical primitives. Differently put, the meaning of a complex term is a syntax-dependent function of the meanings of its syntactic parts. More recently, Planer (2021, 237) expresses this as follows:

> On this view, to token a thought is to token a mental expression—a representation couched in "mentalese"—where the content of the whole is determined by the content of the expression's parts, together with their mode of composition.

This may sound intuitive but when it comes to natural languages it is wrong. Few would dispute that the meaning of a sentence is determined in a much richer way than by the meanings of the atomic terms and the syntax. Context, common sense, culturally established norms and knowledge, not to mention illocutionary forces and other speech acts (Austin, 1975), together determine the meaning of a sentence. The list of meaning-determinants does not end here when one considers spoken languages. One should add gestures, the pitch and loudness of the voice. Later on, I shall claim that grammar acts as a semantic filter that co-determines, in addition to the meanings of atomic constituents, the meaning of a complex expression, in accordance with semantic compositionality. In natural languages, however, culture is another semantic filter on what the meaning of an expression is.

Putting this reservation aside, and assuming that one is concerned with formal systems or with languages that allow for no ambiguity, metaphors etc., in which semantic compositionality holds, the semantic compositionality of symbolic representations entails that their truth values, are determined by the way their constituents are combined together according to their syntax, and the truth values of the constituents. This is usually defined as the "principle of semantic constituency".

Some philosophers, following Fodor, hold that a representation is digital iff it satisfies this principle (Werning, 2012). The principle states that if a representational vehicle *s* is a syntactic part of a representational vehicle *p* (recall that vehicles are situated at the syntactic level of analysis of a representation), then the meaning of *s* is a semantic part of the meaning of *p*. It follows that the compositional structure on the syntactic level is mirrored on the semantic level, the famous Fodorian parallelism between syntax and semantics that underscores all discussions of the *Language of Thought* (LOT).

As we have seen, symbolic representations are discursive and can be recombined the right sort of way. An important trait of symbolic representations is that their contents allow the representation of logical connectives and quantifiers. This, combined with the fact that symbolic representations have discursive syntactic and semantic structure, entails that states with symbolic contents can recombine in systematic ways to produce other new states. This makes them able to play the role of concepts and, indeed, conceptual contents are digital symbolic contents. Cognition is characterized by the recombinability and systematicity of its contents, which are the traits *par excellence* of thought (Fodor & Pylyshyn, 1988). Thus, the compositional structure of conceptual representations *qua* symbolic representations is the best explanation of the systematicity and recombinability of thought.

Having discussed the main characteristics of digital symbols, we are in position to address a problem that was lurking in our discussion from the beginning. As the reader recalls, it is a standard practice to assume that a symbol is digital if it refers to some entity through a convention. This convention establishes a link between the digital symbol and its referent so that the activation of the former invokes the latter. Thus, the basic idea is that a symbolic association is formed when we learn the convention and reliably pair a syntactic entity, be it a sound or a typed string, with something else. Thus, the relation between the signifier and the signified is a mere correlation between objects. As the reader also recalls, this is what an indexical association between an index and its referent means. Indices are involuntary objects and have a form that bears a physical, factual, or causal relationship with what they indicate. Similarly, when the association between a digital symbol and its referent is learned (notice that indexical relations have to be learned as well), the association becomes involuntary and a factual or causal relation is formed between the digital symbol and what it signifies, in the sense that the occurrence of the symbol causes the occurrence of the referent. In this sense, digital symbols function like indices.

As a matter of course, symbolic or digital representations and indexical representational systems are not equivalent and agents that process symbols have much more representational and expressive capabilities that agents that simply use indices (think of the cognitive differences between humans and many primates, for example). What is, then, their differences? I think that the main difference is that symbols form a system in which its members are interrelated. Eventually some symbols become concepts and concepts do not stand alone (barring Fodor, of course) but are engaged in various inferential and associative relationships. Indices do not have this property. This explains a significant difference between symbols and indices. For an indexical relationship to hold, there must be a correlation in time and place of the sign and its object. If the correlation breaks down, the association is eventually forgotten, and the indexical power of that sign to refer is lost. However, symbolic reference remains stable nearly independent of any such correlations. The reason is that the referential relation is sustained through the interconnections with the other symbols in the system. A symbol can refer to an object even if the physical association between the symbol and the object is very rare, or even an impossibility, as with unicorns. No indexical association could have been developed with so little correlation, or no correlation at all.

As a consequence of this difference, indexical associations are largely independent of one another. If one is lost, this has no, or almost no, repercussions for the other indexical associations; they are essentially independent. This is not the case with symbols, however. Symbols are incorporated into quite specific individual relationships to all other symbols in the system. Change in the use of one may require a restructuring of the whole system, as it happens in cases of strong conceptual changes. This is why symbols need to be in context with other symbols/words (a predicate term, for example, must have an argument and both together have a truth-value) in order to have any truth value and determinate reference.

Symbolic systems, thus, differ, and emerge, from a ground of indexical relations between objects when these indexical relations are organized so as to form a set of interconnected, through logical/conceptual connections, symbols. Unlike indices, digital symbols do not merely accumulate in unstructured sets of indexical relations, but they form a nexus of interrelated objects that is a hierarchic network of nodes and connections that defines a vast and constantly changing semantic space.

All these are closely related to Evans' (1982) *Generality Constraint* (GC), according to which mental states that have conceptual contents are systematic and recombinable in the sense that if one thinks that a is F and one possesses the concept for G, one can think that a is G. Or, if one thinks that a is F and one possesses the concept for object b, one can think that b is F. Finally, since thought is characterized by the systematicity and recombinability of its contents and since these two traits characterize conceptual/symbolic contents, the contents of thought are conceptual (Toribio, 2011). Note that GC may apply either to the mental representational states or (inclusive) to their contents. Fodor (1998, 2004, 2007), for example, who construes concepts as mental representations that are nomically related to items in the environment, and for whom concept possession has no epistemic consequences but only signifies that a person who possesses a certain concept is just having a mental representation of what that concept refers to, is committed to the view that GC applies only to representations (that is, the representational vehicles) that are syntactic entities physically realized in the brain, and not to their contents, which they do not have. Or, if one wishes to claim that such a representation has a content, but this content is not representational but some item (entity or event) in the world to which the representation refers, GC obviously is not applicable to these contents.

However, if one thinks, a view with which I agree, that concept possession has important epistemic consequences for the cognitive economy of those who possess them, then concepts are tied to the way representations present a worldly state of affairs as being, that is, to the modes of presentation of the representations, which constitute the representational contents of the representations. This applies clearly to perceptual representational states that present a worldly state of affairs as being such and such as well. Concerning symbolic representations, most notably sentential representations, the cognitive significance of the representation is determined by the proposition the sentence expresses, that is, the Fregean sense of the representation. The discussion concerning the satisfaction of the GC by symbolic representational vehicles that we discussed above should be extended, thus, to the digital representational contents of the representations.

The recombinability and the systematicity of concepts to form new thoughts (Beck, 2017; Peacocke, 1992) results from their being symbolic representations. The fact that digital or propositional representational systems are recombinable, productive, systematic, and satisfy GC, is made possible by two factors. First, the fact that symbols are context independent. To be able to ascertain that if one thinks that a is F and one possesses the concept for G, one can think that a is G, or, that if one thinks that a is F and one possesses the concept for object b, one can think that b is F, "a", "b", "F", and "G" must be definable independently of one another, and applied across different contexts. In other words, these symbols should have their reference values fixed in a context-independent way. Second, and relatedly since the second fact requires the first, the fact that if a representation is digital, it satisfies the principle of "semantic constituency."

As a foreword of what is to come, I argue in the next chapter that some sort of systematicity, recombinability, and productivity can also be found in iconic representations, but these properties differ significantly from those that characterize symbolic representations, owing to the fact that iconic representations do not compose in the discursive, formal way of the digital representations. Iconic representations are inherently compositional in that parts are combined or compiled to form a whole, and the representational value of the whole is determined by the representational value of its parts and the way these parts compose to form the whole. This presupposes that a semantics is possible for iconic representations, but this semantics is not determined by a set of syntactic rules of good formation, which means that iconic representations are not underwritten by a set of syntactic rules that determine well-formedness, but, rather, by the significance of the representational parts for the representing organism and the way they should be combined to form natural wholes. This is a matter which, as we shall see, is studied by Mereotopoloy and not by formal logic.

Complex symbolic representations consist of distinguishable discrete parts that are composed according to some rules of good formation, as dictated by some formal logic or some set of rules. Consequently, for any decomposition to yield constituents, it must follow the rules. As we saw, the syntactic structure of a complex symbolic representation is determined by the syntactic structure of its parts and the syntactic features that are used in the composition. A kind of composition is concatenative composition in which various symbols are put together to form a more complex symbol. The symbols 'C', 'A', and 'T', for example, are concatenated to form the more complex symbol 'CAT'. Another example of composition is the complex symbol p&q, which is syntactically composed, and whose truth (semantic) value depends on the truth values of p and q, and the way the logical connective '&' functions. This entails that not only should the constituent parts/types of a complex digital representation be differentiated, but that their modes of combination should themselves be digital types, which may be a set of logical rules, some formal grammar, a set of mathematical instructions as to how compose complex numbers or perform arithmetic acts, or any set of instructions, such as those that regulate game-playing. The demand that the modes of combination be of digital types means that the instructions to carry out the combinations do not allow any equivocation, and are both clearly readable and produced (to use Haugeland's (1998) account of digital systems.

Having discursive or logical syntactic structure means that some parts of the symbolic representation are constituents and others parts are not, depending on whether the decomposition that yields the parts follows the rules of the system. Since complex discursive/digital representations consist of distinguishable discrete parts that are composed according to some rules of good formation, as dictated by some formal logic or some set of rules, for any decomposition to yield constituents, it must follow the rules. 'Φ' is a constituent of the representation 'Φ(a)' but 'Φ ('is not a constituent. This is so for two reasons, one pertaining to syntax, and the other to semantics. 'Φ ('is not a constituent of 'Φ(a)' because it is not formed according to the rules of good formation as determined by formal logic (which dictate that there should be no unmatched parentheses). 'Φ ('is not a constituent because it does not have a semantic interpretation that contributes to the semantic interpretation of 'Φ(a)'. This shows that discursive representations are subject to canonical decomposition.

All these have repercussions concerning the compositionality of symbolic representations, but, importantly, they mark some basic distinctions between symbolic and iconic representations. First, through the use of discrete symbols, any discursive representation can be analyzed into distinct parts each of which can represent a unique object or property without representing anything else. Through concatenation, such representations can be combined to represent, say, a whole visual scene. Iconic representations do not have such a property, although they consist of parts too. Second, symbolic but not iconic representations support logical connectives. Third, symbolic but not iconic representations satisfy the *Generality Constraint*.

Some of the properties of symbols, as applied to language, are summarized by Deacon (1997, 32–33)

> A language-like signal would exhibit a combinatorial form in which distinguishable elements are able to recur in different combinations. It would exhibit a creative productivity of diverse outputs and a rather limited amount of large-scale redundancy. And although there would be a high degree of variety in the possible combinations of elements, the majority of combinatorial possibilities would be systematically excluded. In terrestrial examples, where it would be possible to observe the correlations between signals and contextual events, there should be another rather striking and counterintuitive feature. The correlations between individual signals and the objects and events that form the context of their production should not exhibit a simple one-to-one mapping. The correlations between sign elements and their pragmatic context should differ radically yet systematically from occasion to occasion, depending on how the signals are arranged in combination with respect to one another. These, of course, are all general features associated with syntax, though not just language syntax. Human games, mathematics, and even cultural customs exhibit these features.

Let me close the discussion of symbolic representations by returning to mental propositional representations to highlight their main characteristics, as they emerge from our discussion thus far, adding a new twist. Mental representational states have representational contents. Those mental states that are sentences or sentence-like entities have propositional contents that form propositional structures. Propositional structures are complexes of representational contents that are formed by means of various combinations of atomic propositional units that are performed by means of specific compositional mechanisms. The simplest and most fundamental such mechanism is predication, where an unsaturated, predicative part of thought is completed by a term that denotes some entity in the world (such as a proper name, for example) and becomes a complete part of thought, that is, a proposition that has some truth value (Frege, 1892, 34). (Recall that according to Frege predicates or concepts are functions that take as inputs objects or denoting terms and output a truth value). Rescola (2009, 177) defines predication as 'a compositional mechanism whereby terms fill the argument-places of a predicate that carries their denotation into a truth-value.' For example, the proper names "John" and "Mary" are denoting terms that when they fill the two argument-places of the 2nd order (relational) predicate "Love" form the complete term or proposition 'John loves Mary', which is true or false depending on whether John loves Mary or not. The basic structure of symbolic representations, therefore, is that of subject-predicate or noun-phrase, which I call the "R-relation". Building complex symbolic representations consists in the compilation of various R-relations binding objects to predicates, a compilation that composes the complex symbol out of its parts (the R-relations) according to some formal rules of good formation.

Propositional structures are, thus, characterized by the kinds of operations that can be performed on them. Camp (2018, 25–26) argues that the operations that underwrite propositional structures share four basic features.

1. The operations are highly digital, which means that they take a small number of discrete elements as inputs.
2. They are universal in that they can combine a wide variety of inputs. Predication, for instance, is a highly general operation taking many kinds of objects and predicates and delivering truth-values;
3. The combinatorial operations are asymmetrical. One element must be of a type that enables it to serve as input for the other; only a denoting term can fill the argument-place of a predicate but not the other way around, and this makes the predicate and the denoting term asymmetrical. This asymmetry means that the same terms combined differently yield different propositions, and also entails that a hierarchical structure is created.
4. The operations are recursive, that is, its' outputs can serve as inputs to the same operation. Thus, one can have sentences or phrases within sentences or phrases, or, in case of cognitive abilities, one can have thoughts within thoughts and, thus, one can think thoughts about thoughts. The recursive application of an asymmetrical relation produces an intricate hierarchical structure that contains

nested iterations of the same type of representation. For many researchers, deep hierarchical recursion is a defining trait of propositional structures.

Hierarchy is a fundamental characteristic of G2 and G3 languages and shows in the multi-tier organization of sentences in these languages. (There are languages, the G1 languages, that do not have this property.) Finally, hierarchical structure makes embeddings possible (embedding is placing one element inside another). For example, on may take 'my mother' and place it inside the phrase 'my mother's second cousin'.

It is undoubtedly true that formal systems such as symbolic logic and most modern languages have hierarchical and recursive constructions. However, to claim that all propositional structures display these properties is to claim that all languages have this complicated syntactic structure. This is clearly Chomsky's idea (for a recent statement see Hauser et al. (2002)) according to which recursion is the only specific pre-wired linguistic mechanism, and, therefore, all languages are by definition recursive. In Hauser's et al. (2002) paper, the recursive operation characterizing languages is *Merge*, which is a binary operation that combines two linguistic objects to form a larger object. It is a recursive operation because its output can be used as input to the operation.

However, this view is most likely wrong, as there are languages that do not have a recursive structure, since they do not allow for hierarchical and recursive constructions, and, in fact, do not have any syntactic rules (Deutscher, 2005; Everett, 2017); these are the 'Me Tarzan, you Jane" languages, the G1 languages in linguistic theory (G2 being the languages with hierarchical constructions but no recursive operations, and G3 being languages that display both and which are languages properly speaking if one follows Chomsky), in which sentences are merely conjoined together to yield complex stories without the use of any complex syntactic rules. '[S]ubortinating clauses, complex noun phrases, word-compounding and others, are not crucial for language and are later additions made for cultural reasons.' (Everett, 2017, 74). In addition, there are expressions in modern languages that lack syntactic structure and, yet, are perfectly intelligible because it is the context and common sense that determine the meaning (recall that semantic compositionality fails in natural languages); A typical example is the expression "you do the crime, you do the time."

One might object that even in these cases, the ordering of symbols is important, and this ordering surely must be determined by some sort of syntax. This is not necessarily so, however. Recall that according to Peirce there is a historical evolution from iconic signs to symbolic signs, or, in the vocabulary of this book, form iconic symbols to digital symbols. This means that at the interface of the transition some of the properties of iconic representations may have been retained in the digital representational system. Since iconic representations in their original function were resembling their representatum in some respects, they were reflecting reality. It is plausible that such a reflection of reality passed onto languages, and it also plausible that the languages at their first stages of development relied more on reflections of reality than on syntactic rules that took many years to develop. Linguists call reflections of reality in language "iconicity". Thus, at the initial stages of languages, or even in some modern languages that lack intricate syntactic structure, it is reality rather than a set of syntactic rules that determine the ordering of words and the ordering of sentences.

Considering that the sentences produced this way purport to represent reality, it follows that meaning determines the form of the representation, that is, its syntax. At the interface of iconic and digital representations one finds that digital representations display a basic property of iconic representations (which we discuss in the next section), namely that it is meaning that constrains the form and not the other way around, as is the case in mature digital representational systems. This means that the way a word behaves in a sentence, its syntactic properties, are determined by its meaning, which is determined in turn by the way the world is. Deutscher (2005, 214–224) argues that the ordering of the words was initially determined by some principles that echo reality and are ingrained in Homos' cognitive mechanisms that allow them to perceive the world. He proposes three basic mechanisms. Things that belong together in reality ought to appear side by side in language. The order in which events occur should be mirrored in the order in which events occur in language. The words that are included in a narrative are those that cannot be inferred from the context, and that the actor of an act comes first as s/he is the focus of the narrative.

The relation between 'symbols' and 'concepts' allows another way of characterizing symbols, by borrowing from the properties that concepts in their classical or standard form are thought to have. A concept, whatever entity this may be, is abstract in the sense that allows the categorization under its scope of a wide range of items that otherwise can be very different; the concept abstracts away all differences among some items and categorizes them together as belonging to the same class of items on account of some defining core characteristic that they all share (whether it be a chemical composition, a range of spectral reflectance's, etc.) The concept 'apple', for example, applies to many different sorts of fruits that may differ in shape, size, color, texture, taste etc., but owing to their chemical composition belong to the category of apples. Let us assume that at the syntactic level of analysis, the symbol 'A' stands for apples. Each time an apple is encountered, it is designated by A, and whenever an A appears, an apple is being designated. Suppose that one wants to represent a rotten apple. To do that, one should use A and a symbol that designates 'rotten', say 'R'. A rotten apple is RA. If one wishes to add more properties to the apple, one should use other symbols all built around A, because it is A that grounds the representation on apples, that is, allows us to understand that rotten and all the other properties are about apples. This means that the context in which an apple appears is built around A, by using symbols for the other items or properties that constitute the context that surrounds, as it were, A. In this context, in any context in fact, the category/concept 'apple' retains its autonomy and no matter the context, A always represents 'apple' and 'apple' is represented by A. Applying this to the symbols involved, one could say that the context in which a symbol appears is built around the symbol, and the context neither affects what the symbol denotes, nor does it change the symbol to ensure successful reference; symbols are context independent entities. We shall see that this does not apply to iconic representations, which are context dependent.

However, one must also note that a symbol X represents, or stands for, Y in a certain context C. "Cat" refers to cats in English (C1) but in French, a different context (C2), it is the term "Chat" that refers to cats. This follows directly from that fact that X represents Y by fiat, that is, through some sort of convention, and conventions may differ from one context to another. This seems to create a paradox. In the previous paragraph, I pointed out that symbols are context independent and now I just argued that digital symbols represent something within a given context. However, the paradox is dispelled if one distinguishes between the context in which a symbol acquires its meaning, by referring to some item, and the contexts in which this symbol is used after it has been determined what the symbol stands for. The first context is a constitutive fact because it determines what the symbol stands for, while the other contexts are the contexts in which the symbol is used. The first, constitutive context is not truth-evaluable, while the others are because the attribution of the symbol in them may be correct or wrong.

References

Austin, J. L. (1975). *How to do things with words* (2nd ed.). Harvard University Press.

Beck, J. (2017). Marking the perception-cognition boundary: The criterion of stimulus-dependence. *Australasian Journal of Philosophy, 96*(2), 319–334.

Blachowicz, J. (1997). Analog representation beyond mental imagery. *Journal of Philosophy, 94*(2), 55–84.

Camp, E. (2018). Why maps are not propositional. In A. Grzankowski & M. Montague (Eds.), *Non-propositional Intentionality* (pp. 19–45). Oxford University Press.

Carey, S. (2009). *The origin of concepts*. Oxford University Press.

Deacon, T. W. (1997). *The symbolic species: The co-evolution of language and the brain*. W. W. Norton & Company.

Deutscher, G. (2005). *The unfolding of language: An evolutionary tour of mankind's greatest invention*. Picador.

Evans, G. (1982). *The varieties of reference*. New York, N.Y: Oxford University Press.

Everett, D. L. (2017). *How language began*. Liverlight.

Fodor, J. A. (1998). *Concepts: Where cognitive science went wrong*. Oxford University Press.

Fodor, J. A. (2004). Having concepts: A brief refutation of the 20th century. *Mind & Language, 19*, 29–47.

Fodor, J. (2007). The revenge of the given. In B. P. McLaughlin & J. Cohen (Eds.), *Contemporary debates in the philosophy of mind* (pp. 105–117). Blackwell.

Fodor, J., & Pylyshyn, Z. (1988). Connectionism and cognitive architecture: A critical analysis. *Cognition, 28*, 3–71.

Frege, G. (1962), (1892). Begriff und gegenstand. In G. Patzig (Ed.), *Funktion, Begriff, Bedeutung*. Vandenhoek and Ruprecht.

Goodman, N. (1978). *Ways of worldmaking*. Hackett Publishing Company.

Goodman, N. (1976). *Languages of art* (2nd ed.). Hackett.

Goodman, N. (1984). *Of mind and other matters*. Harvard University Press.

Haugeland, J. (1987). An overview of the frame problem. In Z. Pylyshyn (Ed.), *The robot's dilemma: The frame problem and artificial intelligence*. Ablex Publishing Company.

Haugeland, J. (1998). Representational genera. In J. Haugeland (Ed.), *Having thought* (pp. 171–207). Harvard University Press.

Hauser, M. D., Chomsky, N., & Fitch, W. T. (2002). The faculty of language: What is it, who has it, and how did it evolve? *Science, 298*, 1569–1570. https://doi.org/10.1126/science.298.5598.1569

Jackendoff, R. (1987). *Consciousness and the computational mind*. MIT Press.

Peacocke, Cr. (1992). *A study of concepts*. Oxford University Press.

Planer, R. J. (2021). What is symbolic cognition. *Topoi, 40*, 233–244.

Pylyshyn, Z. (1984). *Computation and cognition*. MIT Press.

Rescorla, M. (2009). Predication and cartographic representation. *Synthese, 169*, 175–200.

Toribio, J. (2011). Compositionality, iconicity, and perceptual nonconceptualism. *Philosophical Psychology, 24*(2), 177–193.

Werning, M. (2012). Non-symbolic compositional representation and its neuronal foundation. In M. Werning, W. Hinzen, & E. Machery (Eds.), *The Oxford handbook of compositionality* (pp. 633–724). Oxford University Press.

Chapter 4
Analog or Iconic Representations

In this chapter I turn to discussing the properties of icons. I examine iconic or analog representations highlighting their main properties. I concentrate on their significant difference from symbolic representations concerning both the way icons refer, and the way iconic representations compose to form more complex iconic representation. I also attempt to elucidate the notion of iconic representations because the discussions in the literature tend to be, to say the least, chaotic and perplexing. Researchers usually fail to distinguish between external and mental representations, between representational vehicles and contents, and between individual representations and representational systems, Lyons (2022) *being a notable exception. Before proceeding, I discuss the reasons why icons, qua representational signs, succeeded the non-representational indices, that is, what made possible the emergence of representations in the first place. Then, having examined the properties of icons, I turn to addressing the question of when hominins acquired the cognitive arsenal that allowed them to acquire the capacity to understand and employ iconic representations, and also the question why was the development of these capacities necessary, evolutionary speaking. This will allow us to pinpoint the cognitive capacities that agents must possess in order to be able to form iconic representations of their environment.*

Among representations, symbols are contrasted to icons. In this chapter, I analyze the notion of "iconic representation" and brings to the fore the main properties of this kind of representations. Determining and understanding these properties serves two purposes. First, it allows us to pinpoint the cognitive capacities that organisms must possess in order to be able to form iconic representations of their environment. Following this, second, one can turn to addressing the question of when hominins acquired such cognitive capacities and why, and the question concerning the reasons that representational icons succeeded the non-representational indices, that is, what made possible the emergence of representations in the first place. In view of the fact that symbolic representations succeeded icons, one can also discuss, bearing in mind

A. Raftopoulos, *The Origins of Representations: Cognitive and Brain Development*, Studies in Applied Philosophy, Epistemology and Rational Ethics 73,
https://doi.org/10.1007/978-3-031-91213-9_4

the properties of symbolic representations, what is required for symbolic abilities to emerge.

Cognitive and perceptual mental and, thus, internal states are held by many (Ayers, 2019; Beck, 2017, 2018, 2019; Burge, 2010, 2018, 2022; Burnston, 2017; Carey, 2009; Crane, 2003; Cummins & Roth, 2012; Dretske, 1981; Fodor, 2007; Goodman, 1976; Heck, 2007; Haugeland, 1987; Jackendoff, 1987; Kosslyn, 1994; Peacocke, 1986, 2019; Sainsbury, 2005) to be cast in different representational formats, namely, symbolic and analog or iconic formats respectively. Icons also include external, representations, such as pictures, paintings, drawings, maps, analog thermometers, etc., while external symbolic representations include digital measuring devices, languages, etc.

Among the first iconic representation were the Paleolithic cave drawings, and, later, pictograms, in the form of the Neolithic Sumerian cuneiform and Egyptian hieroglyphics. Sumerian writing was first developed as a means of accounting material goods, such as sheep or grain. The quantity or number of sheep were incised in clay tablets or seals. This presupposes two sorts of signs; one for the number and the other for the sort of goods that the number counts. Since these seals were also used for foreign commercial exchanges along vast regions (Sumerians, for example, had trade outposts in middle modern Turkey) with peoples speaking different languages, the signs used should be such that people who did not speak, say Sumerian, could understand what goods were referred to in the seals or tablets. Sumerian may be different from Elamite, but the cows were the same and this is why the symbols for cows and other goods resembled the physical form of the cow so that everyone could understand the significance of the seal independent of the linguistic conventions of each civilization or local community. The pictograms resembled the physical form of their referents. Pictograms, and the signs for numerals, belong to the family of logograms, which are signs that represent a significant element of speech, such as a word or a part of a word that has a meaning on its own. This means that a logogram represents a part of speech that has a semantic value, that is, a referent.

This is in contradistinction to phonographs that represent nonsignificant elements of speech that, have, however a phonetic value. A typical example is alphabetic writing where each sign, a letter, has a phonetic value but no reference. A logogram signifies or refers to whatever object the word refers to and, thus, a logogram is a symbol of an identifiable object. The significance of the sign is independent of the phonetic value of the word and, thus, one can understand what a symbol means without knowing the underlying language.

Interestingly, the pictograms were not initially related to the names of the goods they referred to, the way 'Cat" is associated with the animal it refers to through the name of the animal and, thus, through the concept *CAT*. In this sense, pictograms referred directly, that is, without any conceptual intermediary, to some physical entity. This is why iconic representations were first associated with the notion of physical resemblance, which is grounded in qualities that representations and their representata share. Even recently, Planer and Kalkman (2019, 20) define icons as follows: A structure S is an icon for content C if and only if: (i) S is a signal for C; (ii) S stands in some physical resemblance relation R to C; and (iii) that S stands in R to

C is manifest to receivers. For well known, good, reasons (Crane, 2003; Goodman, 1972, 1976; Greenberg, 2013; Newall, 2011), the view that resemblance is necessary for iconicity no longer prevails. On the one hand, one may talk of iconic represenations even when these do not resemble their representata, and, more importantly, it is the consensus that resemblance is not even a sufficient condition for representation, since, for example, it is required that receivers be able to detect perceptually the resemblance in the first place. However, as we shall see, even in modern notions of "iconicity" there remains a grain of the older view since an iconic representation is iconic precisely because it captures in a natural way, that is, through some natural correspondence (Abell, 2009; Beck, 2017, 2018, 2019; Blachowicz, 1997; Burge, 2010, 2022; French, 2003; Haugeland, 1987; Heck, 2007; Jackendoff, 1987; Kulvicki, 2006, 2014; Maley, 2011; Peirce, 1906/1977) some aspect of the physical constitution of the items or arrangements of items it purports to represent, even if this correspondence is restricted to the spatial arrangement of these items or to some other relational property. This way, a faint trace of the resemblance requirement between the referring sign and the referred entity remains.[1] As we shall see when we discuss visual perception, the notion of "resemblance" comes back with some force, because even if it is true that visual perceptual representations are iconic in virtue of an appropriate natural mapping onto the represented visual scene, which is some form of isomorphism or homomorphism, as Haugeland (1998, 174) puts it, this isomorphism in perception is obvious enough to strike the viewer as resemblance. The representational content of a visual perceptual state resembles a visual scene in many respects pertaining to the qualitative features of the objects in the scene, their spatial arrangements, their relations, etc.

Even though pictograms are often referred to as iconic representations since they represent through some physical resemblance to the form of the representatum, there are some notable differences between pictograms and icons. The Sumerian pictogram for fish in its first form resembles schematically the shape of a fish, that is, it leaves out most of the details pertaining to fishes except its generic shape. In other words, the pictogram represents fishes with a high degree of abstract ion. This is to be expected, since it is supposed to represents all sorts of fishes and, to be able to do that, a high degree of abstraction from most of the fishes' various other attributes is required. An icon of a fish, say a picture of a fish, on the other hand, is a detailed (depending on the level of analysis of the picture, of course), representation of the fish in the sense that it captures several of its features other than the generic shape. For this reason, many think that pictograms are simplified versions of icons. One could also say that while an icon can represent anything (a person, for example) on some material, pictograms are used to represent words or ideas by illustration. This is why there are no pictograms of persons but pictograms of fishes.

[1] I do not have the space to elaborate on the problems associated with the resemblance case. Those who are interested in an illuminating discussion of resemblance and its role if defining pictorial representations should read Kulvicki (2014), who gives a very thoughtful consideration of Goodman's objections to resemblance as a necessary trait of pictorial representation.

At the same time, an icon can be used as a picture or as a pictogram. Consider a sign that shows an old-fashioned bus, a vehicle that is shortnosed, longish, with four doors, two sets of wheels on each side, etc.) This may be taken to be an icon of an old-fashioned bus, but it could also be taken to mean that there is a bus stops at this location, a bus that need not resemble the bus in the picture. In this latter case, the sign is not, strictly speaking, an icon but a pictogram since it represents the idea that this is a bust stop and does not represent a bus. This is the difference between a "pictorial" and a "pictogrammatic" reading of the sign (Kjørup, 2004). The difference comes down to the fact that the pictorial reading simply reads off the physical similarity between the icon and the bus, while the pictogrammatic reading also requires the use of certain conventions that go beyond physical similarities. For example, it requires knowing that there is public transport in the area, that bus stops are indicated by certain signs, etc.

Think of a sign consisting of a red circle with a diagonal red line that crosses an arrow pointing right. This sign, which is a pictogram, represents the instruction 'no turn right' in the driving code. One need not know some symbolic language to understand what the sign means; one gets its meaning by seeing the red diagonal line crossing an arrow; so, the meaning can come across irrespective of what language the driver speaks or knows. Still, to be able to read the sign, knowledge of some conventions is needed. One must know what a red circle and a red line crossing some symbol means. One must also know what an arrow pointing right in a highway means. Once one knows these things, the pictogram conveys the same information as a sentence reading 'no right turn is allowed', except that to understand the sentence one must speak English. Reading the sign goes beyond using one's visual system and grasping physical similarities, as one does when seeing a drawing of a cat.

There are, thus, pictograms, such as the cuneiform sign for fish, that are more like icons, albeit with some sort of abstraction, which do not require any knowledge of some convention(s) to be understood, and pictograms that are so abstract that understanding them presuppose knowledge of a set of conventions. However, if conventions are needed for understanding some pictograms or, even, some abstract icons in some artistic tradition, what is the difference between these icons and symbols that are traditionally distinguished from icons on account of the fact that they are conventional? The difference lies, as we shall see, in the way even very abstract icons are composed as compared to the way symbols compose. We shall discuss this in Sects. 4.8 and 4.9.

In the modern literature one finds discussions of iconic representations, of analog representations, and of pictorial representations. The relations between these forms of representations are intricate and depend on the perspective of the writer, which leads to a confusion as to the scope of each form of representation. Analog representations are ascribed many properties, such as density, continuity, undifferentiation (Goodman, 1976). They are also thought to carry many messages if they carry any, which means that they contain information packages rather than a single source of information (Burge, 2014; Dretske, 1981, 73; Quilty-Dunn, 2019), to be unit-free (Jackendoff, 1987; Peacocke, 1986, 2015, 2019).

Some of these properties, if not all of them, have been disputed as to whether they constitute necessary conditions for a representation to be analog. Even though, uncontroversially, many analog representations exhibit all these properties, a case has been made that many representations that should be construed as analog do not exhibit some or even all of these properties but they share another property which should be thus considered as the defining characteristic of analog representations (Beck, 2018, 2019; Kulvicki, 2015; Maley, 2011), this property being 'iconicity', whose definition will be discussed shortly. Others (Burge, 2010, 305; Blachowicz, 1997) claim that when representations do not exhibit all the traditional properties of analogicity it is because they are hybrid analog and digital representations, or, as Blachowicz (1997, 69) expresses it, they have both digital and analog elements. There are others who refuse to consider as analog representations those that are not dense and continuous even if they satisfy the condition for iconicity (Schonbein, 2014). In most accounts of analogicity, however, there is a property that characterizes all analog representations, a common denominator if you will, irrespective of which other properties these representations have, which is that analog representations exhibit some right sort of natural mapping (whose account may differ from author to author) to the represented domain; it is this natural mapping that renders a representation analog and none of the other properties that analog representation may or may not have.

Here, I employ a minimal conception of analog representations that equates them with iconic representations; it is a minimal sense because it posits iconicity as an essential, defining characteristic of analog representations independent of what other properties analog representations may have. A representation is iconic or analog if it has an inherent structure that maps naturally onto the structure of the represented entity (note that physical resemblance satisfies this condition) in such a way that some part, at least, of the semantic structure (that is, its properties and relations) in the represented domain that are relevant to the representing task, and the spatial structure of the represented domain (the way it is arrayed in space), are preserved in the representation. Iconic representations, thus, represent structural and semantic relations in the represented domain by instantiating these relations, rather than merely describing them the way symbolic representations represent a domain.

This instantiation can take place either in a physical, external medium that is itself arrayed in space and therefore can use spatial structure to represent structures arrayed in space, or in an internal brain representation that is not arrayed in space but has what is called functional spatiality, that we will shortly discuss. The represented entity can be an object, in which case the iconic representation may represent it through some mapping of shapes (similarity between the shape of the representing entity and the shape of the represented entity), or some other characteristic feature including its spatial location (as when a simple dot may stand for an object in a setting). Or, the represented entity can be a visual scene in which case the scene is represented (a) through some spatial mapping from the spatial structure of the representational content to the spatial structure of the visual scene, (b) through mappings from the attributives (that is, the representations of the features of objects in the visual scene) to the attributes in the visual scene (although it is not necessary that the mapping contain all the features in the visual scene), and (c) through a mapping of particulars in the representation to the objects in the scene.

4.1 Iconicity and Resemblance

It is undoubtful that the first iconic representations (paintings in the walls of caves from 67,000 to 40,000 years ago made by Neanderthals and *Homo sapiens* respectively) did resemble their referents in this way, so that by viewing them, perceivers brought into their mind the represented animal or scene. Many icons do resemble their referents; some pictures, some photographs, some paintings do exactly that. Visual perceptual representations, as a matter of course, do resemble aspects of the environment, namely those that are significant for the interactions of the perceiver with the environment. Thus, let us examine the notion of resemblance a bit closer.

Planer and Kalkman (2019, 20) define resemblance as follows:

> (Icon) A structure S is an icon for content C if and only if: (i) S is a signal for C; (ii) S stands in some physical resemblance relation R to C; and (iii) that S stands in R to C is manifest to receivers.

The motive behind clause (iii) is manifest. S is an icon for C only if the receivers can register in some perceptual manner the presence of S. It is also worth emphasizing that Planer and Kalkman (2019, 20) correctly note that the "manifestness" of the resemblance does not require that receiver forms any actual mental representations or ascends to the resemblance. "This allows for a resemblance relation between a signal and its content to be manifest to a receiver without him actually making use of this resemblance in interpretation or even noticing it." The detection at issue, in other words, can be purely perceptual and non-conceptual. If it were otherwise, only organisms with conceptual capacities could perceive icons and retrieve information from them.

If iconic representations are representations that map on to their respective domains in a way that preserves spatial or semantic structure, iconic representations resemble in some respects what they purport to represent. This resemblance ranges from strong visual resemblance to resemblance with respect to some abstract structure. This idea gains tract if one considers realistic pictorial representations, but, also cubic paintings; in both cases the drawing resembles what it represents but the resemblance ranges from faithful depiction to capturing abstract structure in the represented domain. The idea that underlying iconic representations is some sort of similarity between some aspects of the picture and the represented item that is experienced by the viewer, which entails that the representation resembles in some respects what it represents, is found in many accounts of pictorial or iconic representations.

This is clearly stated in Kosslyn's (1994, 5) discussion of depictive (remember depictive representations are a subset of iconic representations) representations.

> In a depictive representation, each part of an object is represented by a pattern of points, and the spatial relations among those patterns in the functional space correspond to the

> relations among the parts themselves. Depictive representations convey meaning via their *resemblance* (my emphasis) to an object, with parts of the representation corresponding to parts of the object.

Kosslyn (1980, 33) is more specific as to what the resemblance amounts:

> Images occur in a spatial medium ... the pattern formed in the spatial medium is a topographic mapping from the represented object such that (a) each local portion (set of contiguous points) of the image corresponds to a portion of the represented object as seen from a particular point of view, and (b) the interval [that is, the space between two points, my comment] relations among the positions of the image implicitly represent ... the interval distances among the corresponding portions of the represented object.

What is this functional space Kosslyn refers to? The general problem that led to the need to introduce the notion of functional space is that the demand that an iconic representation be mapped appropriately onto its represented domain so that it captures at least its spatial structure applies to mental images as well, but mental images are quasi-pictures because they are not arrayed in space, as literal pictures are, and, in addition, it is not quite clear in which sense one could talk of parts in non-spatial representations. Thus, one must find a way to introduce spatial structure to mental representations, or, to understand how mental representations could have spatial properties despite the fact that the representational vehicles are not arrayed in space.

Kosslyn's (1994, 5) answer is to introduce the notion of "functional space", according to which the space in which mental images appear need not be physical but can be like an array in a computer that specifies spatial relations purely functionally. This means that the physical locations of each element in a representation that represents a spatial array are not themselves arranged into a spatial array and it is only 'in virtue of how this information is read and processed that it comes to function as it were arranged into an array.' In such a functional spatial medium, distances are not defined as actual physical distances but are expressed in terms of the number of locations intervening between two locations. More about that in a while.

Jackendoff (1987, 181–182) explicates "resemblance" in iconic representations as follows 'In a geometric representation [Jackendoff's term for 'iconic representations'] objects are necessarily represented in terms of their shapes and apparent sizes... In a geometric representation multiple objects under simultaneous consideration are necessarily spatially related in distance and orientation.' Jackendoff also points out that in geometrical representations distances can be measured directly off the image without any specification of units. To the extent that objects in iconic representations are necessarily represented in terms of their shapes and apparent sizes, it follows that the relations among aspects within the representation are not arbitrary but depend on the relations between the aspects of the visual scene onto which they map. In this sense, an iconic (or an analog in the minimal sense) representation may be said to approximate the representatum (Peacocke, 1986).

Hyman (2006) and Hopkins (1995, 1998) make this proposal more specific. Hyman (2006, 148–149) thinks that a pictorial iconic representation represents some object by eliciting the recognition of this object on account of the fact that the picture allows viewers to get in touch with the occlusion shapes of the depicted objects through the shapes of the parts of the picture that purports to represent the object; 'If O is a depicted object and P is the smallest part of a picture that depicts O, the general principle can be stated as follows: the occlusion shape of O and the shape of P must be identical.' (Hyman, 2006, 81). Hopkins (1995, 443) uses the notion of outline shape and argues that if a part P of a surface resembles an object O in outline shape, appropriate observers see O at P; thus, pictures are experienced as resembling things in outline shape to various degrees of determinateness.[2] The reader should be reminded, however, that the sort of similarity required by iconic representations in general may concern purely relational attributes. An arrangement of matches in a sketch may represent a column of tanks if the spatial arrangement of the matches maps one to one (and in this sense it similar) to the spatial arrangement of the tanks. Since, in general, resemblance is always defined relative to some quality or set of qualities, if a picture P represents O with respect to quality Q, this means that P depicts or represents O as being Q, which, in the case of the tank column, means that the matches represent this column as having the spatial arrangements that the set of matches does.

The discussion on experienced similarities as underlying the property of iconic representations to depict their objects is in line with Ayers' (2019, 77–78) view that pictorial representations, which are a subset of iconic representations, convey information about what is represented 'by exploiting the content of the experience of seeing it, that is, how the object or scene looks from a particular point of view'. Note that this view of iconicity imposes a restriction on the resemblance relation between an icon and its representational content; this relation must have caused in the past, or cause in the present, the iconic representation to be a sign of its content. If a pictorial representation conveys information about what is represented by exploiting the experience of seeing the representatum, the resemblance between the pictorial representation and its content had played a causal role in establishing the representation as a sign for its content. Perceptual experience works by exploiting the resemblance, seen from a certain perspective and point of view, of course, between the perceptual representation and its representatum, since visual perception, in order to guide perceiver-environment interactions successfully, retrieves information from a visual scene in a way that preserves the main properties and relations, that is, the semantic structure, of the visual scene.

Almost fifty years earlier, Gombrich (1951, 220) had claimed that pictorial representations are records of visual experiences that act as substitutes for one's perceptual encounters with the depicted items. Many authors that work on imagistic representations agree with this view. Recognitional accounts of pictorial representation, for example, hold that pictures represent their objects the way they do because they enable viewers to recognize the objects the pictures depict in a way similar to the

[2] Kulvicki (2014, 57) defines *Outline shape* and *Occlusion shape* as follows: *Outline shape* is the pattern of angles that the rays emanating from an observer (considered as a point source) make as they touch the outlines of the object from that point in space. *Occlusion shape* is the flat pattern that the object forms on a planar surface that is placed perpendicularly to some line of sight cutting the rays that were passing through the space that the plane occupies.

one the viewers would use to recognize these same objects in ordinary perception; in other words, a picture P depicts an object O if P elicits viewers' capacities to recognize O. This is a clear case in which iconic representations convey information about their objects by exploiting the content of the visual perceptual experience, which renders them imagistic iconic representations. It is also true, however, that on several occasions one can see what a picture depicts without necessarily being able to recognize the *depictum*. It has been proposed, therefore, that a picture P depicts O if P elicits in viewers a visual experience as of O (which differs from recognizing O). Thus, imagistic representations are directly tied the experience of seeing. Even in the case of the set of matches that represents iconically a column of tanks by preserving the spatial arrangement of the tanks, this spatial arrangement is visually perceived.

However, this is not generally true. The perceptual processes at work in normal visual perception do, indeed, parallel the cognitive processes governing the construction and interpretation of pictures, paintings, and drawings that use the rules of linear perspective and linear geometrical projection. However, linear perspective in picture, painting, and drawing construction is not the only game in town. One could equally use, and for certain purposes it is advisable to use, parallel perspective and parallel geometrical projection (Giardino & Greenberg, 2015, 20). In a nutshell, in linear perspective, the projection lines (that is, the lines that start from the depicted object and terminate to the viewpoint) converge on a single viewpoint. In parallel perspective, the projection lines are all perpendicular to a single plane, and are, thus, parallel to each other, rather than converging to a single point. (It is evident now why visual perception uses linear perspective; the fact that the eye is at a distance from the source from which the light emanates, and is, usually, smaller in size that the items in the visual scene entails that the projection lines terminate to a single point rather than on a plane.)

The existence of two different perspectives, namely linear and parallel, which also corresponds to two sorts of geometrical projections from 3D surfaces to the 2D surface of the eye, namely the linear and the parallel projection, is important for another reason too. The content of a picture may vary depending on the perspective one adopted in drawing the picture or taking a photograph (using the appropriate lenses for each occasion). A picture of an irregular solid may represent a cube if linear perspective is used, but it represents an irregular solid if parallel perspective is used. The reason is that with linear perspective parallel lines tend to converge to the viewing point and, thus, the fact that the edges of the solid are not really parallel does not clash with the fact that cubes have parallel edges; simply put, parallel edges seem to converge. However, had one employed a parallel perspective, in which parallelism is preserved, the fact that the depicted solid does not have parallel edges entails that it is an irregular solid.

This has an important repercussion for the representative function of iconic representations, even for those that are purely iconic (like pictures, drawings, and paintings). If the content of a pictorial representation may vary depending on the way the representation is constructed, this entails that pictorial representations have determinate rules of interpretation and construction (in our example, the rules underlying linear perspective and projection, on the one hand, and the rules underlying parallel perspective and projection, on the other hand).

This means, in turn, and this is the upshot of this discussion, that iconic representations are relative to a system of rules, exactly like symbolic, and most notably, linguistic representations. This is very important because if iconic representations are relative to a system of rules, they also have (against, as we shall see, Fodor (2007)) syntactic and semantic structure. One might think that the dependence of iconic representations on a set of determined rules of construction and interpretation undermines the condition that purely iconic representations map on to their represented domains in a way that does not depend on any conventions because, one might think, these rules constitute a system of conventions. However, these rules (the rules of projective geometry, of perspective, and the rules underlying the state transformations that characterize perceptual processing) are rules determined by the geometry and physical constitution of our environment, and the physiology of our perceptual systems fixed by the dynamic interaction between human anatomy and our environment. There is nothing conventional about them.

Emphasis on the role of some system of rules for iconic representations is also put by Haugeland (1998, 174). Haugeland thinks (although he will revise this view later) that usually iconic representations are distinguished by the fact that they are somehow and to a certain extent isomorphic to their contents, that is, to the items they purport to represent. He adds that 'which isomorphism a given scheme employs is entirely arbitrary or conventional, but when the isomorphism is fixed, the content of a representational token is not arbitrary.' Thus, the choice of a parallel or a linear perspective is arbitrary (it depends on the use of the representation), but once a perspective is chosen the content of the representation is not arbitrary.

One could generalize from this to argue that in some sense and for a certain class of iconic representations conventions are necessary to fix the representational content of the representation. Consider for example an officer drawing on the sand a map in which the officer uses line segments to represent the arrangement of the enemy tanks in the field. For this map to work as a representation of the enemy tanks disposition, one must establish by convention that each line segment stands for an enemy tank. Once this isomorphism is fixed, however, the content of the representation is not arbitrary because the number of the tanks (line segments) and their spatial disposition must reflect (if the map is accurate, of course) the actual number and disposition of the enemy tank column. It follows that if one thinks that a representation is iconic if it maps in an appropriate natural and non-conventional way on its intended referent, the presence of conventions in securing understanding the representation does not undermine the characterization of the representation as iconic, in the same way that the non-dense character of a representation does not entail that the representation is not iconic. Suppose now that the sketch of the tank column contained instead of lines icons of tanks to represent the enemy tanks. In this case, no convention would be necessary to interpret the sketch because the natural, physical resemblance between the iconic of a tank and the actual tank would suffice to fox the referent of the sign. This means that conventions become necessary as the mapping between the icon and the referent becomes more abstract. When conventions intervene to allow understanding the icon, this means that conceptual elements intervene too and that purely perceptual processes do not suffice to secure understanding. This means, in turn, that the representation involves both purely iconic and symbolic elements (because concepts are symbols) and, thus, is a hybrid representation. The question, the, arises, as to the way the iconic and symbolic formats combine to form the hybrid representation. I shall address this problem when I discuss the compositionality of iconic representations in Sect. 4.8.

If one restricts pictorial representations to those that rely on linear perspective and linear geometrical projections (because our visual perceptual system employs them), the iconic or pictorial character of a representation is grounded in its adherence to the character of visual perceptual representations. One could take an opposite approach, however, and try to define iconic content independently of the format of perceptual content and then examine whether perceptual visual content meets the criteria of iconic content. Be that as it may, if a representation is pictorial because it exploits some relevant perceptual content, it follows that the components of a pictorial representation represent components of the represented domain because components of perceptual contents represent parts of the represented visual scene. One might go further and extend this requirement to all iconic representations.

According to Carey (2009), analog representations are iconic in the sense that their parts represent parts of what the representation as a whole represents, a condition that is analysed by Fodor (2007, 173) as the *Picture Principle*, which states that if P is a picture of X, then parts of P are pictures of parts of X. Kosslyn (1994, 5), as we saw, holds the same view about depictive representations. In all these views, iconic representations have a whole/part structure, that replaces the compositional/combinatorial structure of digital representations (I will return to the whole/part structure of iconic representations later). Quilty-Dunn (forthcoming, 4) calls it the 'Parts principle', according to which 'parts of icons correspond to parts of what they represent', whereas elsewhere (Quilty-Dunn, 2016, 47) calls it 'Iconicity'.

A similar view is held by Tye (1991, 44), who defines a representation R of an object O as an image if and only if it satisfies the following conditions.

> (i) every part of R that represents anything represents a part of O visible from V; (ii) a sufficient number of adjacency relationships among parts of O visible from V are represented in R: (iii) any apparent adjacency relationship among parts of O that is represented in R is represented in such manner that the parts of R representing those O parts are (literally) adjacent to one another.

Note that both Kosslyn (1980, 1984) and Tye (1991) think of the resemblance between the iconic representation and the represented item in terms of contiguity and interval/connection relations, which are topological rather than metric factors. Topological properties of a spatial order are its non-metric properties, and concern neighborhood and connection relations. As such, they do not include distances (which are metric relations).

This is the most natural reading of iconicity as resemblance between the representation and what is represented. However, the way I explained iconic representations above entails that a representation is iconic when there is a natural, structure preserving mapping between the icon and the represented object. Physical resemblance is a first-order similarity in which representational contents map through resemblance to physical properties in the represented domain, but the similarity may also be between structures (second order similarity). An arrangement of matches in a sketch may represent a column of tanks if the spatial arrangement of the matches maps one to one, and in this sense is structurally similar, to the spatial arrangement of the tanks. Since, in general, resemblance is always defined relative to some quality or set of qualities, if a picture P represents O with respect to quality Q, this means that P depicts or represents O as being Q, which, in the case of the tank column, means that the matches represent this column as having the spatial arrangements that the set of matches does. This sketch is considered to be an iconic representation of the column because it satisfies the sense of (structural) mapping that is required for iconicity.

4.2 Iconicity and Mapping

I referred to some sort of appropriate mapping from the representation onto the represented entity as the defining characteristic of iconic representations. This mapping is usually described as an isomorphism or homomorhism between the representation and the represented domain. An isomorphism between two domains A and B holds in general if the structure of A corresponds to the structure of B, which means that the relations among the elements of domain A can be put in a one-to-one correspondence to relations among elements in the domain B (in case of a homomorphism the relation can be one-to many). Mapping two structures does not require that the elements in the two domains be the same or similar, in which case the two structures share some abstract property (think of the points in the number line and the natural numbers the line represents). In other cases, the two structures may share elements, in which case the isomorphism also concerns concrete elements (think of visual representations).

Although detailed accounts of the exact nature of the sort of mapping that renders a representation iconic differ from one author to another, as we shall see all accounts more or less agree on the following: an individual iconic representation satisfies the following conditions:

(i) the representation maps onto the entity it represents in a natural, non-arbitrary way; what "natural" means has many interpretations that we shall address when discussing the various accounts of iconicity or analogicity, with a special emphasis on what this natural, non-arbitrary is when it comes to perceptual iconic representations;

(ii) the spatial and semantic structure of the represention maps onto, or corresponds to, the internal structure of the represented entity in such a way that the representation preserves (at least functionally) the spatial structure (that is, spatial relationships) in the represented domain. This means that the components of the representation are spatially arranged so that this arrangement mirrors the spatial arrangement of the components of the represented entity. For some representations the semantic structure, that is, the semantic relationships among the parts of the represented entity (such as relations of size or orientation) in the represented domain are preserved in the representation. In other iconic representations, some of the surface properties of the components of the representation (such as shapes or colors) correspond to the represented properties on the represented entity.

The second condition needs further elaboration. While mapping of spatial structure is clear, one need explain what is the semantic structure of the represented domain that the representation must preserve and why is the qualification "to a certain extent" needed. By semantic structure I mean, first, the distribution of qualitative features of items in the represented domain. If the represented domain is a visual scene, these items are the objects in the scene and the qualitative features are their properties. A perceptual representation of this scene, if accurate, preserves this aspect of the semantic structure to a certain extent. This last qualification is needed for the following reasons. The representation represents only those properties that are suitable for the representing organism's interaction with the environment. Humans and bats, for example, differ with respect to some of the properties of objects they extract from the environment. Moreover, a visual perceptual representation is from the perspective of the viewer and, thus, the properties represented are as perceived from that perspective. Thus, a visual representation of a visual scene from one perspective may be different from a representation of the same visual scene from another perspective. In addition, representations of the same scene, say drawings, may omit several properties in the scene depending on the intentions of the artist. Finally, other representations, say a sketch of the arrangement of the tanks in a tank column on a piece of paper may omit all the qualitative features of the tanks and preserve only their spatial arrangement with respect to each other and the elements of the terrain and, perhaps, the number of the tasks.

Semantic structure, second, comprises the relations among the features of the items in the represented domain, such as relative sizes, for instance. On certain occasions, the semantic structure also includes, third, the numerosity of the items in the represented domain, even in the form of a one-to-one correspondence between these items and the objects files that perceivers create for the items in a perceived visual scene. This makes sense in the case of perceptual representations, but, obviously, is irrelevant in other iconic representations that represent magnitudes, such as temperatures or time. Fourth, and this is important for characterizing the representation as iconic, the representation preserves the semantic structure of the represented domain in the sense that state transformations in the latter (say, an increase in size or in temperature) must be captured by analogous transformations in the representing medium (say, an increase in the height of mercury in regular thermometers), so that the mappings between elements in the representation (at the syntactic/vehicle level) and elements in the represented domain be preserved after the change in the represented domain. As we shall see, this demand imposes restrictions on the syntax, that is, the physical properties, of the representation, since they must allow state transformations that map onto transformations in the represented domain.

I have pointed out that the representations whose iconicity is discussed are individual representations and not representational systems. This is important because isomorphism at the systemic level does not warrant the conclusion that the individual representations involved are iconic. Recall our discussion of the digital thermometer; it is a representational system whose structure is isomorphic to the structure of temperatures in the environment since it preserves the ordering and arithmetic relations among the temperatures. This isomorphism withstanding, for reasons discussed in the preceding chapter, the representational system is a digital representation. It is individual representations that must have the requisite structure to map appropriately onto their *representanta* so that they may be deemed to be iconic representations. Depending on whether the representation is physical or mental, this structure may be at the level of the representational vehicle or representational content. In physical representations whose contents are the represented domains, the structure at issue concerns the representational vehicle. An iconic individual representational vehicle (whether it be a map or a picture) is isomorphic to the entity it represents; Lyons (2022) calls it "vehicle isomorphism."

A representation is isomorphic to what it represents if both the representation and the representatum instantiate relational structures that are isomorphic to each other. The representation has a number of components and those components bear various relations to each other, and all these relations together specify a relational structure. A similar situation holds for the representantum. The representation and the represented entity are isomorphic if and only if their relational structures are isomorphic. Since it is not required that every part of the *representatum* has to map to every part of the representation, *or vice* versa, in order for *the* representation to represent the entity, a homomorphism may also suffice for iconicity. On certain occasions, both the representation and the representatum have common some of their elements (as in photos where the colors and shapes in the photo may be the colors in the pictured entity), but this is not required, as in black and white photos.

In mental representations, in order for the representation to be iconic in the mapping sense, it has to have the requisite structure both at the vehicle and representational content level. Both the representational vehicle and the representational content must map appropriately onto the represented domain. This means that in iconic mental representations both the vehicle and the content must have internal structure. I expand on this problem when I discuss the problem of the internal structure of perceptual iconic representations.

Independently of whether an iconic representation is external or mental, the following are true. First, each individual iconic representation maps naturally on its represented item. Second, a system of such representations forms a representational scheme that appropriately and naturally maps onto the represented domain that consists of the individual entities that the individual representations represent. As we shall see, an iconic perceptual state maps onto a visual scene both at the level of representational vehicle and the level of representational content. It does that because both vehicle and content have an internal structure that is mappable onto the internal structure of the represented visual scene. In this vein, individual iconic perceptual representations naturally onto their represented entities (through, say, similarity of shapes or colors, etc.). Moreover, when they form a system, this system is an iconic representations that iconically represents worldly states of affairs, say visual scenes. Perception is such a system or rather capacity that is a representational determinable in the sense that its values, that is, each individual perceptual representational state (a determinate), represents a specific visual scene.

A related situation holds for external iconic representations. An amount of liquid (a representational vehicle) in an analog thermometer iconically represents a certain temperature because it is a magnitude that can be naturally mapped onto a temperature *qua* magnitude in the sense that when the temperature increases so does the amount of liquid in the thermometer in a systematic way. Let me expound on this sort of systematic relationship between the iconic representation and the represented domain. Maley (2011, 122) offers a covariational account in which a representational medium R of a domain Q (Maley talks about representations of numbers but his account can be extended to other domains as well) is analog just in case there is some property P of R such that the quantity of P determines Q and as Q increases or decreases by an amount d, P increases or decreases as a monotonic function of Q + d or Q − d.

This demand is clearly exemplified in Kulvicki's (2015) account of analogicity. Kulvicki also eschews the requirement of density and continuity for analogicity and argues that analog representations are those that bear a certain mirroring relationship to the domain they represent, a requirement that may be satisfied by non-dense representational schemes. Analog representations require structure preserving syntactic-semantic links (syntactic refers to the representing medium, while semantic refers to the represented domain) that result in representations with vertically articulated content. A representation has a vertically articulated content (Kulvicki, 2015, 173) when it represents objects as being P but also represents them as being Q, where Q is an abstraction from P. A mercury thermometer is such a representation because it represents temperatures and when it designates a certain temperature T1 through the measurement of some measured height of the mercury, it also represents indefinitely many abstractions from T1, that is, other temperatures that correspond to heights that are very close to the measured type that, as such cannot be discriminated from that height. Recall that measurements in such thermometers rely on estimations of the height of the magnitude of mercury and each specific height of mercury is a syntactic type. Since magnitudes are dense there is always a value of the magnitude between any two other values of the magnitude, and this means that the noise inherent in any measurement system renders precise readings, which would make the set of possible abstractions from a reading empty, impossible. No matter how careful a measurement of a magnitude is, it will always be an approximation, and, therefore, compatible with many values that fall within a range (modulo noise). This means that "no matter how careful a measurement is, there will always be indefinitely many syntactic types under which any specific height might fall." Kulvicki (2015, 168).

Kulvicki explicates this as follows:

> What is the content of the thermometer representation? One might just tease as much information as possible from it. Measure the determinate height, as carefully as possible, and state its associated temperature. This suggests that the thermometer's content is fully captured by

> a careful description of temperature, or even a name which indicates just that value. But stating the most specific temperature does not give voice to all of what the thermometer represents. The indeterminate temperatures—those ranges corresponding to abstractions over the specific heights—are also represented. The motivation for this claim is that there are syntactic features of the thermometer responsible for representing just such features. The thermometer's content spans levels of abstraction, and in that sense is highly vertically articulate.

Abstractions over syntactic properties (the properties of the representing medium, the mercury, for example) map readily onto abstractions over semantic contents (the temperatures) and so "the content is only articulable at spanning levels of abstraction." The existence of such syntactic-semantic links is what makes a representation analog. Kulvicki rightly points out that dense, smooth representational systems, to the extent that they make many abstractions available, exemplify analogicity; these systems are necessarily analog. At the same time, Kulvicki (2015, 176) points out that discrete representational systems are analog when abstractions over the syntax map readily onto abstractions over semantic contents, even if the syntactic details are determined, given that the system is discrete; in contradistinction to smooth representational systems that must be analog, discrete representational systems may be analog.

Linguistic expressions, that are typical digital devices, do not have vertically articulate content (Kulvicki, 2015, 173).

> Contrast this with a linguistic expression that represents a very specific temperature: 'the freezing point of water.' The temperature described is, in fact, colder than 40° and warmer than −40, but the description does not represent the freezing point of water in that way. Facts about abstractions from the most informative thing a linguistic expression conveys are rarely part of the representational contents of such expressions. Why? There are precious few abstractions over syntactic features of the description that are responsible for doing interesting semantic work. Isolate 'the freezing point of' from the rest and you find a part of the expression responsible for representing the freezing point, just as isolating 'water' gets you the water-representing bit. That's pretty much all you get by playing the abstraction game with the description. It represents no temperatures less determinate than the freezing point of water.

Having explained the "co-variation" relation, let us return to discuss the relation between iconic individual representations and the systems that consist of a set of such individual representations. A system of such vehicles, an analog thermometer, iconically represents the domain of temperatures of some entity. It may seem to strain the meaning of "iconic representation", but an individual magnitude of liquid in an analog thermometer is an iconic representation of some temperature. An analog thermometer is a system (and not a mere collection, which means that the components of the system are interrelated in a systematic way) of such individual representations. Being a representational system and not an individual representation, an analog thermometer does not represent anything but it is a representational determinable. The specific volumes of Hg comprised in the system represent specific temperatures; they are determinate representations. As I argued above, a specific volume of Hg is an iconic representational vehicle. The thermometer is an analog representational system on account of the fact that the individual representations of which it consists are iconic (remember that mappings and even isomorphisms at the scheme level do not suffice to establish the iconicity of the representation), and on account of the way the individual representations are interrelated in the representational system. In this particular case, the individual representations together form a dense system that can represent temperatures without any loss of information. (In contradistinction, symbolic representations suffer from loss of information because, as we saw in the previous chapter, they are syntactically and semantically disjoined, which means that there is no syntactic entity, that is, a symbol, in the representation to refer to some temperature that lies between two consecutive symbols in the representational system. This entails that this temperature cannot be represented in the system and this information is lost). This way ensures that the representational system is isomorphic to the domain of temperatures that it purports to represent.

The mapping account of iconicity presented in the preceding paragraphs is unproblematic with physical or external representations that are instantiated in some publicly accessible physical medium because such representations are spatially arrayed and, thus, have a spatial structure that can be mapped onto the spatial structure of the represented entity. Venn diagrams, for example, are semi-iconic representations of logical relationships among classes of objects, because by using them one exploits the spatial relationships between shapes, in this case, circles, to represent logical relationships between classes of objects. Let me first explain the qualification "semi-iconic". Venn diagrams include a non-iconic representative element, namely the shading of an area, which signifies that that area is empty. In order to be able to understand and use Venn diagrams to represent Aristotelian logic, one must know the convention that a shaded area is empty; this is a convention as there is no natural correspondence between a shaded area and an empty area and, thus, shading is not an iconic representation of emptiness. It follows that Venn diagrams are hybrid representations that involve both iconic and digital symbols. In fact, the shading of an area, or a cross that indicates that there exists an item in some area, are not the only digital elements in Venn diagrams; each circle must be accompanied by a label indicated the type of class the circle represents and this label bears a conventional relation to the indicated class.

Physical, non-mental iconic representation, such as photographs, pictures, paintings, etc., involve space essentially, since that they are themselves spatially arrayed. This does not mean, of course, that there is a spatial-structure preserving mapping from the representation to the represented entity concerning all spatial properties involved (objects in pictures, for example, are smaller than the objects in the pictured scene, and the distances between the objects in the picture are scaled down with respect to the actual distances), but, still, the spatial arrangement of the objects in the picture is the same (and does not merely correspond to) the spatial arrangement of the objects in the pictured scene. This means that the isomorphism involved is a "strong isomorphism" since some of the spatial relations among the components of a representation will not be merely isomorphic to the spatial relations among the components in the pictured scene, but the spatial relations among the components in the representation are literally shared with the components of the pictured scene. Since physical representations are typed and individuated only by their vehicles, it follows that being spatially arrayed is a property of representational vehicles.

Significantly, this also applies to neural states, which are the vehicles of mental representations, although the spatial relations involved are, as we shall see in the next paragraphs, functionally defined.

However, the spatial-mapping condition in the definition of iconicity poses problems for mental representations, such as perceptual representations because these representations purport to represent spatial arrangements (such as visual scenes) and, yet, they are not spatially arrayed, being mental entities. We shall return to see how theoreticians attempt to account for, and surmount, this problem.

The first clause of the mapping condition requires that the mapping between the symbol and the represented entity be *natural*. Although I shall explain in which way the mapping is natural with respect to visual perceptual representations, let me say here that, in general, a mapping is natural if understanding the representation does not rely on some set of conventional rules that are learned and are thus, the result of enculturation (such as linguistic systems), but on 'the degree to which human nature—including relatively universal aspects of cognition, physiology, social behavior, and environmental interaction—rather than enculturation, makes that system easy to internalize and use.' (Giardino & Greenberg, 2015, 8).

When, later on, I explain the naturalness of perceptual iconic mappings, I ground my arguments in the physiology of the human visual perceptual system and the interaction with our environment to claim that the mappings between the constructs of our perceptual systems and the visual scenes in the world are natural because of the aforementioned two factors, and do not rely on any conventions. The mapping between a perceptual iconic representation and the represented visual scene is natural because it does not rely on some learned convention but on the way our perceptual systems are structured and on the interactions with our environment.

Some iconic representations, however, do require some sort of enculturation in order to be understood and used. A sign with an airplane on the road may signify the direction to the airport, but this is understood only by those viewers who have experience of airplanes or have learned about airplanes even though they have never seen one. The relation between the sign and airplanes is natural in that the sign depicts the shape of an airplane, while, in contrast, the term 'airplane' is conventionally related to airplanes, but understanding and, hence, using, the sign is learned. It is true that by depicting an airplane, or, rather, the shape of a typical airplane, the sign is easy to internalize and use. Accounts of iconic representations that are based on some notion of a natural mapping between the representation and the representatum, therefore, should consider the role of convention in establishing the correspondence between the sign and the signified item in parallel with the role of enculturation in understanding and using the sign. In view of the trade-off between "naturalness" and "enculturation" one should talk of degrees of iconicity; some representations are more iconic than others. In view of the fact that several iconic representations, such as diagrams, are in reality hybrid representations because they necessarily involve digital information elements to denote relations, types of magnitudes etc., it makes sense to claim that a diagram with less digital elements is more iconic than a diagram with more digital components. Talk of degrees of iconicity, on the other hand, does not make sense in the case of pure iconic representations where all information is conveyed solely by the mapping of the representational vehicle and content onto the represented domain.

Returning to the view that a representation is iconic owing to the way it maps onto its represented domain, a way to discriminate between iconic representations and symbolic or notational representations, such as sentential representations, that is based on the mapping of the representational scheme onto the represented domain is offered by Haugeland (1987, 88–91). Haugeland assesses Palmer's (1978) distinction between notational systems (such as set of propositions) and iconic representations, according to which iconic representations represent properties and relations intrinsically, whereas digital systems represent them extrinsically. For example, in a digital representation of a domain in which an object is taller than another, the relation 'taller' must be explicitly represented by a distinct symbol. Imagistic representations (the term Haugeland uses for pictorial or quasi-pictorial representations), on the other hand, just show this relation and they do not need to represent it by importing an extrinsic element, such as a digital symbol. The reason is that in imagistic representations 'the representing relation has the same inherent constraints as the represented relation. That is, the logical structure required of the representing relation is intrinsic to the relation itself rather than imposed from outside.' (Palmer, 1978, 271). One could say that the intrinsic logical structure of the representing relations in iconic representations, which mirrors the structure of the represented relations, entails that all representationally salient information in the represented structure is explicitly shown in the representation. This is not the case with quasi-linguistic, sentential representations in which the fact that relations are represented extrinsically entails that there is a clear distinction between what is represented explicitly and what is represented implicitly, since inference is needed for that information to be extracted from the explicitly represented information.

Consider a map that depicts the spatial relations among three cities A, B, and, C and shows their distance sin terms of magnitudes. Now think of a set of sentences that express the same information, namely, A is x miles north of B and y miles northwest of C. In this system, the relation between B and C can also be determined but this would take some geometrical calculations and in this sense the relation between B and C is implicit in the system and must be extracted from the explicit information that A is x miles north of B and y miles northwest of C. In the map, however, no calculations are necessary; the relation between B and C is shown as a result of representing in the map the relation between A and B, and A and C. The distinction between explicit and implicit representations in the map collapses because just looking at the map one cannot tell which are the original representations and which is the resulting one. Haugeland (1987, 90) calls this property of imagistic representations 'complicity'.

For this reason, iconic representations are characterized by a unique property that radically distinguishes them from digital representations, namely, that in the former, representations of the side effects of a change are side effects of the representation of that change and, hence, are automatic and free (Haugeland, 1987, 91). Returning to the map example, if the represented relation between A and B changes, the whole map should be automatically updated and all the representing relations change as well. In fact, in perception or if one was taking a video of the scene, the update would take place instantaneously. This entails that there is a mapping relation between spatial relations of the elements in the map and the spatial relations between the represented cities, a mapping relation that allows for spatial changes in the represented domain to be automatically spatial changes in the representing medium. In the notational representational system consisting of individual digital representations, on the other hand, the explicit representations are independent one from the other and, thus, a change in any of them does not entail any change in the other; any such change must come from the outside, as it were, because someone should rewrite the sentences to express the new spatial configuration that resulted from the change.

Heck (2007, 127) points to the same direction noting that in iconic representations a change in the representation is automatically a change in what is represented, and that the effects of changes in the represented domain automatically show as changes in the representation, whereas in propositionally structured representations one has to make these changes manually. Underlying this difference is the fact that in a propositional representation of a map one first forms the propositions that describe parts of the map, by assigning properties and objects that the map purports to represent to specified locations in the map that denote locations of the mapped entity, and then one conjoins these propositions to get the propositional representation of the entire map. The conjunction of the propositions, however, fails to show the spatial relations among the assigned properties and objects. These relations must be added by using additional representational resources and by inferential derivation. Rescola (2009) calls the process by means of which propositional representations of maps are constructed 'predicative analysis' and uses the failure of propositional representations to show automatically the spatial relations among the elements of the map to argue that maps do not have predicative structure. Kulvicki (2015) thinks that maps predicate properties and objects to locations, but maps organize those predicates in a fundamentally different way than sentential vehicles, partly because maps exploit spatial structure and, thus, show the spatial relations among the representational elements of the map and there is no need to add distinct representations of these relations, as the predicative analysis of propositional representations of maps requires. I revisit the distinction concerning compositionality and predication in digital representations and iconic representations later.

The map in the example that Haugeland uses is iconic because it satisfies the mapping or mirroring criterion, since it mirrors what it represents in that it bears a structure-preserving relation toward what it represents, which in our case is the spatial structure of the cities A, B, C. Haugeland shows that a representation such as a map that bears a structure preserving relation to the represented domain differs from a series of descriptions that expresses the structure preserving relation because the representational format of the map allows it to update automatically its structure when a change occurs. Rescola (2009, 197–198), echoing Palmer (1978) relates the property of automatic updating to the fact that iconic representations show directly the relations between the elements of the representation and do not need a distinct representational symbol to represent the relation as propositional representations do; 'A map's geometric structure is not just another element to be listed alongside its markers and coordinates… the markers and coordinates *stand in* [italics in the text] geometric relations, and those relations have representational import.' So, when these relations change in the map, which is the representational vehicle and, thus, these changes are syntactic changes, they automatically correspond to the relevant change to how the mapped entity is represented as being. Therefore, there is no reason to perform some sort of inference to arrive at the conclusion that describes the change occurred: an active inference from premises to a conclusion in a predicate calculus is transformed to a "free ride" in iconic representations such as maps.

These considerations suggest that Lyons (2022, 6) is likely wrong to claim that since the truth, or satisfaction-condition, of a subway map can be given without loss or addition of information by a complex verbal description, this refutes the view that iconic representations have distinctive contents. The verbal descriptions do no satisfy complicity. Lyons confuses expressing a content and having this content. Iconic representations instantiate (functionally or actually) the spatial relations in the representatum and not merely describe them, as digital representations do. This is why the latter do not update automatically.

Allow me to elaborate on this because this view is important in understanding the cognitive capacities that enabled the emergence of iconic representations from mere non-representational signals, such as indices. This is the demand that the physical processes applicable to the representing medium be isomorphic to the state-transformation processes of the represented domain (Gallistel and Gelman (1992). Gallistel and Gelman (1992) in their discussion of the analog representations involved in various mathematical abilities of infants, a discussion that relates analog representations to the representation of magnitudes. They ask us to think of the formations of ink patterns that form the numerals "1', "2" and "3" with which we represent symbolically the numbers *1, 2, 3*. The physical processes that apply to the formations of ink that represent numbers are not isomorphic to the computational processes of mathematics, the represented domain. That is, the transformations that befit the representing medium are not isomorphic to the transformations in the represented medium. For example, when the addition $1 + 2$ is performed, the mathematical, computational transformations yield *3*, but there are no physical transformations of the ink patterns that correspond to addition and can yield the ink-pattern '3' that represents the number *3*. If someone added the ink used in '1' to the ink used in '2', they would not get the ink pattern '3'.

Consider now iconic representational systems of number, such as histograms, that use magnitudes and not digital symbols to represent numbers. The processes on the histograms, the physical processes applicable to the representing medium, are isomorphic to the computational processes of arithmetic, since the addition of numbers corresponds to the physical superposition of the histograms. To obtain the addition $1 + 2 = 3$, one can put the column representing *1* on the column representing *2*. Representational systems in which the applicable processes are isomorphic to the processes applicable in the represented domain are iconic systems.

In the previous example, we compared, on the one hand, the physical manipulations on the ink patterns of which the digital symbols of numbers consist (in the representing domain) with the computations performed on numbers (in the represented domain), and, one the other hand, the physical manipulations on iconic symbols (histograms) of numbers (the representing domain) with the computations performed on numbers (in the represented domain), and we saw that in the second, but not in the first, case there exists the Shepard-isomorphism required by analog representations.

The view that iconic representations are representations in which the processes that can be applied to their (syntactic) forms are isomorphic to the processes applicable to the transformations in the represented domain can be extended to account for the relationship between pictorial representation and perception. As we have seen, many authors suggest that pictorial representations convey information about what is represented by exploiting the content of the experience of seeing it, and that pictorial representations are records of visual experiences that act as substitutes for one's perceptual encounters with the depicted items. These traits of pictorial representations explain why one can get information from the depiction of an object that one would get from a perceptual encounter with that object.

Comparing pictorial representations of some item with a digital representation of the same item, one should substitute visual actions for the physical manipulations of the previous case. Suppose one perceives a picture of an object O, and, in addition, one reads a description of O that, let us assume, conveys the same amount of information about O. In both case, the viewer performs some visual actions; the viewer scans the image of O and, also, scans the text to read the description of O; both are visual actions. At the same time, the description my engender a vivid visual image of O, which means that imagery may also be involved. There is, however, an important difference that sets the two forms of representation apart. The visual action on the picture of O makes one think that one scans O, itself, or the substitute for O in the picture. When one scans the verbal description, in contradistinction, one scans the page on which the description is written and one thinks that one does so; one does not have the impression of scanning O itself or a substitute of it, even if one has a vivid imagery of O. These are different visual actions even though both are cases of scanning, and each action characterizes a different form of representation.

This last characteristic of iconic representational systems is a direct consequence of the fact that iconic representations have an internal structure that determines both how they represent some domain, and their relations to the represented domain. Palmer (1978), for this reason, calls iconic (Palmer discusses 'analog representations' representations but his views apply to iconic representations in general as well) "intrinsic". In digital symbolic representational systems, in contradistinction, the relation between the system and the represented domain is not determined by the internal structure of the representing domain but is imposed externally, by means of some convention; digital representations are extrinsic representations. We had the opportunity to discuss Palmer's view and Haugeland's (1987) extension of them above. Here, it should be noted that the mapping between physical manipulations of the representing medium onto computational transformations in the represented domain that characterizes icons helps explain Haugeland's (1987) complicity of iconic systems. This is so because the spatial transformations in the represented domain are automatically registered by spatial transformations in the analog representing medium.

In mathematical domains, as a matter of course, one can talk of computations in the represented domain, but not all state transformations in represented systems could be called computations. Many represented domains, such as a visual scene, do not involve computations (at least directly, because one could perceive a scene in which someone performs some computations). The analog of computational transformation in visual scenes is the changes that may occur in the visual scenes, because when things in the represented scene change this amounts to state transformations in the represented domain. Therefore, the mapping between the representation and its representatum should be extended to include changes or state transformations in general. One should say that in iconic representations, albeit not necessarily in a pure form owing to the fact that these representations may involve digital symbols, physical manipulations of the representing medium map onto state transformations in the represented domain. In contradistinction, in a digital representation, a physical change does not entail anything about the represented domain. Increasing the size of the numeral '3' to '3' does not mean that the cardinality of the set represented by '3' is increased. When however, the length of the histogram changes, this necessarily means that cardinality of the set has increased as well. It is this short of mapping on which Haugeland's complicity rests.

As the preceding discussion suggests, the reason underlying "automatic updating" is that the physical transformations of the information-bearing vehicles in the environment must be reflected and matched by the physical transformations of the representational vehicles, whether it be external representations or internal, mental representations.

Kulvicki (2006, 2014) also holds the view that imagistic representations are those representations that preserve the structure of what they represent. They present their *representatum* to us by manifesting aspects of their *representata* structure. He notes that structure-preservation differs from resemblance, even though those representations that systematically resemble in some respect their *representata* make use of isomorphism (Kulvicki, 2014, 105), because structure preservation could concern relations between represented elements and not their qualities, which is along the lines of Shepard's (1978) criterion of 2nd order isomorphism required in analog representational systems. According to it, relations among elements of the computational processes in the represented domain should be mapped onto relations among elements of the physical processes in the representing system, that is, that there must be an isomorphism or a homomorphism between syntax and semantics. We find again the idea that since iconic representations map appropriately onto their represented domain, they must necessarily depict, or correspond to, at least some aspects of this domain. Hence, the form of the iconic representation must enable it to depict these aspects, for example, shapes and apparent sizes, and this demand put restrictions on the form of the representation; semantics constrains syntax. In contrast, in digital representations where digital symbols refer by convention and need not bear any relation to their *representata*, semantics puts no constraints on syntactic forms.

4.3 Iconicity and Context Relativity

Contra the widely held view that is the nature of mapping or isomorphism (homomorphism) that determines iconicity, Haugeland (1998, 182) argues that isomorphism is a common basis for making all kinds records of information expressed by some medium, and not a distinctive mark of iconic representations. We saw above that the number line is isomorphic to the set of natural numbers, or the digital thermometer to the set of temperatures, but no one would call either of them an iconic representation of the natural numbers or temperatures. Moreover, especially since isomorphism may concern abstract properties, by choosing some general abstract property, one can find isomorphism between completely unrelated domains of which no one would think that one represents the other. We also discussed various (and we shall discuss more) attempts to constrain the isomorphism so as to capture the essential sense of mapping characterizing iconicity.

Haugeland thinks that it is the distinctive, context dependent or relative nature of some contents that determines the iconic nature of the representational format of the representations of these contents and not some form of isomorphism. Haugeland (1998, 192–193) defines as relative those elements of the contents of representations (which are environmental entities for external representations) that are defined relative to all other elements along the relevant environmental dimension; temperatures and instances in time are examples of such dimensions, as are the points in a structured space that form a shape insofar as "shape" makes no sense except in terms of relative locations in this structured space. This means that relative elements do not stand alone since they are always individuated in relation to other elements in the environment. By insisting that it is the nature of the contents that determine the sort of the representation that can represent them, Haugeland subscribes to the idea that semantics constrains syntax in all sorts of representations. Hence, the syntactic form of a representation that purports to represent relative entities must be iconic. Similarly, the syntactic forms of representations that purport to represent absolute elements must be digital/discursive.

As we have seen, Haugeland (1998, 182), argues that isomorphism is a common basis for making all kinds records of information expressed by some medium, and not a distinctive mark of iconic representations. He cast doubt, therefore, on the claim that a representation is iconic if it maps in a structure-preserving way on its represented domain, because if a representation is isomorphic to its domain, it preserves the semantic structure in that domain in so far as A and B are isomorphic if and only if the structure of A mirrors the structure of B, which means that the relations among the elements of which A consist can be put into correspondence with the relations among the elements that make up B. Iconic representations are those that represent in virtue of structure isomorphism (Burge, 2018; Kulvicki, 2014, 2015; Maley, 2011), or some other form of structure preserving mapping, such as homomorphism.

To be able to appreciate Haugeland's objection, the reader should recall the distinction between a representational scheme that consists of a series of individual representations, and individual representations. A representational scheme may be isomorphic to its domain even though it is comprised on digital representations. As Lyons (2022, 7) argues,

> The internal states of a calculator can represent the numbers only because the causal relations among those states correspond to mathematical relations among the numbers ... There is therefore an isomorphism between the scheme, consisting of all the possible internal representational states of the calculator, and the domain, consisting of numbers in Plato's heaven. The *system* of representation (the set of internal states of the calculator) mirrors the representandum domain, although there is no interesting sense in which the individual representations mirror their individual representanda ... So the relevant isomorphism here holds between representational scheme and representational domain, not between individual representation and representandum. These schemes are paradigmatically symbolic, despite this scheme isomorphism.

Haugeland's point applies to such cases. At the level of the representational scheme there is an isomorphism between the representational system and the represented domain, but this does not warrant the conclusion that the representation is iconic. Digital representational systems can be isomorphic to their domains. However, if one narrows down the scope to individual representations, a representation that maps in a structure-preserving way onto its domain is, as we have seen, arguably iconic.

Let us concentrate for the moment on, what I hold to be (Raftopoulos, 2009, 2019), purely iconic visual perceptual representations. In Raftopoulos (2009), I argued that the elements of the representations in early vision, that is, the singular elements that stand for objects in the visual scene, and the attributives that stand for properties and relations in the scene are applied always demonstratively and put the viewer is a *de re* perceptual relationship with the contents of the visual scene, a kind of demonstration that on account of the fact that it involves no concepts I have called "nonconceptual demonstration". The demonstrative nature of reference of these elements signifies that they are constitutively dependent on the perceptual context of the specific perceptual encounter and function only in that context. In other words, they are context-dependent elements. The way the elements of the states of early vision refer to objects and properties in the visual scene entails that the contents of the states of early vision do not *encode* environmental information, that is, they do not represent environmental information in terms of context independent symbols that could function as concepts that could apply across different contexts. This is, by the way, why perceptual iconic contents do not satisfy the *Generality Constraint* and do not admit of Boolean operations.

Thus, Haugeland's definition of relative elements in representational contents is coextensive, if not co-intensive, with my definition of context-dependent symbols. Purely iconic perceptual contents, that is, the contents of the states of early vision, are always demonstratively applied and, thus, acquire their meaning only in the context of the specific perceptual act, which means that they cannot represent the same entities in another visual scene. Being constitutively related to the elements of the visual scene, the perceptual contents make no sense except in terms of their *de re* relations to the elements (objects, properties, and relations) in the visual scene. If one thinks of these elements as elements arrayed in a structured space occupying a certain location, the perceptual contents are defined relatively to these locations and are, thus, "relative" in Haugeland's sense.

It follows that the structure of a visual scene as it figures in the purely iconic early vision is a rich topographical array of demonstratively applied singular terms that designate objects, and demonstratively applied attributives that designate visible properties and relations in the visual scene, as they are arrayed in the structured space that contains the visual scene, and which corresponds to a functional, isomorphic or homomorphic mental space. This structure, lacking context independent symbols, is not discursive and does not support discursive operations. I will argue that the structure of the iconic contents of early vision is described by a whole/parts structure that is studied by Topology and Mereology.

These considerations could be extended to other iconic representations; in iconic representations the representational contents are defined relatively, that is, in a context-dependent nature. Haugeland's criterion of relativity of the elements of the contents of iconic representations means that the elements of these contents are defined relatively to other elements in the content of the representation, and this makes them dependent on the representational context. In a graph of temperatures vs. time, which is an iconic representation (albeit not a pure one, since it necessarily involves digital symbols), the elements of the iconic content of the graph, namely the temperatures and instances of time, are relatively defined because they become relevant in the representation only in relation to the other temperatures and instances of time respectively. This makes them context dependent, that is, their representational value and function depends on the specific representational context; if a certain temperature is defined as T1 and to T1 corresponds an instance I1 in time in the graph, T1 and I1 have a significance only in the specific context and cannot be used elsewhere. In contradistinction, were T1 an I1 to figure in the content of a thought, they would be transferable to other contexts in which they would retain their significance.

Think now of an iconic representation, say the sketch of an airplane. When one sees this sign in an airport, it designates the area of the departure gates. This icon is tied to a specific spatio-temporal context, otherwise it would fail to transmit its intended information. When one sees the same sign in the road, it designates the way to the airport but it is not contextually bound. When one sees the sign on a map, it designates the location of the airport on the map. In another context, it may designate travel. The same icon, thus, means different things in differing contexts; what the sign represents is a function of the context. Since iconic representations map on to their representational domains in a way that preserves semantic and spatial structure, the mapping relation is essential to iconicity. Mappings by their nature are sensitive to context since a mapping that works in a certain context might cease to do so in another context. For example, the airplane sign within the airport maps onto the directions to the departure gates but if one applied this mapping to the airplane sign in the highway one would not necessarily get to the departure gates; one could get to the arrival gates, for example.

This extends to the contents of perception. To repeat what I said before, I have argued (Raftopoulos, 2009), the contents of early vision (the perceptual stage that has purely iconic contents, or so I argue) are defined demonstratively, a view shared by Burge (2010, 575) according to whom '[v]isual perception consists in a rich, topographical array of demonstratively applied attributives, at various levels of specificity.' I have argued that the nonconceptual content of the perceptual states induced when one uses a perceptual demonstrative constitutes the mode of presentation of the perceptual demonstrative that individuates, but does not identify, the object of perceptual awareness and allows reference to it. On that account, perceptual demonstratives put us in a *de re* relationship with objects in the world through the nonconceptual information retrieved directly from the objects in the environment. Demonstratives in perception are context dependent in that they work within a certain context; for, example, an object in a perceptual may be individuated on the basis of some of its properties (whether it be spatiotemporal or other visible, features), but in another encounter the same object might be individuated through a different set of properties; so, object individuation takes place always in the context of a specific perceptual encounter. Importantly, this context is not external to the individuation act, but it is constitutive of it. This means that the perceptual representation of the object has the context in which takes place built-in and not around it, as digital representations do.

I have referred to the context sensitivity of the contents of early vision and related this context sensitivity to Haugeland's (1998) view that iconic representations are required to represent relative elements. However, Haugeland restricts his discussion on the relative nature of perceptual contents to patterns of light intensities, whereas early vision contains much more visual information than light intensities and one could question whether it is legitimate to extend context relativity to these richer contents.

To start the discussion on this problem, let us revisit Haugeland's relative contents. Haugeland (1998, 192–193) writes

> On a genuine dimension (in the definition of iconic representations) the relative positions of the content elements are not all arbitrary. Quite the contrary, those elements are always organized relative to one another in some regular structure that every representational token of the pertinent scheme essentially presupposes ... The point is even more conspicuous for maps and pictures: the shapes of the variations are the represented (skeletal) contents—and "shape" makes no sense except in terms of relative locations in structured spaces.

Some explication is required first. Haugeland calls skeletal or bare-bones contents of iconic representations those contents that are formed through mechanisms that do not contaminate their products with digital elements. This is important if one wishes to characterize the essence of some genus of representation, say, iconic representations. Since for Haugeland this can be done only by considering the contents of the various representation *genera,* to accomplish this one must examine contents that are clearly exclusively iconic and do not carry the traits of other representational schemes. Haugeland thinks that the way perceptual systems function (perceptual systems use a host of common-sense information about the world that goes far beyond what is included in the information from the visual scene that impinges the eyes to construct the percept), if one excludes all the additional information that relies on digital representational schemes, what one is left with is certain variations of incident light with respect to direction ... The point is ... to distinguish, within the undeniable contents of everyday representations, a substructure that is skeletal—in the special sense that it does not draw upon the user's antecedent familiarity with the situation and the world, and hence cannot exhibit any admixture of characteristics from other (probably symbiotic) representational schemes. (Haugeland, 1998, 189)

Thus, the "variations" mentioned in the 'shapes of variations' in the quoted passage above are variations in light intensities. Patterns of variations of such intensities form edges and eventually shapes, which are thus skeletal, that is, purely iconic contents. This generalizes to maps and pictures where the patterns of variation form shapes, and "shape" makes no sense except in terms of relative locations in structured spaces, that is, shapes are defined by set of points where each point's location is relative to the locations of the other points.

One should point out, first, that in this passage Haugeland does not discuss relative entities in the environment (the contents of external, non-mental representations), but relative contents in perceptual representations (to the extent that maps and pictures rely on the perceptual capacities of their viewers in order to be understood).

Haugeland's concern that the fully fledged content of visual perception is imbued with non-iconic, digital contents since perceptual systems use information about the world in order to construct the percept and this information is some sort of common knowledge that, as such is represented by digital representational schemes, is shared by Tye (1991), who argues that mental images are "symbol-filled arrays" and, consequently most of what they represent is not represented in a pure iconic way. I have argued (Raftopoulos, 2009, 2019) that the representations of late vision are hybrid (symbiotic, as Haugeland calls them) iconic and digital representations since perceptual processing in late vision is cognitively penetrated. However, I think that early vision, the stage at which most of perceptual information is retrieved from a visual scene is cognitively impenetrable, and its contents contain no digital elements. I have also acknowledged (along with Burge, 2010) that perceptual systems use a host of information that is not contained in the patterns of light intensities registered in the retina to construct the percept; this is the information "expressed" in the operational constraints. However, as I have argued (Raftopoulos, 2009, 2019), this information is not represented anywhere in the brain and it does not constitute some sort of information that is digitally represented in the system (Burge, 2010) agrees with this view) so that its usage would entail that early vision contains digital elements. This has the immediate consequence that Haugeland's skeletal content of visual perception contains much more than patterns of variations of light intensities that form some shapes. It also means that early vision states can have a rich semantic structure without being "symbol-filled arrays", a theme to which I return on many occasions in this book, since it goes against Fodor's claim to the contrary, and has profound implications concerning the representational power of pure iconic representations.

The preceding discussion shows that Haugeland's view that iconicity is tied to context-dependence or relativity is justified in a stage of visual perception, namely, early vision. However, generalizing this view to apply to all iconic representations is problematic. First, one may argue that even if it is true that all representational systems satisfy in some way the isomorphism criterion, one should recall that iconic representations do so in a natural way, while digital representations require some convention. Second, it is not at all clear that Haugeland's criterion of relativity holds for all iconic representations.

Even though this dynamic, context dependent nature of iconic representations has clear advantages it also has a major disadvantage. If each representation may change its meaning owing to the context in which it occurs, binding such representations become problematic, which is why iconic representations do not display the systematicity and recombinability of classical digital representations, which also explains why they do not satisfy the *Generality Constraint*. As we have seen, satisfying GC requires that the symbols used to denote properties and individuals be context independent. As I explained above, this does not apply to perceptual iconic representations, owing to the context sensitivity of the iconic symbols employed to represent properties and individuals in the environment. In addition, logical connectives and quantifiers cannot be among the representational contents of iconic representations, as they can be part of the content of propositional states. 'One important thing that pictures cannot do, then, is represent certain sorts of relations between ideas. They cannot represent, for example, those relations which we express using the words *if... then...* and *not*.' (Crane, 2003, 19). Later on I return to this problem to discuss the main differences between icons and symbols.

This can be inferred from two facts. First, that there are no logical contradictions in perception (illusions as of impossible situations are not logical contradictions), while a proposition whose form is p and – p is a logical contradiction. Second, from the fact that if one tries to take a picture, which is an iconic representation, of a situation expressed by a disjunction, say that O1 is either to the left of O2 or to the right of O2, one will get a picture either of O1 being to the left of O2, or a picture of O1 being to the right of O2, depending on the actual spatial configuration of O1 an O2. This, however, is not a picture that displays the disjunctive fact described above; one cannot iconically express the fact of the occurrence of a logical connective. Or, to use Crane's (2003, 19) example, to represent iconically the negation non-rainy, one would draw a sunny scene, but this represents a sunny day, not a day in which it does not rain (even though this idea can be inferred from the representation, it is not a part of it).

Consider universal quantification and a picture containing Fodor's favorite giraffes, three of them. The picture shows three green giraffes but it does not show, and cannot show owing to representational resources limitations, that all giraffes in the picture are green; in propositional representations one achieves this just by employing the universal quantifier. All the picture can do is to place the three giraffes at certain location and show that they are green. This requires co-locating giraffeness and greenness at some specified location in the picture, which means that giraffeness and greenness are necessarily bound to some location. Using the universal quantifier requires that one abstract from the locations of the animals and convey the information that giraffenes and greenness are co-instantiated in the picture; iconic representations cannot do that because iconic representations are inherently tied to space.

The fact that iconic representations do not support logical operations explains the need for civilizations to move from pictograms to ideograms. Sumerians used a symbol for woman that resembles the female pubic area (an iconic representation), and a symbol for mountain that resembles mountains, these symbols are pictograms since the symbols depict, with some abstraction, what they represent. The symbol of womanhood on the left of the symbol for mountain means a woman slave (since Sumerians got their slaves by attacking mountainous tribes). However, these two symbols put together do not resemble a slave woman and their combination cannot explain the logical relations between 'woman' and 'slave'. For this reason, the combined symbol is not a pictogram but an ideogram.

4.4 Iconicity

To sum up the discussion thus far, a representation is iconic if and only if it maps onto its represented domain in a way that preserves the spatial and semantic structure (properties and relations) of this domain. It is also plausible that in a subclass of iconic representations the elements of the representations are defined relatively. That is, a class of iconic representations is also characterized by the property that its elements are context-sensitive. This definition of iconicity entails that the class of representations that conform to the oldest definition of iconic representations, namely that a representation is iconic if it resembles, that is, is physically similar in some respects, to the represented domain (think of pictures, realistic drawings, or pictograms as opposed to ideograms), are indeed iconic, since, if a representation physically resembles some domain, it maps onto it in a structure preserving way. The resemblance definition curves out some iconic representations, but as a definition of iconicity is inadequate since a representation that maps onto a domain in a way that preserves the semantic and spatial structure of the domain, and, thus is iconic, need not physically resemble the represented domain.

Closing this section, I would like to revisit the problem of what makes an iconic representation to be iconic. Since I have argued that this factor is the proper mapping to the represented entity (where what "proper" means was discussed above), the question amounts to what allows a representation to be mappable to its represented domain in the proper way. In physical, external, non-mental representations that have no representational content the answer is simple: it is the nature of the representational vehicle that makes the proper mapping possible. Recall that the physical medium implementing an iconic external representation must be able to support the same transformations that the represented domain exhibits. For example, the addition of quantities, say numbers, can be represented iconically only by representational vehicles in which the manipulations of their physical properties allow the addition of the representational vehicles themselves, as when one uses histograms to represent numbers. Similarly, mental images *qua* vehicles of mental representations should be susceptible to rotation and translation algorithms that apply to visual scenes, algorithms that symbolically encodings of visual scenes could not realize. This is a clear example in which the semantics of the represented entity constrain the syntax of the representation.

Pictures are external representations too, and, so, one could argue that the iconicity of pictures is a feature of the pictorial vehicle itself; it is the physical constitution of pictures that allows the appropriate mapping to the visual scenes they represent. However, things are complicated by the fact that pictures are understood through their contents insofar as they make use of our perceptual capabilities to portray their representata. In other words, pictures have representational contents that are borrowed from our perception of them. The question, thus, is, in virtue of what are pictures iconic representations?

Lyons (2022, 5) argues that the continuity of pictures, say photos, is likely a feature of the pictorial vehicle itself, not of its content, because we may be agnostic about what kinds of contents pictures have, while still finding it obvious that they are continuous. In this part of his discussion, Lyons treats iconicity in terms of continuity, and so his claim is that pictures are iconic representations owing to their vehicle and not their content. Why does Lyons think that we find it obvious that pictures *qua* vehicles are continuous? Apparently because between any two physical parts of a picture there is always a third. But pictures consist of pixels and between any two consecutive pixels there is not a third one, which means that pictures are not continuous according to all accounts of continuity. Lyons writes that he understands continuity in the sense in which analog recordings are continuous. **Analog sound** depends on a **continuously changing audio signal**, which means that the pressure fluctuations fed to a speaker from its signal is ideally an exact reproduction of the original sound at any time. A continuously changing signal is one in which between any values of the signal there is always a third, which means that Lyons subscribes to the standard definition of continuity. So, in which sense is a picture "obviously continuous"? It is not because is consists of pixels, because then it would not be continuous. Is it because it is extended in space and, thus, contains a continuum of spatial points? But, then, so is the algebraic line of natural numbers and no one would consider it to be continuous. The only possible answer must be that a picture is continuous in a representational sense, that is, that each part of it represents something and that between any two representationally relevant parts there is always a third; between points of a colored space that represent that color in the represented scene in the corresponding spatial area there is always a third point that also represents some color in the represented scene. This is why, according to Lyons, pictures satisfy PP. This is a plausible view, but Lyons and I part company because I think that owing to the fact that pictures in general represent because they use our visual capacities, their continuity hinges on whether visual perceptual representations, or at least some of them, are continuous as well. In the next sections, I argue that owing to their main representational role, which is to represent objects, their properties, and their interrelations, visual perceptual representations contain discrete elements that refer to the objects in visual scene, and, thus, these representations cannot be continuous. However, these discrete elements are not symbols and their presence does not undermine the purely iconic character of the representations of early vision.

I shall argue that pictures, to the extent that they rely on the perceptual representations of their represented domains formed by the viewers of the pictures, are iconic owing both to their representational vehicles and contents.

4.5 Iconic Representations in Perception

Iconic representations were defined as those that map onto their representatum in a spatial and semantic structure-preserving way and, additionally, instantiate physically or functionally the spatial structure of the represented entity. It is easy to see that this is plausible in the case of iconic representations such as pictures or maps because they are spatially arrayed and, thus, have parts that are also arrayed in space, and which could be mapped onto parts of the represented entity. The same can be said for the *Picture Principle* (PP) that for some is the main trait of iconic representations; according to PP (Fodor, 2007, 108) if P is a picture of X, then physical parts of P are pictures of parts of X. When one considers perceptual representations, however a problem emerges since mental perceptual representations *qua* vehicles are not arrayed in space, unlike neural representational vehicles that are arrayed in space occupying locations in the brain. Even though it makes sense to talk about parts when the representation itself has spatial properties (as pictures, maps, or number lines have), there are mental iconic representational vehicles that do not involve parts properly speaking because they are not arrayed in space and, therefore, PP does not apply to them. This means that PP cannot be a necessary condition for iconic representational vehicles (Beck, 2019; Blachowicz, 1997; Peacocke, 2019). This has repercussions for the imagery debate, which is very important because mental images are considered to be iconic representations and, in fact, the empirical evidence for their being genuine parts of the cognitive architecture is considered to be evidence for the existence of iconic representations, against Fodor and Pylyshyn's (1981) view that representations are digital and that the analog properties that some representations exhibit should be either delegated to the properties of the physical medium of the sensory detection systems, or be explained away by digital representations.

As we have discussed, the problem that PP, or more generally, the demand that an iconic representation be mapped appropriately onto its represented domain, poses for mental imagery is that according to Kosslyn (1994), PP applies to mental images as well, but mental images are quasi-pictures because they are not arrayed in space, as literal pictures are and it is not quite clear in which sense one could talk of parts in non-spatial representations. We saw that Kosslyn's (1994, 5) answer was to introduce the notion of 'functional space', according to which the space in which mental images appear need not be physical but can be like an array in a computer that specifies spatial relations purely functionally. This means that the physical locations of each element in a spatial array are not themselves arranged into an array and it is only 'in virtue of how this information is read and processed that it comes to function as it were arranged into an array.' In such a functional spatial medium, distances are not defined as actual physical distances but are expressed in terms of the number of locations intervening between two locations. '[E]ach part of the depiction must correspond to a visible part of the object such that the distances among the points on the object are preserved by the corresponding "distances" among their representations' (Kosslyn, 1984, 107). Thus, the implementation of a mental representation in the brain that can represent spatial relations does not require that the brain have a particular spatial physical organization. It suffices that the brain realize 'a system of representational states with the same expressive power as, the same systematicities as, the same analyticities as, and is subject to the same algorithms and same mistakes as, some essentially spatial representational scheme' Lyons (2022, 19).

Heck (2007, 127) remarks that in perceptual representations, as in cognitive maps that are also iconic representations, the represented entities are 'spatial distributions', which are geometrical entities. The spatial relations represented in perception are local in the sense that objects are located in perception only relative to nearby objects and not relative to all the objects in the representation. This entails that distances are not defined as actual physical distances, but in terms of the number of locations intervening between two locations. It follows that the claim that images resemble their objects with respect to contiguity and interval relations that was explicitly made by Kosslyn and Tye should be read to apply to 'contiguity' and "interval relations' as they are functionally defined in a topological manner that does not require metric considerations.

It is important to note that talk about images preserving contiguity and interval relations (or, more generally, spatial structure) in the represented item concerns the syntactic, physical features of the representation and not, at least immediately, their representational contents. (In a while, I will expand the claim to include representational contents as well.) When, for example, Kosslyn asserts that distances among parts of the represented object are preserved by the "distances" in the representation, he means the distances between the neuronal assemblies in the brain that represent the parts of the object. Kulvicki (2014, 163) hastens to note that this means that representations need not resemble even qualities like distances and contiguity in the represented item in order to be mental images. It suffices that there be a structural isomorphism between the mental state and the represented item, and, so, mental states need not share any qualities with the represented items but only structure. Even though Kulvicki is correct about structural isomorphism when it comes to the class of iconic representations in general, there is still a sense in which brain representational states do resemble their representata with respect to contiguity and distances. As we shall see, the visual representations of items from the retina to the higher parts of the brain have wholly, or retain to a certain degree, a retinotopic character. Retinotopic is a form of neuronal organization that preserves the structure of the retina. Since the retina mirrors the spatial structure of the visual field, a retinotopic organization is also a spatiotopic organization. It follows that objects that are closer in the visual scene are represented by neuronal assemblies that are closer, and in this sense, the brain states that represent a visual scene are similar with respect to distances to the visual scene they represent.

This is important for the thesis that visual representations preserve the structure of the represented items and, in this sense, resemble their *representata*. The resemblance concerns spatial structure, most notably contiguity and interval relations, and sometimes those visible features, such as shapes and colors between objects, but not necessarily all visible qualities of the represented item. Thus, the form of a pictorial representation of an item need not share with the items it represents qualities, which is as it should be because brain states represent shapes, shades and hues, but brain states do not have these shapes (which explains the qualification "sometimes" above), shades and hues. In other iconic representations, however, such as drawings or pictures, parts of the forms of the representation (that is, some of their syntactic features) do resemble some of the qualitative features of their *representata*, shape being the most uncontroversial example. Moreover, the fact that interval relations and contiguity in the brain states are functionally defined does not negate the fact that the retinotopic mappings preserve, in a brain scale if you wish, contiguity and distance relations in the represented visual scene. Thus, visual representational vehicles, the forms of the representation, resemble the items they represent in a limited sense.

In the introduction to this chapter, I noted that what determines the sort of symbol a symbol is, is the interpretant used on a specific occasion and that the same sign could be an icon, an index, or a digital symbol. In visual perception, the interpretive process that generates iconic reference is none other than visual perceptual recognition. Representation is to present something again (re-present). Perceptual iconic relationships are the most basic means by which things can be re-presented. As we shall see, they are the basis on which digital symbolic representations are built, in the sense that iconic representations ultimately ground digital representations into reality.

Going back to visual perceptual representations and mental images, these need to be 'physically instantiated in some way or other such that they can "act" or "function" like images, in an appropriate "processing" environment' (Haugeland, 1987, 91). This view has its own problems (whether they are insurmountable or not is another issue), however, especially since assimilating mental images in the brain to the functional spatiality of digital computers renders mental image susceptible to Pylyshyn's (1984) criticism as to whether they are genuine elements of the cognitive architecture (see Beck (2018, 4–5) for a discussion). Regardless, it is important to remember that an iconic representation need not be arrayed in space itself. A representation could be iconic even if it is not arrayed in space, and, thus, does not have spatial features, since it suffices that it determine a geometrical structure that maps naturally onto some spatial structure. To put it in topological terms (I discuss topology next), the structure of the an iconic representation that is not arrayed in space emphasizes topology but it is not an abstraction from the geometry of the representatum, in which case it should consist of spatial points that would involve metric considerations and that would presuppose an actual arrayment in space, but, rather, the representational structure encodes relations of similarity (in terms of contiguity and interval relations, as we saw above) among the elements of the representation.

Recall that according to Kosslyn (1994, 5), in functional spatial mediums, such as those related to mental images or visual perception, distances are not defined as actual physical distances but are expressed in terms of the number of locations intervening between two locations. In other words, distances in a functional sense do not require a metric but are determined through relations of contiguity and neighbourhood. Thus, what is important in functional spatial mediums is this non-metric notion of distance between locations or points. These mediums are defined in purely topological terms and constitute topological spaces that are characterized by a closeness relation (neigborhood) rather than by a metric relation, as metric spaces are. Of course, distances suffice to determine closeness relations, but they are not necessary to determine whether two things or events are close to each other.

In Algebra, a metric space is a set together with a metric on the set. The metric is a function that defines a concept of *distance* between any two members of the set, which are usually called points. Any metric space can be viewed as a topological space, because a metric on a space induces topological properties like open and closed sets, which lead to the definition of more abstract topological spaces. A classic example is the topology of the set of real numbers R where T is the collection of all subsets of R. However, not all topologies need a metric.

A topological space is, in general, a set of points together with a topology defined on them, that is, any set of points that satisfies a set of postulates of some kind. Formally, if X is a set, a topology on X is a collection T of subsets of X satisfying certain conditions. X along with the topology T defined on X is the topological space (X, T). Thus, a topological space, also called an abstract topological space, is a set X together with a collection of open subsets T that satisfies these conditions: (1) The empty set is in T; (2) X is in T; (3) The intersection of a finite number of sets in T is also in T; and (4) The union of an arbitrary number of sets in T is also in T. Simply put, a topology is the way in which the constituent parts of the set are interrelated or arranged, a way that is determined by the set of postulates that the topology satisfies. This general definition applies both to metric spaces and non-metric spaces. In the latter case, the purely topological, non-metric properties of the set are brought forth (recall that these properties are contiguity and relations of interval) and this makes the notion of "neighborhood" central in defining a non-metric topology and substitutes for "subsets: in the preceding definition of a topological space.

Accordingly (Hausdorff (1914), as adopted by Russell (1927)), a "topological" space is a manifold [set] whose elements x are associated with sub-classes [neighborhoods] Ux of the manifold [set] such that:

(1) To every x corresponds at least one Ux, and every Ux contains x;
(2) If Ux, Vx are both neighborhoods of x, there is a neighborhood of x, say Wx, which is contained in the common part of Ux and Vx;
(3) If y is a member of Ux, there is a neighborhood of y which is contained in Ux;
(4) Given any two distinct points, there is a neighborhood of the one and there is a neighborhood of the other such that the two have no common point.

It is not difficult to see that the two definitions are equivalent if one substitutes relations between neigborhoods (such as containment in the sense of belonging and having common parts) for relations between subsets, (such as union and intersection). The difference is that in the latter definition there are no metric components but only topological properties. This renders this definition of non-metric-geometrical topology very suitable for the purposes of defining iconic mental representations in view of the fact that these representations are not arrayed in space and, thus, have no metric properties.

In view of these, Kosslyn's and Haugeland's idea that iconic representations need not be physically instantiated in space acquires more credit. In fact, the general idea that representations in perception or imagery even though not literally arrayed in space as their representata are, still can be iconic representations of spatially arrayed properties, is further supported from our knowledge of the brain. Let me elaborate on this point. Consider a perceptual representation that presents a visual scene as being a certain way. Let us examine its representational content. Its content presents objects, properties, and relations arrayed in space, that is, it presents objects, properties, and relations as filling some space in a particular way. Let the objects, properties, and relations that figure in the representation constitute the set X1. The topology on X1, T1, is the collection of subsets of these objects, properties, and relations. These subsets, as it is easy to verify, satisfy the four conditions stated above. For example, the intersection of a finite number of subsets of X1 each consisting in certain objects, properties, and relations, obviously is a set of objects, properties, and relations that belongs to X1. Similarly, the union of an arbitrary number of sets in T1, that is, of subsets of X1, is again a set of objects, properties, and relations, that also belong to X1. Thus, any perceptual representational content is associated with an abstract topological space and its topology, abstract in that the represented relations are not out there in the real world but in the mind as contents of neural representational vehicles. This topology determines the way the elements in the representational content are arranged to form the representational content as a whole. The definition of topological spaces in terms of neighborhood relations is also satisfied. If, for, example, an object O is contained in two neighborhoods V1 and V2, O is the intersection of V1 and V2. The intersection defines a neighborhood V3 in which O belongs and which is contained in the common part of V1 and v2, being in their intersection.

Let us turn now to the neural representational vehicle of that content. It consists in a set of neurons across the brain that fire when a viewer perceives the visual scene; let us call this set X2. A topology on X2, T2, is similarly a collection of subsets of these neurons, their properties (activations, activation thresholds, etc.), and relations (excitatory or inhibitory connections, connection weights, etc.) that satisfy the four conditions mentioned above. T2 defines a metric concerning the way the relevant neurons are interrelated, for example, which neurons are connected, how they are connected with other neurons, the distances in the brain of the connected neurons, etc.

Finally, consider the visual scene, which consists in a set of objects, their properties, and relations. Let us call this set X3. As before, let us define a topology T3 on X3, which determines the way the objects, the properties, and relations in the visual scene are arranged. That is, the relevant metric concerns the spatial relations between the entities (be it objects, properties, or relations) in the visual scene that satisfy the postulates of Euclidean geometry.

The claim that visual perceptual representations are iconic even though they are not arrayed in space is grounded in the fact that the abstract topology of the perceptual content (T1) is mapped onto literal topology in the brain (T2), which is mapped onto the literal topology in the visual scene (T3). Let us examine first the mapping between the visual brain (T2) and the visual scene (T3). The optic tectum is a structure in the midbrain to which the optic nerves project. The cells in the tectum are arranged in layers called the retinotopicmaps in which the positions of optic tectum cells relative to one another correspond to the relative positions of their visual fields on the retinal surface. Thus, the topological structure in the tectum maps onto the topological structure of the retinal neurons, which maps onto the to the structure of spatial positions of areas of surfaces in the environment.

Let me give an example by using the well-known neural feature maps involved in visual neural processing (Hubel & Wiesel, 1968). These neuronal assemblies are characterized by both a receptor topology and a feature topology. Each neuron in a feature map corresponds to a specific receptive field on the retina. As the cells in the retina are connected to the neurons in feature maps, the pattern of synaptic connectivity of neurons within a feature map reflects the geometry of the receptor. Thus, the topology of the feature maps reflects the topology of the retina (receptor topology). Accordingly, neurons closer in a feature map are connected to cells in the retina that are close to each other. Since the cells in the retinal are connected to neurons in feature maps through synaptic projections, the density of synaptic connectivity introduces topological relations, such as contiguity and closeness, among the neurons of a feature map. A neuron counts as closer to a given neuron as compared to a third neuron if the connectivity density between the first and the given neuron is greater than that between the third and the given neuron. In humans (and many other mammals too) this pattern of connectivity is (roughly) congruent to the topography of the feature maps given by the actual physical distances between neurons. In vision, these feature maps are retinotopic. So, in this case, topological and metric properties are congruent.

Each neuron has a specific feature selectivity, that is, its activation is a function of a specific feature so that the neuron fires more when the specific feature is in the retinal receptive field (feature topology). A specific neuron in a color map, for example, fires if a stimulus projected on the retina is located in the receptive field of the neuron (that is, if the neuron corresponds to the receptive field on the retina on which the stimulus projects), and its color has a value in an interval of angles for which the neuron is selective. The topology of the thin strips of V2 that form the color feature map in that area, for example, is such that the neurons of neighboring strips have neibhboring or overlapping receptive fields.

The combination of receptor and feature topology of feature maps yields some remarkably interesting properties. First, having a receptive field entail that there is a topologically continuous mapping from the connectivity topology of the feature map into the geometry of the receptor. That is, given a neuron in the feature map, all neurons close to it have receptive fields in the retina that are close to the receptive field of the neuron. Second, having a certain feature selectivity is a topologically continuous mapping from the connectivity topology of the feature map into the topology of the attribute space. Given a certain neuron in the feature map, all neurons close to it have feature selectivities, that is, respond preferentially to properties in the environment, that are similar to the feature selectivity of the given neuron.

Similar retinotopic maps that preserve topological structure of retinal receptors are found in LGN cells. Optic nerves terminate at LGN cells that are arrayed in layers and each layer 'contains a retinotopic map of half of the visual scene, those in the right LGN having maps on the left side of the visual field and those in the left LGN maps on the right side.' (Bruce & Green, 1990, 50) This is called retinotopic or topographic mapping since it preserves the relative location of cells from retina to LGN; 'nearby positions on the retina project to adjacent regions of the LGN, thus preserving qualitatively (not quantitively) spatial relations in the retinotopic map.' (Palmer, 1999, 149–150).

This retinotopic mapping extends to the entire brain. There is

> An orderly retinotopic mapping of the visual world onto the surface of the cortex, just like that in the laminae of the LGN, with the left and right halves of the visual field mapped onto the right and left cortices respectively. Cortical cells, therefore, have the same topographic relation to each other as their receptive fields have in the visual field. (Bruce & Green, 1990, 54)

Palmer (1999, 150) agrees; 'This kind of spatial mapping is a common feature of higher levels of the visual nervous system, having been found in many areas in the visual cortex.'

Thus far, we have established a spatial mapping between the topology in the retinal cells and the topology of higher brain areas; relative positions on the retina cells are preserved so that they are mapped onto the relative positions of the cells in most of the brain. To establish, however, the spatial mapping of brain structures to the actual physical spatial relations in the visual scene, one need establish the topological mapping between the topology of retinal cells, the entry point of the incoming environmental information, and the topology of the visual scene. In other words, one needs to establish that the spatial relations between the entities in the visual scene are preserved in the retinal registration of the information from the visual scene as it impinges on the retina. This is a problem concerning mappings from 3D space to 2D space and, specifically, the problem of mapping spatial relations in 3D space onto spatial relations in 2D space, and of mapping 3D shapes to 2D surfaces. I shall discuss this projection shortly. In addition, and foremost, if visual perception is to be successful, there must be a consistent relationship between the geometry of surfaces in the visual scene and the light that enters the eye at a particular observation point. This relationship indeed exists and is governed by the laws of geometrical optics.

Let us turn to the mapping between T1 and T2, that is, the mapping between abstract, or functional, topology of the perceptual representational content of a visual scene and the literal topology of brain states *qua* representational vehicles. This can be put more specifically as follows: what is the relation between the contents of perceptual states in general and of conscious experience in particular and the brain areas that implement the representational states/vehicles whose content is the visual experience? We know, for example, that neural activity in the cortical areas that connect to retinal cells on which light impinges from the upper right corner of the retinal image correlates with aspects of conscious perception that represent entities in the upper right corner of the visual scene. Similar mappings between conscious experience and other, higher, brain areas are also well documented.

The mapping relation is transitive, which means that T1 maps onto T3 as well. In other words, the abstract topology of our perceptual experience maps onto the literal topology of the represented visual scene through the topology of the brain states. This can be surmised from the nature of the experiential contents themselves. These contents preserve the spatial relations of the objects in the environment, that is, the spatial relations between objects in our experience (ceteris paribus) correspond to the actual relations of the real objects in the visual scene. The mapping between T1 and T3 is not confined only to the preservation of spatial relations. There is also a mapping, which sometimes assumes the form of similarity, between the attributives in our experience and the actual properties of the objects in the visual scene.

Additional evidence that the contents of our visual states, starting from the sensory non-representational contents of the states that consist of the activation of retinal cells (in other words starting from the retinal image, map on to the spatial structure of the perceived visual scenes comes from the way the brain solves the projection problem. A visual scene consists of three-dimensional surfaces with an overlying reflectance function at each point on these surfaces. The three-dimensional information is determined by the spatial structure of the visual scene and by the shapes and textures of the overlying objects. The reflectance function at any given point on the surfaces is simply a measure of how efficiently the surface at that point reflects light of various wavelengths. The information that reaches the eye from this stimulus is a complicated function of the three-dimensional structure of the visual scene, the reflectance function of the surfaces of the objects at each point, and the illumination and viewpoint conditions. Despite the complicated nature of the information in visual scenes and the complexity of the tasks required to achieve some constancy in representing this information, human observers are remarkably good at recognizing and categorizing visual scenes and their objects. It is known that to achieve this task, the brain uses both 3D shape, and 2D surface reflectance information, which means that the brain representations must include both a structural coding of viewing object faces, and an image-based coding. However, the 3D shape information itself must be reconstructed from the information impinging on the 2D retinal surface. That is, the brain has to get from the retinal images of scenes back to the objects in the visual scene that gave rise to the image in the first place (the *inverse problem*).

Going from the 3D visual scene to the 2D retinal image of it, a vital piece of information, depth, is lost. In order for the 3D representation of the visual scene to emerge, this information must be recovered from the information in the 2D optical image formed in the retina. The problem, thus, is to recover 3D structure from its projection onto 2D space. To solve this problem, independently of the solutions one proposes, it is required that one start with the mapping of three-dimensional shapes on to a two-dimensional plane. This mapping must satisfy some constraints, for example, that the non-accidental properties of the objects in the visual scene, that is, those properties whose form is not substantially changed through rotations of the objects (such properties are parallel lines, and collinearity (Biederman, 1985)) are preserved in the retinal image and the brain representation. Other constraints include the preservation of other nonaccidental properties such as symmetries (Leyton, 1992), and the defining major axes and radii of objects (Marr, 1982).

The problem of 3D to 2D projection is known as the 3D projection problem; the processes that map the 3D visual scene to the 2D image can be analyzed mathematically through projective geometry. Projective geometry specifies where each point in the 3D visual scene projects onto the 2D retinal image and which properties of these retinal images will be invariant over different projections. Projective geometry by itself, however, cannot solve the problem of image formation because it cannot model reflection, absorption, or refraction of light. A consideration of these factors brings into the picture the perspective from which a visual scene is viewed which, in turn, determines which objects are closer to the eye hiding thus other objects by preventing light emanating from them to reach the retina. This means that in addition to projective geometry, one must factor in the way light from a 3D visual scene forms an image when is projected onto a 2D plane surface. Fortunately, there exists a consistent relationship between the geometry of surfaces in the visual scene and the light that enters the eye at a particular observation point. To factor in perspective, perspective projection is used, which determines the process of image formation in which light converges to a single focal point. There are two main sorts of projection that can be employed in that regard. The first is linear perspective in which the projection lines converge to a single point; this is the linear perspective projection. This corresponds to the situation of a human viewer and, as I said before, this is why the contents of linear projection images are closer so the contents of visual perception.

However, even then, the problem is complex, and visual scientists use orthographic or parallel projection, which is the second kind of geometric projection, to simplify the overly complex processes involved in the 3D to 2D projection. In orthographic or parallel projection, one assumes that the retinal image is formed by rays that travel parallel to each other and perpendicular to the retinal plane, rather than rays that converge at a single point, a situation which obtains when the distance from the image to the object is relatively large compared to the depth of the object. This simplification entails that depth is simply ignored but all spatial information in the plane perpendicular to the viewing direction is preserved without a change.

Since in visual perception we do not see 2D images but 3D visual scenes, the visual system must somehow reconstruct the 3D scene by recovering from the retinal image all necessary information (the inverse problem that we encountered before). It is not the aim of this book to address this problem. What suffices for our purposes is that the way 3D objects in a visual scene project on the 2D retinal plane to produce the retinal image preserves (ceteris paribus) the spatial information in the visual scene. This supports the view that even though perceptual states are not literally arrayed in space, there is a structure-preserving spatial mapping between the contents of perceptual representational states of spatial structures (T1) and aspects of the represented spatial structures (T3). This is compatible with the fact that the reconstruction of the visual scene from its 2D projection is not uniquely determined. The content of a visual perceptual state is fixed by the scene from which the 2D image on the retina could have been projected, plus illumination conditions and viewer perspective. To put it differently, the percept, the interpretation of the visual content, depends on the sort of the visual scene recovered from the projection of the scene on the 2D retinal surface. However, depending on the nature of the projection, that is, on whether the projection is from a linear perspective or is a parallel perspective projection, the reconstructed scene may be, at least slightly, different. In other words, different perspectives may result in different recovered scenes and thus, percepts. The interpretation, therefore, of a pictorial representation depends on specific assumptions about the method of 3D projection used to reconstruct the visual scene. Thus, we come across one more time to the view that iconic representations are relative to a construction and interpretation system of rules.

I said that scientists usually employ the parallel or orthographic projection to solve the 3D projection problem because in this way depth information can be ignored. It is advisable to ignore this information because working the precise distances of the objects in a visual scene from one another and from the viewer is, computationally speaking, a very demanding task. Since in linear perspective projection depth information is somewhat preserved, it must be worked out and this is why parallel projection is preferred. However, as we have seen, the visual perceptual system most likely uses linear perspective and linear perspective projection, which means that it has to work the complicated computational processes involved in scene recovery from a linear perspective projection. An efficient and quick solution to this problem requires that the perceptual system use a host of heuristic assumptions about the depicted environment and, specifically, its geometry and physical properties. I have argued (Raftopoulos, 2009, 2019) that this sort of heuristic assumptions, along with a host of other assumption required to solve other underdetermination problems of the percept form the retinal projection of the depicted scene, constitute the operational constraints that are at work in perception.

I think that the mappings from the world onto the neural brain circuits, and through them to the contents of our visual perceptual experience are sustained by the way our neural circuits are causally related to and have been related through our ontogenetic and phylogenetic developmental history to, the surrounding world. To explain how this mapping could have occurred and why it did occur is not the subject matter of this book. Suffice it to say that any such account should involve our success (in fact, the success of any creature) in interacting with the environment (see Raftopoulos, 2008). It should also involve the way our causal interactions, through perception, with the world have shaped our neural circuits partly forming the neural structure of the brain in order to hardwire in the brain many realities about the geometry and physics of our world so as to facilitate or even render possible perceptual processing, by imposing on the brain a set of rules that constrain the sorts of processes that can take place when the brain processes incoming information from the environment; these principles I have called *operational constraints* (Raftopoulos, 2009). I revisit this topic in a while. Let me stress that these constraints are necessary to solve the underdetermination of the inverse problem by the sensory data in the image; the 3D objects that give rise to 2D projections in the retinal image cannot be uniquely recovered from this image because many different objects could have produced the same projections.

The main conclusion to draw from this discussion about the topologies involved in perceptual representations and the inter-mappings is the following. Recall, first, that a representation is iconic if it has an inherent syntactic and semantic structure that maps naturally onto the structure of the represented entity. The iconic nature of perceptual representations is grounded successively in the layout of the retinal cells that maps onto the spatial layout of the environment, and in the orderly retinotopic mapping of the visual world onto the surface of the cortex through the retinotopic mapping of the surface of the cortex onto the retinal cells. Thus, the physical layout of the retinal cells and the other visual receptors higher in the hierarchy of the brain renders registration of information from the retinal image iconic. The iconic registration of the retinal image maps onto representational states in the brain rendering them in turn iconic, and both map onto to visual perceptual representation in experience rendering perceptual content iconic as well. Thus, all these mappings are grounded in the mapping of the topology of information registration in the retina cells onto the topology of spatial and featural structures in the environment and this results in perceptual representations whose (syntactic) form preserves both the spatial structure of the represented visual scene, that is, the arrangement of the represented objects and properties of the objects in the visual scene.

In addition, the representational contents of the representations retain the spatial relations in the visual scene and also some of the qualitative features of the objects in it, namely those features retrieved by early vision. In this sense, perceptual contents resemble to a certain degree their representata with respect to some qualitative features and not just with respect to some abstract structure. Thus, the demand that iconic representations map in an appropriate way onto their representata concerns both the vehicles and the contents of the representations.

In view of these, Kosslyn's (1994) thesis that distance relations in the visual representational perceptual state preserve distance relations in the represented visual scene is correct, and, as I have argued, this holds true for both representational vehicles and their contents. This is not an essential property of iconic representations, however, because one can imagine pictures taken under such conditions so that the spatial relations in the representation do not preserve the spatial relations in the environment. Think of the 3D to 2D projection problem not in perception but in photography. In a picture two points may be close in the 2D spatial frame of the picture even though they are far apart in the third dimension lacking from the photograph.

If the above account is correct, this should alleviate Kulvicki's (2014, 106) worry that if one understands imagistic representations in terms of experienced contents—as when one defines a pictorial representation as one that manifests its content to a viewer, in the sense that it leads the viewer to experience visually the picture's content (along the lines of Ayer's (2019) view that we discussed above)—then one could not extend this account to explain the imagistic character of mental visual representations because, apparently, even though Kulvicki does not explain his reasons, in the brain there is only neural activity and talk of experienced contents is not readily available. Kulvicki's worry can be met by arguing, as I did above, that there is a structure preserving relation between brain states and experienced contents. Since Kulvicki (2014, 104) agrees that images preserve the structure of what they represent and in that sense they present things to us, he should agree that the structure preserving relationship between activity in the brain and the contents of experience suggests that the perceptual brain states can form an imagistic representational system.

Closing this part of the discussion on the iconic nature of visual perceptual representations, let us note that, if my views are correct, perceptual representations are iconic representations not simply because they map appropriately on to their domains but also because they resemble in some respects the visual scenes they represent. The resemblance is twofold. It concerns both the spatial structure and the relations between the objects in the represented visual scene, because a perceptual state (barring illusions and hallucinations) preserves or instantiates the spatial and semantic structure of the visual scene it purports to represent, and also preserves at least some of the qualitative features of the objects in the visual scene (otherwise, as I argued in Raftopoulos (2009), object segregation both from the background and other objects would not be possible).

It should also be noted that thus far I have claimed that a representation is iconic if it maps onto its represented domain in a way that preserves the spatial and semantic structure in this domain. I have not elaborated on the form of this mapping, however. If the structure of the representation maps onto that of the represented domain, does this mean that elements in the representation covary with the elements in the represented domain? And what is the precise form of this covariation? Moreover, how are changes in the represented domain tracked by the representation? With respect to the second point, I have said, and will elaborate in a while, that since iconic representations instantiate spatial and semantic relations in the represented domain, any changes in the latter instigate an automatic update in the former.

4.6 The Picture Principle and Iconic Representations

We saw that for several researchers what distinguishes at a last analysis iconic from digital representations is the *Picture Principle* (PP). Sometimes, PP is thought to apply to all iconic representations, but the discussion of PP by Fodor and others show that they mean to apply it to a specific subset of iconic representations, namely pictorial representations. Therefore, in the discussion about PP that follows, the reader should bear in mind that by "iconic representations" one intends "pictorial representations", and this applies more forcefully to visual iconic representations that are pictorial representations *par excellence*, as they present the world as being a certain way. The putative satisfaction of the *Picture Principle* by iconic representations has some immediate consequences for the way they represent a domain.

At a first pass, if by "parts of an iconic representation" one intends parts that are arbitrarily extracted from the iconic representation by cutting it in arbitrary ways, if PP holds it follows that no matter how one cuts an iconic representation the resulting part represents some part of the represented domain. Hence, all parts of an iconic representation contribute in the same way to the representation, in the sense that (i) any part represents by showing, or my being mapped onto, its *representatum*, and (b) that any part represents a part of the represented entity no matter how one cuts the representation. This means that iconic representations are homogeneous; any way you cut them, you get a representation. It also means that there is not a way determined by some rules (whether it be logical or syntactic in general) in which to decompose an iconic representation in order to get representing components, in contradistinction to digital representations where not any arbitrary part of a complex representation is a representation and where, in order to get a representing part, one must decompose the representation according to an established set of rules. Iconic representations, unlike digital representations, lack canonical decomposition (Fodor, 2007).

Iconic representations, as we shall see, have a structure and have parts but these parts do not have discrete representational properties, as in discursive representations in which some parts refer to individuals, other parts to attributives, and other parts to the relations holding among the individuals. These distinct representing functions are not applicable to iconic representations owing to the fact that they lack discursive structure but, instead, have a whole/part compositionality that does not support these distinctions.

The fact that iconic representations are homogeneous in the way explained above, in addition to the fact they cannot support and represent logical relations, entail that they cannot satisfy the *Generality Constraint* (which is the main point of Fodor (2007) and Heck (2007)), according to which if one thinks that a is F and one possesses the concept for G, one can think that a is G, or, if one thinks that a is F and one possesses the concept for object b, one can think that b is F. To satisfy this constraint, a representation must have contents whose constituents are systematic and recombinable according to some rules of good formation, and iconic representations do not have these properties. Fodor (2007) goes further to claim that iconic representations cannot distinguish between individuals and attributes because this requires that individuals and attributes are represented as individuals and attributes so that they can be separated, but iconic representations cannot do that because that would require the use of discrete symbols and having a discursive structure. By not having discursive structure, and, therefore, by lacking discursive compositionality, iconic representations lack syntactic and semantic structure.

Fodor joins a long tradition that Burge (2010) admirably assessed and criticized as *individualism*. Joining forces with Quine (1960), Fodor argues that individuation presupposes the capability to quantify over domains and pictorial representations do not offer this capability. Thus, 'there is no right answer to the question which things (how many things?) does this iconic symbol represent.' (Fodor, 2007, 110). Fodor things that the lack of the capability to individuate is the hallmark of iconic representations that sets them apart from digital, discursive representations. Fodor (2007, 110) understands that this may sound counterintuitive and this is why he hastens to add that a photograph may show three giraffes in the field, but, he adds, it also shows a family of giraffes, an odd number of Granny's favorite creatures, or a number of Granny's favorite creatures, and the picture itself provides no means to determine which one is the correct interpretation. Fodor's view that perceptual representations, by themselves, cannot support object individuation because they lack symbols/concepts, and object individuation requires some sort of conceptual apparatus has a long history in the Analytic tradition. The idea that object individuation requires more than mere perception may be traced back, in modern philosophy at least, to Frege (1892) and from there to Quine (1960), Strawson (1959), and Evans (1982).

Frege thought that to represent the world perception should be complemented with something non-perceptual, to wit, the capability to understand abstract, structured, propositional thoughts. Only when perception is supplemented with judgments does the representation of the world become possible. Frege does not seem to argue for this thesis, but Quine, Strawson, and Evans provide such arguments.

Strawson's (1959) holds that reference to objects requires more than a description of these objects, irrespective of the accuracy of the description. It requires that one be in a demonstrative perceptual relation with them. Demonstrative reference cannot be reduced to descriptions that do not contain indexicals because it is essentially contextual. The proposition expressed by a perceptual belief is not detachable from the perceptual context in which is believed, and cannot be reduced to another belief in which some objective content from a third person perspective is substituted for the indexicals that figure in the thought, because the belief is tied to an idiosyncratic viewpoint of the viewer by making use of the viewer's physical presence and occupation of a certain location. Thus, the singular element in the perceptual content "is an occurrent context-bound application of 'that' referring to a non-repeatable property-instance such as an object or event or a trope." (Block, 2014, 560, ft. 1).

According to Strawson (1959), reference to worldly objects presupposes a minimum number of constitutive conditions. Without them we perceive just bundles of features and the perception of a bundle of features does not necessarily mean perception of a thing that carries these features. One may respond to the presence of certain features that one has encountered before without being able to single out an object as an object that carries these features. To be able to refer to particulars, viewers should have the conceptual ability to distinguish between the way things seem to be and the way they really are, the conceptual ability to distinguish the subject from the predicate (which means that a viewer should be able to apply sortal concepts). They should possess the concept of causality, and they should be able to represent spatial relations within a spatiotemporal framework. In the absence of these conceptual capabilities, a perceptual experience is not the experience of particulars, individuated objects, but the experience of feature-placing universals that represent that certain conditions are instantiated somewhere in the environment. To pass from the stage of pure perception to the stage of experiential representation of particulars, the viewer must have the conceptual capacity to individuate objects, to recognize or re-recognize the objects in a visual scene and this requires that the viewer apply certain criteria.

For Evans (1982), the representation of particulars in perception and in perceptual beliefs presupposes that one know which is the object to which they refer and can provide a definite description that could individuate this object from other objects, that one can determine what category the object belongs to and place it in a determined place in space and time. Viewers should be able to individuate and categorize the object, which means that they can conceptualize the reasons that set an object apart from others. In addition, they should have propositional knowledge of the concepts involved in these capabilities, that is, they must know and represent what differentiates an object from another, and what makes this object to be the object that it is. It follows that perception has a representational content that refers to particular objects only if perception is related to certain concepts that allow demonstrative propositional thought that is required for an objective perceptual representation.

Evans, however, endorses two further theses that pave the way to a different consideration of what perception could do. He introduces the notion of 'nonconceptual content' even though this content cannot secure reference to particulars. Evans underlines the role of perception as a demonstrative act whose content is determined within a specified spatio-temporal framework. He introduces the idea that perception is inherently 'centered'. Evans thinks that a perceptual concept F refers through the demonstration "That O is F" while one points to the object O. A perceptual concept is about an object or of some of its properties if the attitudes of the viewer with respect to the contents that contain the concept are sensitive in the appropriate way to perceptual information about O. Evans thinks that this perceptual information cannot be conceptual because then we could not explain how the concept connects to the world. Thus, the essential perceptual causal relation has as *relata* the visual objects and their properties, and the nonconceptual information of the viewer.

Discussion of reference brings us to Quine's (1960) *Gavagai* problem that purports to show the indeterminate character of translation from a set of speech or mental acts. Quine's conclusion is that the reference of the terms in an unknown language is underdetermined by the set of linguistic expressions, since the observers of the behavior of the members of the tribe that use the language, even if they possess the entire set of the natives' linguistic expressions, cannot know whether "gavagai" that the natives use each time they see what the observers perceive as a hare refers to the hare or to a set of undetached parts of the hare that always move together. The observers' experience cannot settle this issue because each time a hare is present so is the set of undetached hare-parts. The only way to solve the problem is for the observer to have decided beforehand the applicable ontological framework.

According to Quine, the determination of the ontological framework that fixes the referents of the terms of mental states is related to the possession and application of criteria that individuate and categorize objects allowing object identification. These criteria are expressed in a language and, thus, language possession is a necessary condition for reference. Without these criteria, the reification problem, that is, how one passes from the feature-placing level of experience to the particular involving level of experience, could not be solved. Since the expression of the criteria of individuation and identification of objects in a language presupposes the existence of existential and universal quantifiers, the ability to refer to objects presupposes a complex logical arsenal.

Underlying all these views is the thesis that in order to individuate an object one must be able to refer to it, which, following Russell's principle that one can refer to O only if one knows what O is, and this presupposes that one know both what the object is and how it differs from other objects. Since at the nonconceptual level at which the contents of the perceptual representations of early vision reside (recall that early vision representations being purely iconic cannot contain concepts because concepts are digital elements), neither knowledge of an object's identity, nor knowledge of what sets this particular object apart from others is possible, Fodor's view that purely iconic perceptual content does not allow object individuation follows naturally.

To support the claim that iconic representations do not allow object individuation, Fodor says that a picture that depicts a number of giraffes shows three giraffes, or a family of giraffes, or an odd number of giraffes and which one of this is correct is undeterminable. His point seems to be that picture parts cannot fix any referents, which is why they cannot individuate the objects in the picture; individuation takes place just in case the parts of the representational vehicle refer to elements in the represented scene; hence, if reference fails there is no object individuation. Thus, according to some philosophers, iconic representations in general, and perception in particular lack discursive semantic and syntactic structure that would allow them to individuate objects and attributes and distinguish between them.

There are at least two problems with this set of views. The first concerns the *Picture Principle* itself, and the second concerns the putative lack of syntactic and semantic structure of iconic representations and the ensuing inability of these representations to individuate particulars and attributes in a visual scene. In the next section, I argue, first, that there are certain restrictions in the way the *Picture Principle* applies to iconic representations in general and perceptual representations in particular, which cast doubt on the claim that iconic representations are homogeneous in the sense that any way one cuts them one gets a representation, and, second, that the view that iconic representations do not have semantic structure and cannot refer to particulars and attributes is false.

4.7 Picture Principle, Semantic Structure, and Perception

The first problem concerns the *Picture Principle* itself, especially as it applies to iconic representations in general and visual perception in particular. Visual perception is a candidate for an iconic representational scheme that satisfies PP, because perceptual representations are functionally arrayed in space and, thus, amenable to the part/whole composition and decomposition that PP requires. *Pace* Fodor (2007), there are some limitations as to what constitutes a proper part of an iconic representation, such as a picture, a map, or a perceptual state. Not all parts and not all combinations of features of these iconic representations could be considered genuine representational parts of the image; it is not true, for example, that any part of a picture P that represents X is a picture of some part of X. First, not all parts of an image are representational. Think, for example, of the molecules of which a picture ultimately consists; even though they are parts of the image, they do not represent anything.

This is not to say that there are not representational primitives in iconic representations; the pixels in an image are primitives in that they cannot be further analysed as representational units, but they do represent location and color information.

In addition, there are some limitations concerning admissibility as a genuine representational part even for representational constituents of a map or a picture. A circle in a map represents a city, but half of this circle does not represent anything. This problem of PP shows more clearly in perception. Consider the back part of an object and a part of the immediate background in a visual scene and combine them. In perceptual terms, that is, in terms of what is computationally relevant in perception, it is highly unlikely that this complex part of the visual scene is represented by anything in the perceptual state representing this visual scene. One should bear in mind that X is a representation if it fulfils certain epistemological criteria (Bermudez, 1995) and is employed by some process that interprets the representation and uses it to achieve some purposes (later on I will connect this consideration with Pierce's "interpretant"). This latter factor introduces a pragmatical ingredient that concerns the reasons a representation is used for and the ways the representing system has been shaped or developed to achieve this purpose.

In the case of an organism's visual perceptual system, this means that what counts as a representation should factor in the way the organism interacts with its environment and the way this interaction has shaped its perceptual system. I have argued that the operational constraints inbuilt in our perceptual systems, by hardwiring in the brain regularities in our physical environment and its geometry, allow it to interact successfully with the environment. They constitute, thus, the "pragmatical baggage" of our visual representational system and co-determine what counts as a visual representation. This means that they co-determine in a principled way how our perceptual system curves the world for its representational purposes. In particular, the operational constraints guide the perceptual processes to reconstruct in perception the visual scene by individuating objects from the background and parsing them from other objects. In this perceptual setting, there is no room for representations of parts of objects conjoined to parts of the background. Burge (2018) offers a similar reasoning when he claims that iconic representations have canonical or natural decompositions due to our visual processing of them.

It is not true, thus, that any way one cuts a perceptual iconic representation one gets an iconic representation that depicts a part of the represented scene; the operational constraints limit the set of acceptable partitions of the representation; only parts that are admissible as components of visual processes are admitted, and which parts are these is an empirical issue. There is a similarity with the analogue situation in digital representations. As 'Φ('is not a constituent of '(Φa)' because it does not have a semantic interpretation that contributes to the semantic interpretation of '(Φa)', on account of the fact that it is not a syntactically well-formed expression, so some parts of a visual perceptual representation do not have any functional representational value, given the computational requirements of the perceptual system involved, and, thus, do not have a semantic interpretation that contributes to the semantic interpretation of the whole picture. Note that this is not done by applying some formal rules in a language, that is, a set of syntactic rules that are paralleled by semantic restrictions. There are no syntactic restrictions that preclude conjoining the representation of back part of an object with a representation of a part of its background. The fact that the visual system does not do that is due to the semantics and pragmatics of the representational system that constrain the coherence that results from the combination of the predicates and objects in the visual scene. In contradistinction, in a digital, propositional representation of the scene, it is the syntactic, formal rules that determine well-formedness.

To repeat the main point because it is very important, it is not true that in iconic representations anything goes, unlike in digital representations. There are rules of well-formedness in iconicity, except that they are not syntactic. In fact, the semantics of perceptual representations constrain the syntax or form of the representations, that is, the neuronal configurations that subserve the representation, in the sense that it is the semantics of the visual scene (the semantic structure and spatial relations of the scene) that determine which neuronal configurations may serve as representational states; recall, for example, that parts of the visual scene that are contiguous are represented by contiguous cells in the retina and from there on in the visual cortex, which means that the spatial relations in the scene co-determine which neurons will represent what. It follows that PP holds only for admissible parts of the image. The reverse of PP is unconditionally true; pictures of parts of X are parts of the picture of X because they are representations of parts of the existing natural whole and they cannot fail to be parts of the representation of the whole.

Fodor's account of PP and his talk about iconic representations being homogeneous and lacking semantic structure suggests that this is how he perceives PP. Tye, on the other hand, is more careful. As we saw, Tye in offering his version of PP, stipulates that 'every part of R that represents anything represents a part of O visible from the viewer. This immediately excludes all the counterexamples to PP mentioned above concerning the non-representational parts of iconic representations. Kulvicki (2014, 169) argues that Fodor, in discussing PP, should have focused his attention to the syntactic features of a representation, that is, those features that do some representational work, and not on the incidental features of an image (such as the back parts of a painting or the mass of the material used to make the painting) that do not represent anything. This is correct, but Kulvicki (2014, 159) also remarks that any disjoint part of an image represents a part of the represented item. I think, for the reasons explained above, that, at least, as far as the visual perceptual system is concerned, this disregards the way visual perceptual systems function to segregate objects from background and from each other.

One could, thus, hold onto some limited version of PP without necessarily agreeing with Fodor (2007) that one can arbitrarily segment icons into spatial parts that represent parts of the represented scene and are, thus, semantically significant for the representation, and that, in this sense, iconic representations do not have any sort of a canonical decompositions. Quilty-Dunn (2019, 25, ft. 7) argues that iconic representations satisfy PP since parts in the representation represent parts of the represented domain, but he specifies that this does not mean that one can freely segment icons spatially and preserve semantic significance, which means that it is not true that icons have no canonical decompositions at all. This raises the problem of how exactly PP applies to visual representations, and in particular which segments of the representation are not amenable to free segmentation. This is the problem that I tried to address by claiming that PP applies to those parts of the representation that have functional significance for the perceptual system.

I have argued that maps and visual perceptual representations do not satisfy PP. What about photographs? To answer that one must take recourse to the reasons maps and perceptual representations fail to satisfy PP. Our discussion above suggests that, in the case of perceptual representations, PP fails because perceptual representations of visual scenes are designed to represent objects and their properties, and the relations among the objects in the scene. Objects are discrete entities and, thus, their representations must involve discrete components too. Any decomposition of the representation in order to be able to represent anything must leave the components that represent objects [illegible] and cannot split them up, as it were. Of course, objects can be divided and thus, cut in half, which means that the corresponding representational component must be decomposed too. However, this decomposition is not free but must follow the rules that ensure that the resulting object is still a natural whole with the appropriate properties (for example, one cannot take half of an object and assign to it a color which that that part of the object did not have before but belonged to another part of the object). Thus, even in this case, it is not true that anything goes.

I argued that perceptual states, including states of early vision, contain discrete elements and, thus, are not continuous representations. I will discuss the nature of perceptual representations in detail in the following chapters, and I will claim that the states of early vision are purely iconic, which means that they do not have digital components, despite the fact that they involve discrete elements, the so-called object-files (see Raftopoulos, 2009 for an extended discussion of object-files). This means that I disagree with Lyons (2022, 6) who thinks that the reason some iconic representations do not satisfy PP is because they contain digital elements. After he has argued, rightly, that subway maps do not satisfy PP because they have discrete constituents, he concludes:

> Subway maps violate Fodor's picture principle. The left half of a given circle doesn't represent half—or any part—of the stop that the whole circle represents. Parts of the circles and the lines don't represent anything at all, contrary to (PP); they function similarly to sub-morphemic parts of words. Not because subway maps represent arbitrarily, as words do, however, because subway maps are discrete (/digital), as words are. Fodor thinks (PP) shows that iconic representations lack canonical decompositions into constituents; but this is simply mistaken. It's only continuous iconic representations that lack canonical decompositions; subway maps have clearly defined, canonical, constituent structure: only some cuts result in a smaller map. The same is true for other discrete iconic representations.

As we shall see, the reason I disagree with Lyons, as far as the iconic representations of early vision are concerned, is that the discrete elements in the representations of early vision that refer to objects in a visual scene are not digital components, their discreteness notwithstanding.

However, if PP holds under this proviso, the same could be said of digital representations, since it holds true for them too that admissible parts of digital representations represent parts of the representatum of the whole digital representation. There is one crucial difference, however, between the two cases. In perceptual representations, which parts are admissible is determined by the make-up of the perceptual system and the constitution of the world; thus, what is a part properly speaking of a perceptual/iconic representation is determined naturally and not through some convention. In digital representations, in contradistinction, what constitutes a proper part is determined by the conventions governing the relevant grammar.

Even if PP is understood in the constrained way I discussed above, it is still true that from (i) the fact that iconic representations do not have syntactic rules that determine well-formedness; (ii) the fact that they have a part/whole compositionality (more about this in Sect. 4.9), and (iii) the fact they cannot support and represent logical relations, it follows that they cannot satisfy the *Generality Constraint* (Fodor, 2007; Heck, 2007). To satisfy this constraint, a representation must have contents whose constituents are systematic and recombinable according to some rules dictated by logic of a syntactic theory (in cases of languages), and iconic representations do not have these properties, because well-formedness is not determined syntactically.

According to a simple application of the *Generality Constraint*, if one thinks that a is F, one should be able to think that a is not F. Perception lacks the apparatus to express negation, as all other logical connectives, and thus if one perceives a as being F it does not follow that one should be also able to perceive a as not being F; in fact, the latter does not even make sense in perception. Moreover, the fact that one perceives a as being F and b as being G does not entail that one could perceive a as being G or b as being F, for the simple reason that a may not admit the attribute-type G as a possible attributive. Or, to borrow, and expand on, an example from Heck (2007), from the fact that one perceives object O1 as closer than O2 and from the fact that O2 is viewed as closer than O3, it does not follow that one perceives O1 as closer than O3, because the relation "closer than" is not transitive in perception, as is in discursive structures. The cause of this lies in the local character of visual representations in which one can perceive the relation between O1 and O2 and between O2 and O3, even though one may not be able to perceive the relation between O1 and O3.

I have said that the second problem with the Fodor's account of PP is the claim that iconic representations, by not involving digital symbols and by satisfying PP, fail to refer and, thus, cannot individuate individuals and attributes in the represented domain; iconic representations lack essentially semantic and syntactic structure. Recall that a digital representation is semantically structured since its semantic interpretation is exhaustively determined by its syntax together with the semantic interpretations of its lexical primitives (for the reasons explained above, I am restricting discussion to formal systems and languages that do not allow context and other factors except the meanings of the lexical constituents of a complex representation to determine the meaning of the whole). I have argued (Raftopoulos, 2009, 2019) that perception acts so as to individuate objects in the visual scene by parsing them from other objects and by segmenting them from the background. It does that by demonstratively referring to them in a contextual, *de re*, way. Object individuation occurs through the processing of the object-features (primarily spatio-temporal information) even though these features are not permanently assigned to the objects. The individuated objects are assigned object-files that initially contain only transient featural information that may change; these files are accessed by pointers or labels. It follows that the representational states of early vision consist of objects that have some properties, namely those properties used in the process of object-individuation. The representations of a visual scene in early vision, therefore, represent a manifold of objects, properties and events.

According to Crane (1992), even though the nonconceptual content of perception does not have the structure of judgeable content, it still represents a manifold of objects, properties, and events. Peacocke (2001, 241) argues that the nonconceptual content of perceptual experience represents things, events, or places and times in a certain way, as having certain properties or standing in certain relations, "also given in a certain way." I have argued, (Raftopoulos (2019)) argues that early vision, the stage of perception whose states have nonconceptual contents, represents a structural description of the visual scene that contains information about 3D shapes as viewed from the perceiver, spatio-temporal and surface properties, color, texture, orientation, motion, and affordances of objects, in addition to the representations of objects as bounded, solid entities that persist in space and time (although later on, in Sect. 6.1, I qualify this last thought). The numerosity of objects is also represented. The contents of early vision are context-bound and demonstratively applied (Raftopoulos, 2009; Raftopoulos & Muller, 2006) in the sense that these contents are inextricably tied to the context of the perceptual encounter and to their *de re* application to the objects and properties in the visual scene.

Burge (2014, 574–575) argues for the existence of 'preconceptual cognitive representations', where 'pre-conceptual' intends to convey that these representations are not propositionally structured, and also claims that these pre-cognitive, perceptual as we shall see, representations have the same structure as noun phrases, the difference being that they do not employ concepts as linguistic noun phrases do.

> Another type of (cognitive) pre-conceptual representation is formed through learning or other processing in long-term memory (modal or amodal). Even certain operations on working visual memory count as cognitive. Like perceptual-level representation, the foregoing cognitive representations are pre-conceptual in lacking propositional representational format.

These pre-conceptual cognitive representations are perceptual because

> Like perception, these types of pre-conceptual cognitive representation have the same structure as noun phrases constituted of contextual-determiner-dominated attributives—the structure of that F or those Fs. When representation occurs, the representational types are applied in a demonstrative-like manner. In all perceptual-level and most pre-conceptual cognitive level representations, such determiner-governed attributions are part of a complex iconic array. Visual perception consists in a rich, topographical array of demonstratively applied attributives, at various levels of specificity.

The demand that iconic representations do not have a discursive structure is equivalent to the thesis that the contents of iconic representations are not sets of structured propositions. Therefore, iconic representations do not support Boolean combinations (Crane, 2003, 2009; Heck, 2007; Kulvicki, 2006; Sainsbury, 2005). If iconic contents were sets of structured propositions, there is no reason why they should not be able to represent Boolean combinations (including negations) of the propositions that figure in their contents, as Heck (2007, 126) justifiably argues. A second reason why iconic contents cannot be sets of structured propositions is adduced by Heck (2007, 126), when he argues that there is no unique description is the content of an iconic representation in terms of propositions because an iconic representation (in Heck's case a cognitive map) lacks the syntactic structure of the set of structured propositions that purport to capture, or verbally describe, the content that the iconic representation represents. As I noted above, this should not be taken to mean that iconic representations lack any sort of syntactic and semantic structure, since they have the structure expressed by the basic relation R. The point is, rather, that this structure should be expressed in terms of part/whole compositionality rather than in discursive compositionality. This view is partially shared by Camps (2018), who argues that, for the reasons discussed above, iconic representations do not have a predicate structure as digital representations do. However, whereas Camps holds that the way iconic representations compose entails that they do not have a predicate structure, I shall argue in the next section that iconic representations have the predicate structure of the R-relation, as this relation applies to iconic representations, and that this sort of predication functions differently in iconic representations than in digital representations.

Contra to the thesis defended here, Grzankowski (2015) aims at showing that pictures have propositional and not iconic contents, and starts by rejecting the view that the fact that pictorial representations cannot represent negations or other Boolean operations entails that they do not have propositional contents. Grzankowski (2015, 159) accepts that one cannot purely pictorially express a negated content. However, he argues correctly, this does not entail that iconic representations do not have propositional contents but only that they cannot express some sorts of propositional contents, namely, negated contents or other Boolean combinations of contents.

Grzankowski then proceeds to offer a positive argument that pictures have propositional content. He starts (ibid., 160) by arguing that when one denies an iconic representation such as a picture (when, for example, one holds a picture and utters 'this is not how it was'), one denies in fact a set of propositions, which means that the picture expresses a set of propositions. Grzankowski (ibid. 160–161) notices immediately that there is an obvious reply to this objection, a reply that is based on the distinction between the content expressed by a representation and the content that can describe the expressive content of a representation. These two need not coincide; an iconic representation could have, and, thus, express, iconic, non-propositional content, but obviously there is some propositional content that could describe, at least partially, the non-propositional content expressed by the iconic representation. Thus, by denying a picture one does not deny its content but the set of descriptions (or a subset of them) of the picture's content, and, thus, the fact that one can deny a picture does not entail that the picture has propositional content.

Grzankowski's (ibid. 161) rejoinder to this objection is that the fact that one can use modal operators with pictures (when, for example, one utters by looking at a picture 'this is possible', where "this" refers to the picture's content) entails that the picture has or expresses some propositional content because modal operators are sentential operators and apply to propositional contents. This holds true on account of the fact that modal operators refer to, or make use of, situations in some possible worlds that might actually obtain and in which the proposition is true. It follows that the possibility of applying modal operators to pictures entails that pictures have necessarily propositional contents. I think that the argument is weak, to say the least. I fail to see why the possibility of applying modal operators cannot be treated the same way one dealt with the previous objection concerning truth-functional operators, that is, by claiming that the operator applies to the content that purports to describe the non-propositional content expressed by the picture. Involving possible worlds does not preclude this possibility, one could very well attach some description to a picture and proceed to think in terms of possible worlds. Grzankowski offers no argument against this possibility.

Though Grzankowski fails to offer a positive argument as to why iconic representations have propositional content, his initial objection is valid. From the fact that a picture cannot express a negated content or a Boolean combination of contents, it does not follow that pictures cannot have propositional contents. The problem with Grzankowski's dialectic, however, is that no one argued that pictures have non-propositional contents simply because they cannot express, say, negated contents. This is a consequence of the fact that they do not have propositional contents, rather the reason that explains why they do not have propositional contents. The reason that iconic representations do not have propositional contents is that they, unlike propositional, or digital in general, representations do not compose discursively (Camps, 2018; Crane, 2003; Fodor, 2007; Haugeland, 1987; Heck, 2007; Kulvicki, 2006, 2015).

4.8 Compositionality in Iconic Representations

If perception represents the environment in such a way, the accuracy conditions of a perceptual representation are fixed by the way the properties and objects in the represented scene are bound together to form the parts of the perceptual representation of the scene, and by the way the parts compose to form the whole scene. The problem is that if this composition is not discursively determined by some syntax, as in the case of symbolic representations, one should explain how the parts of iconic representations compose to form the whole. Merely conjoining parts does not suffice to yield the whole, because the atomic parts must be placed together in a way that the spatial and semantic relations in the represented scene be preserved in the perceptual representation. In iconic representations semantics constrains syntax.

The accuracy conditions of a perceptual representation are determined by the accuracy of the representation of its parts and by the accuracy of the way these parts compose to form the whole representation. The difference from digital representations consists in that the composition of iconic parts is not determined by a set of formal rules that constitute the syntax of the representational system, but, as we shall, see by Mereotopology; iconic representations do not compose discursively but holistically.

Let us start by discussing the structure of the content of the visual perceptual representations of objects and of their properties in a visual scene. Since this content consists in objects and their properties arrayed in space, one could render this representational content in the form of a set of subject-predicate structures. If this is correct, it follows that perceptual iconic representations have a rich semantic structure that is to a certain extent similar to the subject-predicate structure of linguistic representation, or of all representations. I refer to this structure as the basic relation R of noun phrases. Let me elaborate here on R. The rich semantic structure of an iconic representation of a visual scene (be it a perceptual representation or a picture or a drawing of the scene insofar as the latter exploits perceptual capacities to interpret the scene) consists in the compilation of various R-relations binding objects to predicates, a compilation that composes the whole out of its parts (the R-relations). The compositionality involved in visual iconic representations is not the discursive compositionality grounded in hierarchical relations and recursive structures that characterizes digital representations, but the compositionality befitting part-whole structures that involve different kinds of restrictions as to what counts as a proper part of the representation, and engage different combinatory mechanisms that underwrite the compilation of parts to form wholes, mechanisms described by Mereotopology.

Accordingly, what counts as a well-formed construction in digital representations is determined by the formal rules of the relevant grammar; the restrictions are at the syntactic level, and they fix in turn the semantics of the representation, that is, they determine which parts are meaningful or not. Things are different in iconic representations, especially visual icons, where what counts as a well-formed part is determined by semantics and pragmatics, in the form of the operational constraints at work in perception, and not by syntax. Moreover, the compilations of parts to form wholes is fixed by formal rules in propositional representations. In iconic representations, in contrast, the compilation of parts to form a whole is not determined syntactically either. As we shall see when we discuss Mereotopolgy, what counts in iconic representations as a well-formed, natural whole is determined geometrically and not formally. The compilation, thus, is not the discursively analyzed composition of digital representations.

These have implications for the nature of R-relation as it is figures in iconic representations and distinguish it from the R-relation in digital representations. First, predication works differently in the two sorts of representation. Predication is a combinatorial operation or compositional mechanism that connects predicates to subjects in language, or to objects in perception. The objects fill the argument-places of a predicate to form a full or saturated expression that has a truth-value. In a propositional representation of a visual scene, one forms atomic sentences that have the R structure that represent parts of the scene, and conjoins them to form the representation of the whole scene. The construction of such "atomic" sentences is done as follows: one identifies an object and predicates to it the predicates that represent the properties of the object in the visual scene. Then, one conjoins the atomic propositions to get the propositional representation of the entire scene. In this construction, the spatial relations between the objects in the scene are not represented directly, but they can be derived if the properties predicated to the objects include spatial information. This entails that changes in the spatial relations in the represented visual scene are not automatically instantiated in the sentential vehicles of the digital representation. In an iconic representation, in contrast, the spatial relations among the objects in the scene are instantiated in the iconic representation (both vehicle and content in perceptual representations, and in the vehicle in maps, pictures or drawings), so that any spatial change in the scene automatically updates the representation.

Iconic representations organize their predicative structures (the R-relation) in fundamentally different ways than sentential representations, exactly because they exploit spatial structure and place objects and properties at locations. Not only in the sense that in an iconic representation an object and its properties as they are perceptually represented are tied to a specific location (and one could add time), which explains the well-known situatedness of perceptual representations and ties them to a specific spatio-temporal context, but also in the sense that spatial information is primarily used to individuate objects in the scene (Raftopoulos, 2009) and is part and parcel of the perceptual representations of the object. In addition, iconic representations do not allow the hierarchical, asymmetrical structure of propositional representations; there are, for example, no nested perceptual representations within other perceptual representations. In this specific sense, iconic representations have a flat structure (Camp, 2018; Deutscher, 2005; Everett, 2017). This means that the combinatory mechanism binding atomic predicative representational components differ between iconic and digital representations. In digital representations, we have a digital, hierarchical, universal, and asymmetrical combinatory mechanism, while in iconic representations the mechanism is holistic, spatially bound, and allows automatic updating in the representational vehicle when spatial relations in the represented domain change (Camp, 2018, 41).

Sterelny (2020, 778), starting from a distinct perspective reaches the same conclusion, arguing that

> Iconic representations, such, maps, photos, diagrams, and the like … can convey a lot of complex information asynchronously and in a more or less language-independent ways … Moreover, in contrast to other powerful graphic codes, the expressive power of these depictions does not depend on the recombination and recursive interpretation of a much smaller number of standardized signals.

Let me emphasize that in the preceding paragraph I wrote 'in this specific sense', because there is another sense of "flat" that does not apply to iconic representations. There are wholes that are flat in the sense that they have no internal organization. Mere sums of elements are a typical example. There are also whole that have an internal, syntactic or semantic, structure and, thus, there is a division into levels of organization. Iconic representations such as images of any kind are such examples. The representation of a house clearly consists of parts that are hierarchically organized into parts of parts, etc.

The semantic interpretation of an iconic representation of a visual scene is exhaustively determined by its whole/part relations together with the semantic interpretations of its parts, that is, the objects and properties to which the analogues of singular terms in perceptual contents, and the attributives in the representation refer. Thus, perceptual, iconic representations are semantically structured. They have the syntactic structure of the basic relation R cast not in a discursive form but in a whole/part form that constraints the sorts of relations that hold between the syntactic parts of the representations.

On a more general note, and returning to the idea that iconic representations have the syntactic structure of the basic R relation that they share with noun phrases, and thus, that iconic representations can express semantic relations that R makes possible, it should be added that perceptual iconic representations in particular and iconic representations in general share two characteristics with linguistic representations that use digital symbols, on account of the fact that they both exemplify the basic R relation. The first is that iconic representations are constructed on the basis of a set of construction rules. This is readily shown in the case of perception; the perceptual system constructs the percept, which is an iconic representation at the level of representational content, first, on the basis of a set of transformations that gradually transform the light pattern impinging onto the retina to the full-blown percept. Marr (1982), for example, was the first to give an account of such transformations from light patterns to primal sketches, 21/2D, and finally 3D visual representations. This set is essentially a set of mathematical transformations that allow the perceptual system to extract from the pattern of light that is the input to the visual system the spatial and semantic configuration of the entities in the visual scene from which the light emanates.

This does not suffice, however, for the construction of the percept. As many have argued, these transformations must be supplemented by a second set of rules that express both the geometry of the environmental space and the basic physical regularities of the interaction between light and our perceptual systems, which I have called operational constraints (Raftopoulos, 2009). These two sets of rules that operate in, and render possible, visual perception are construction rules that parallel the syntactic rules of linguistic systems with the major difference, however, that the latter are conventionally established and culturally learned systems, while the former are neither of these. Thus, perceptual representations are syntactically decomposable, albeit in a non-discursive way, and have a syntactic, albeit non-discursive structure. It is important to note that the operational constraints function by restricting the solutions available to perceptual processing, solving thus the problem of the underdetermination of perceptual processing from the input. Their role is to ensure that perception retrieves form the environment and represents information that is useful for the organism in its interactions with the environment. In other words, these rules contribute to retrieving from the environment and representing (when representations are required) information that is meaningful for the organism, by hardwiring in the perceptual system of the organism the physical and geometrical regularities in the environment that are useful for its survival. In this sense, it is semantics that drives the form of the representation rather than the other way around as in digital representations.

This reintroduces in the discussion the notion of the interpretant that we first encountered in Peirce's taxonomy of representations, In the case of perceptual representations, the interpretant is hardwired in the perceptual system and determines what information is selected to be represented and how these representations are used by the perceptual system. As the reader recalls, the interpretant uses only certain aspects of the representation to process it and interpret it. In the case of the perceptual system, these are the aspects that are representationally relevant, the relevance being determined by the sort of the organism and of the environment in which it lives. Thus, only these aspects are taken into account and the others are irrelevant. This explains the restrictions imposed on the way an iconic representation can be decomposed and falsifies Fodor's view that anything goes. It follows that in iconic representations, the semantics and the pragmatics of the representation determine the way a representation is composed by, or decomposed from, its parts and not a set of syntactic rules.

In Chap. 3, we discussed the discursive compositionality of symbolic representations, and, in this section, we started examining the whole/part compositionality of iconic representations. Let us turn now to discuss the representational format of the hybrid representations, that is, the representations that consist of both iconic and symbolic elements, such as the representations of late vision, the representation in VWM, or the tank sketch that we have discussed. If both symbolic and iconic elements are involved, what is the main format of the representation, or, how are the two formats combine? This is important in understanding how the hybrid representations functions in cognitive economy.

I have argued that the representations higher in the hierarchy of the perceptual brain, such as the representation in late vision, are hybrid representations with both iconic and symbolic components. The iconic ingredient may contain discrete elements (recall that iconic contents can have discrete components and even digital symbols/concepts (that refer by convention) since iconicity is defined by the existence of a natural appropriate mapping between the representation and the represented domain, and as we have seen this may occur even if a representation contains symbols/concepts. Conceptual contents and iconicity are not incompatible, since symbols may be parts of hybrid iconic representations. Hybrid iconic representations that contain concepts/symbols may still be iconic since they preserve the spatial or semantic structure of the represented domain, and since the elements in the representation map onto elements in the represented scene. This is so because the main representational frame is iconic and spatially arrayed and the symbols piggyback on this frame.

In discussing VWM, we saw that VWM representations involve both icons and symbols. We have also seen that the representations in late vision are also hybrid representations. Haugeland (1998, 187) argues that there are hybrid representations, such as maps, in which "piggyback" devices (the symbols) systematically belong, and contribute, to the iconic representations that carry them. This means that the main representational format that does much of the work is the iconic format and the symbols just add information. In the hybrid representations of VWM, late vision, and maps, the discrete differentiated elements that constitute the symbolic component of the hybrid representation are embedded in the continuous, spatial format of perception. Visual perception, as we have seen, consists in a rich, topographical array of demonstratively applied attributives, at various levels of specificity. In general, in representations whose medium format is spatial, the discrete elements, such as symbols, are embedded in a spatial, continuous medium. The numbers ordered serially around a clock face, for example, are embedded in a circular spatial continuum; although the continuum is not part of the notation itself the serial ordering of numbers is taken to be an analog of the serial flow of time as displayed by the continuous spatial medium. Or consider the case of diagrams that, in order to be comprehensible, use of symbols. These symbols are embedded in the spatial array of the diagram. One could say that the main format of perceptual representations of this sort is iconic, and that the symbols they contain are arrayed in space, as the labels with the names of the cities on a map are embedded in the iconic representation and piggyback on it; the symbols are superimposed on the iconic representational format.

Which format is the main should determine the way the hybrid representations compose. Thus, if the main format of VWM representations, late vision representations, and maps, *qua* hybrid representations is iconic and the symbols piggyback on it, these representations should compose according to the parts/whole manner that signals iconic representations, and not discursively, which is the way symbolic representations compose. The same distinctions hold with respect to the ways these representations decompose, because the decomposition of representations is determined by the way the representations compose in the first place. Let us consider a map that portrays the locations and distances of cities in a country, in which each city is marked by a point and overlaid by its name, which is a digital component embedded in an iconic representation. How are the various parts of the map composed to form the map as a whole and, equivalently, how does the map decompose? Suppose that when decomposing the map, we take care to avoid those decompositions that are not amenable to the natural function of our visual perceptual system (such as conjoining a part of an object with a part of the background) and which restrict the application of PP in perception. Suppose that we cut the map into half and we disregard for the time being the digital components of the representation, namely the names of the cities. The two resulting parts are clearly iconic representations (they both map naturally onto their represented domains, and, in addition, both parts satisfy complicity), each one representing the distances and locations of roughly half of the cities in the country. Thus, PP holds true, and the same holds for any way in which one could cut the map. In other words, the half map consists of well-formed parts. In view of this, one has to conclude that this representation is composed in a whole/parts manner. Let us now bring back the digital components of the representation, that is, the names of the cities. If the decomposition into half is such that the names remain intact, there is no problem and the two parts of the divided maps continue to provide information about half the cities each.

However, what happens if the decompositions cuts into half the name of a city, while the point representing the city falls into one of the two halves? In this case, the city cannot be identified and, thus, cutting the map results in some loss of information. More importantly, this decomposition results in a representation that has gaps, as it were, in the sense that one of the cities in not named and, thus, that part of the representation does not represent its intended domain. To determine whether the decomposition results in well-formed parts, it is not enough to examine whether the mapping between points on the map and cities is preserved. One should also determine whether the digital components that result from the decomposition are well-formed. This, in turn, requires that one should check whether the resulting parts satisfy the rules of some grammar or logic, and this is discursive procedure. One could substitute the name by looking at another map, or by inferring its name on the basis of knowledge of the country, but this would involve discursive inferences. In this sense, PP does not strictly apply. The upshot of this discussion is that the iconic representations that are not purely iconic (recall that Fodor's (2008) original discussion of PP concerned pictures that are pure conic representations) do not compose in a whole/part way only, but discursive compositionality (compositionality that conforms to some rules of good formation) must be taken into account as well. Thus, there is not a strict parallelism between iconicity defined in terms of mappings and iconicity defined in terms of compositionality, since iconic representations that involve digital elements and which map appropriately onto their domains do not compose only in a whole/part manner.

In contradistinction to the representations discussed above, the main format of VLTM hybrid representations is sentential-like and the iconic pictorial elements may occur as slots in the sentential structures, as when one reads a sentence and some of the words in it have links to pictures or drawings; the representations compose discursively and carry the iconic representations along. Finally, some representations in VLTM, such as phenomenal concepts, are symbolic, which, when activated, recruit perceptual representations, but they are themselves quasi-iconic in Lyon's (2022) sense because they form a representational iconic system.

4.9 Whole/Part Compositionality in Perception

I have argued that iconic representations do not have the canonical compositionality of discursive representations, but are characterized by a part/whole compositionality. I will elaborate now on this sort of compositionality, starting from an example of what I mean by part/whole compositionality, as it reveals itself in the phenomenology of visual perception. 2-D surfaces are composed of edges that, in turn, are composed of line-segments. When one perceives in realistic conditions a line segment, one sees it as part of an edge, which is also a part of a surface, despite the fact that the perceptual system constructs first line segments, then edges, and then 2-D surfaces. In other words, segments of an edge are perceived by perceiving the edge and, thus, are perceived as edge parts and not as independent perceptual units. The same holds for properties such as size and brightness. Burns (1987) shows that size and brightness are not encoded at first as independent features of objects but are perceived in terms of holistic objects.

This bears directly on why iconic symbolic representations having whole/parts compositionality do not satisfy in general the principle of semantic constituency that characterizes, or even defines, digital symbolic representations. According to this principle, if a representational vehicle *s* is a syntactic part of a representational vehicle *p*, the meaning of *s* is a semantic part of the meaning of *p*. Even though syntactically line segments are parts of edges, which in turn are parts of 2D surfaces, because this is how the neural system constructs the representations, at the semantic, phenomenological level, line segments are not perceived as semantic parts of edges in the sense that to perceive phenomenologically an edge one need not phenomenologically perceive a line segment; in fact, one does not. To put it metaphorically in conceptual terms, one understands an edge without understanding line segments, even though the latter are syntactic constituents of the former. As a matter of course, iconic representations satisfy semantic constituency at some level. If iconic representations bind holistically parts together, some of these parts, say the objects in a visual scene, are individuated as distinct patterns in the representation, and the meaning of the whole representation depends on the meaning of these parts.

The difference of semantic constituency to the extent that it applies to iconic representations from the same principle as it applies to symbolic representations lies in the fact that in iconic representations the meaning of the whole is not determined by combining parts at the syntactic level following the rules of a formal syntax, as it happens in symbolic representations. In symbolic representations, one composes syntactically well-formed expressions following a set of syntactic rules and it is the syntactic composition along the semantics of the atomic parts that determines the meaning of the whole. In iconic representations, in contrast, one does not rely on some formal rules to compile parts to form a whole. Visual iconic representations, for example, are constructed by the process of the visual system and the visual processes of this system that form parts and combine them to form the whole (that is, the percept) are not guided by, and do not underwrite a, set of formal rules but are driven by the stimulus, and are guided by a set of constraints that express the physical and geometrical regularities of the environment in which the perceiving organism lives. In this sense, it is the semantics and pragmatics of the system formed by the representation and the represented entity that guide the formation of the whole, rather than syntax.

A characteristic of many iconic representations is the use of space or some geometrical structure in representing the representatum; this spatial or geometrical structure maps onto the spatial structure of the representatum. Visual representations, a clear-cut case of iconic representations, have a two-dimensional or three-dimensional form. This is the spatial form that grounds the iconic structure of the perceptual representation; spatial representations preserve the spatial relationships of the represented entities owing to the mapping of the spatial relationships between the elements of the representation onto those in the representatum. Visual perceptual representations, being spatial representations, inherit this property and, thus, preserve the spatial relationships of the entities (be it objects or properties) in the visual scene. This is, of course, one aspect of the iconicity of visual representations, the other one being that some of the elements of the representational content of the visual representation map into the visible features of the objects in the represented visual scene.

The whole/part compositionality constrains the way the parts of an iconic representation are put together to form the whole: 'iconic representations acquire accuracy conditions from the way features are holistically bound in each part, together with the spatial arrangement of those parts.' (Quilty-Dunn, 2020). I said above that line segments are put together to form edges, which in turn, put together in appropriate ways, form 2-D surfaces. Moreover, one does not compile a surface from edges by following some logical, or syntactic in general, rules of construction of complex representations. Owing to the whole/part compositionality of the representation, the parts are perceived as parts of the whole rather than as distinct entities; one does not perceive line segments as distinct entities but as parts of edges. In discursive representations such as '(Φa)', in contrast, 'Φ' retains its autonomy and can be seen independently of the whole (Φa); this is the result of the compositionality of discursive representations.

The part/whole compositionality seems to entail that since line segments are parts of edges and edges are parts of 2-D surfaces, the line segments are parts of the 2-D surface, a conclusion that seems warranted. However, even though the handles of a door are parts of a door, and the door is a part of a house, the handles are not properly speaking parts of the house? How is the latter case different from the former? The answer to questions like these should be searched in the mathematical models of Mereology and Topology, the two disciplines that examine part/whole compositionality.

All models in mereology start by defining at a first pass the 'is part of' relation by means of three axiom, namely, (P1) Pxx; (P2) Pxy ∧ Pyx → x = y; (P3) Pxy ∧ Pyz → Pxz, where 'P' is a two-place predicate to be interpreted as the parthood relation. In other words, the parthood relation is reflexive, antisymmetric, and transitive.

I said in 'a first pass' because this definition has a serious shortcoming. If parthood is transitive, since the handle is a part of door and the door is a part of a house, it follows that the handle is a part of the house too. To avoid this conclusion, the general intended interpretation of "P" by the three axioms should be narrowed by introducing some additional conditions restricting its application. Such a condition involves functionality, which requires that x be a part of y if x makes a direct contribution to the functioning of the whole of which x is a part. In this case, the handle is a functional part of a door and, thus, it is a proper part of the door, and although the door is a functional part of the house, the handle is not a functional part of the house and, thus, the handle is not a proper part of the house. Mathematically put, where 'φ' is any formula in the language, the implication: (1) (Pxy ∧ φ[x, y]) ∧ (Pyz ∧ φ[y, z]) → (Pxz ∧ φ[x, z]) may well fail to be a theorem of the mereological theory if x is not functionally related to z. Note that the functional restriction has its own problems. A spot on the door that is painted differently from the rest of the door is a part of the door, but it adds directly nothing to the functioning of the whole entity of which it is a part, that is, the door. Be that as it may, this discussion clearly bears on the problem of which parts of an iconic representations can be construed as parts of it properly speaking. I have argued that the part consisting of a part of the back of an object conjoined to a part of the background has no functionality in perceptual processing given the nature of our perceptual system; hence it is not a proper representationally speaking part of the representation, according to the restriction introduced in Mereology.

To address the aforementioned, and other problems, the meaning of "P" is further clarified by a set of additional theorems that differ from one Mereology theory to another. Here I consider only those extensions that will be used in defining the sum or fusion of two parts to form a whole. These are the following:

(i) PPxy $=$ df Pxy $\cap \neg$Pyx (Proper Part), that is, x is a proper part of y iff x is a part of y and y is not a part of x;
(ii) Oxy $=$ df $\exists$ z(Pzx $\cap$ Pzy) (Overlap), that is, x and y overlap iff there exists a z such that z is a part of x and z is a part of y; and
(iii) Uxy $=$ df $\exists$ z(Pxz $\cap$ Pyz) (Underlap), that is, x and y underlap iff there exists a z such that x is a part of z and y is a part of z.

To define the sum of two parts the theory is extended so as to constrain further the meaning of parthood. The first extension is:

(P4) $\neg$Pxy $\rightarrow$ $\exists$ z(Pzx $\cap \neg$ Ozy), that is, if an individual has a proper part, it has more than one. (P4) entails the following theorems.
(P4a) |– PPxy $\rightarrow$ $\exists$ z(PPzy $\cap \neg$Ozx), that is, if x is a proper part of y, then there exists a z such that z is a proper part of y and it is not the case that z and x overlap;
(P4b)) EM |– $\exists$ zPPzx $\cap$ $\forall$z(PPzx $\rightarrow$ PPzy) $\rightarrow$ Pxy; and
(P4c)) EM |– $\exists$ zPPzx $\cap$ $\forall$z(PPzx $\leftrightarrow$ PPzy) $\rightarrow$ x $=$ y, which means that no two distinct objects can share the same proper parts.

The second extensions is:

(P5) Uxy $\rightarrow$ $\exists$ z$\forall$w(Owz $\leftrightarrow$ (Owx v Owy)), that is, that if x and y undelap, then there exists a z such that for every w, w and z overlap iff either w and x overlap or w and y overlap.

In a Mereological theory that holds (P1) to (P5), the following "sum" definition can be supported:

(Sum Definition) x $+$ y $=$ df iz $\forall$w(Owz $\leftrightarrow$ (Owx v Owy)), where 'i' is a description operator in the language. This says that the sum of two parts x and y is defined as follows: for every w, w and x overlap iff either w and x overlap or w and y overlap. Adopting the ' $+$ ' sum operator allows us to recast (P5) as follows:
(P5′) Uxy $\rightarrow$ $\exists$ z(z $=$ x $+$ y), that is, that if x and y undelap, then there exists a z such that z is the sum of x and y.

The sum definition determines the sum of two parts that are combined or fused to form a whole. This definition can be generalized to yield the notion of Unrestricted sum: $\exists w\varphi_w \rightarrow \exists zSiz\varphi_w$, where φ is any formula in a language (which picks out the components or parts that form the sum, say the grains of sand in a pile of sand). This definition stipulates that if there are parts φ_i, then there is an object z that is the sum of φ_i.

The most pressing problem with Mereological theories, a problem that pertains directly to the compositionality of iconic representations, is that the way the "sum" operation is defined, Mereology by itself cannot capture some of the basic properties we attribute to wholes (for example, a whole is a one-piece, self-connected entity, such as an object in a visual scene, as opposed to a scattered entity made up of several disconnected parts, such as parts of different objects). Parthood is a relational concept, whereas wholeness a global property (Varzi, 2016, 269), and, thus, the former, and the ensuing definition of "sum", cannot capture completely the meaning of "wholeness". This shows immediately if one considers that in Mereology for every whole there is a set of parts, and that for every specifiable set of parts (for example, arbitrary objects) there is in principle a complete whole, i.e., its mereological sum, or fusion. As the "sum definition" shows, the only restrictions concern the overlapping relations between the combined parts, but this is hardly enough to constrain the definition so that only combinations of parts that result in a whole *qua* object that satisfies our basic understanding of what should count as an object. Thus, within Mereology itself, there is no way to draw a distinction between "good" and "bad" wholes, and, thus, there is no way to distinguish between an integral whole and a scattered sum of disparate parts (yes, this is the *Gavagai* problem).

The problem is that it is not possible, on pure Mereological grounds, to determine the appropriate restrictions that would permit the fusion of parts in ways that allow only the formation of integral or natural wholes (such as objects or whole visual scenes), and would exclude the formation of sums of disparate parts or concrete heap-like composites (such as a pile of bricks, or sums of disparate parts of objects or background). Mereology does not say what constitutes a natural whole, since the existence of a sum x + y is conditional to the existence of an object z containing both x and y, in the sense that if x and y undelap, there exists a z such that z is the sum of x and y. This allows that parts of different objects be combined to form a whole, which thus consists in a conglomeration of disparate parts, and which is, thus scattered all over the place. Consequently, any attempt to account for basic spatio-temporal relations, such as the relationship between an object and its surface, or the relation of something being inside or around something else., which are some among the relations that any theory concerned with spatio-temporal entities should supply, cannot be defined directly in terms of mereological primitives only. This is not the only misgiving with the unrestricted form of fusion, as there are arguments that it

> does not sit well with certain fundamental intuitions about persistence through time ... that it is incompatible with certain plausible theories of space ... or that it leads to paradoxes similar to the ones afflicting naïve set theory. (Varzi, 2016, 38)

Many theoreticians accept this limitation of summation but argue that there is no structured way to restrict the notion of fusion or sum and that the standard definition of fusion in mereology is the only plausible option (Varzi, 2016, Sect. 4.5).

It is possible, of course, to restrict the definition of "sum" so as to output only sums that satisfy certain conditions, such as a sum of concrete parts must be a non-heap like entity, for example. In this case, if φ is any formula in a language (which picks out the components or parts of that form the sum, say the grains of sand in a pile of sand), and ψ is a condition that the sum must satisfy (for example, that the sum of some material parts must be a natural whole), the set of all φ-ers has a sum φ_i if and only if every φ is ψ, that is, it is part of a natural whole. This yields the following definition of restricted sum: $(\exists w\varphi_w \wedge \forall w(\varphi_w \rightarrow \psi_w)) \rightarrow \exists zSiz\varphi_w$.

Accordingly, many theoreticians propose that mereology should be supplemented with some topological theory (Norton, 2011; Pianesi & Varzi, 1996). Pianesi and Varzi (1996) and Fletcher and Lackey (2022, 15) call the resulting amalgam of Mereology and Topology "Mereotopology". A topological theory provides a primary predicate that is essential in solving the abovementioned problems, namely the relationship of "connectedness" by introducing the *Connection* predicate 'C'. According to C, in order for two parts to form a sum it is necessary that they be connected or joined to each other, a demand that introduces the notions of contiguity and of interval relations that promptly show in discussions of iconic representations, as indispensable ingredients for forming natural wholes. 'C' is defined as follows: (C1) Cxx; (C2) Cxy → Cyx; (C3) Pxy → ∀z(Czx → Czy). In other words, x is connected to itself; if x is connected to y, y is connected to x; and if x is a part of y, then for every z, if z is connected to x then z is connected to y. This last axiom ensures that putting together disparate parts does not provide an acceptable sum.

There are three ways to accommodate the fusion of Topology with Mereology. The first is to accept that the two theories together can provide an adequate framework to explain the part/whole compositionality, the second is an attempt to fuse Mereology into Topology by defining the *Parthhood* relation P of Mereology in terms of the connection predicate C of topology, and the third is an attempt to subsume Topology under Mereology by defining the connection predicate C in terms of P and the vocabulary of Mereology. The details need not concern us, because what is important is that to determine the conditions under which parts can be summed or fused to form natural, that is, integral wholes, and exclude the formation of sums of disparate parts require both notions of 'parthood' and connectedness.

My analysis of the whole/parts compositionality befitting iconic representations reflects Werning's (2012) work. Werning employs neurobiological findings concerning topologically structured cortical feature maps and the mechanism of object-related binding by neuronal synchronization to argue that iconic representations do compose, but their compositionality does not follow the principle of semantic constituency. The use of neuronal synchronization mechanisms underlying object-related binding can be used to complement my discussion of the neuronal implementation of the principles of Mereotopology underlying the formation of well-formed whole objects out of parts. Synchronization is the preferable mechanism nowadays invoked to explain how feature binding takes place to produce the representation of whole objects, and is used by a class of models that purport to explain how different attributes that are registered and processed in different visual areas can be bound together to form a visual object. The main characteristic of these models is that

> oscillators with neighbouring receptive fields and similar feature selectivities tend to synchronize … whereas oscillators with neighbouring receptive fields and different feature selectivities tend to desynchronize. As a consequence, oscillators selective for proximal stimulus elements with like properties tend to form a synchronous oscillation when stimulated simultaneously. This oscillation can be regarded as one object representation. In contrast, inputs that contain proximal elements with unlike properties tend to cause anti-synchronous oscillations, that is different object representations. This result is in line with the findings of object-related neural synchronization. (Werning, 2012, 642)

Synchronization, thus, could be the mechanism, or one among others, underlying the compositionality of parts to form whole sin vision. As the quotation makes clear, neigborhood relations are crucial in guiding synchronization, which brings in the role of Mereotopology in explaining the bindings of parts of objects to produce natural objects. Werning shows that the network of the top-down and bottom-up signals as implemented by neuronal synchronization can be modelled by means of neural networks that, famously, do not employ symbolic representations. This means that object-binding may be a case of composing non-symbolic, iconic representations following the principles of Mereotopology.

Before I turn to examine in some detail the way neural synchronization is involved in perceptual groupings, let me say that the discussion of iconic compositionality in terms of Mereotopology is relevant to various problems plugging the application of the *Picture Principle* in visual representations (see Burge (2022) for a thorough discussion). One of the problems of the principle is that it entails that any part of a picture is a representation of a part of what the picture represents. This is a problem for applying the principle to visual perception, because if one combines the back part of an object in a visual image and a part of the immediate background in the image and forms a whole, this whole is irrelevant in terms of what is computationally relevant in perception, and, thus, it is highly unlikely that this complex part of the image represents anything in perception. In terms of Mereology and Topology, this combination does not result in a natural, integral whole. In this sense it is not true that any part of the representation represents a part of the domain that the representation represents; only parts that are admissible as components of perceptual processes are admitted. Which parts are these is an empirical issue, which means that it is the perceptual system itself that solves the problem of which combinations of parts of an image are admissible as natural wholes.

We saw before that oscillating neurons with neighbouring receptive fields and similar feature selectivities tend to synchronize. It is thus plausible to suggest that synchronization is the mechanism underlying the neural implementation of the constraints of "local proximity" and "feature similarity". Since such neurons are usually grouped together in the brain, Local Field Potentials (LFP's) are very useful to study the activity of these neurons and their synchronization, because the LFP's are transient electrical signals generated in neurons by the summed and synchronous electrical activity of the individual neurons. They express the aggregate activity of small populations of neigborhing neurons represented by their extracellular potentials. Unlike action potentials, which are generated by individual neurons, LFPs primarily measure synaptic potentials pooled across groups of neurons near the recording electrode.

Several studies (Baldauf & Desimone, 2014; Bastos et al., 2012; Fries et al., 2001; Gregoriou et al., 2009, 2015) shed light on the role of LFPs in tasks involving top-down attention and VWM tasks, since these two play a pivotal role in grouping behaviorally relevant stimuli. These studies suggest that attention increases gamma frequency synchronization, increases low-frequency alpha-band synchronization for distractors, reduces low-frequency alpha-band synchronization of V4 neurons representing behaviorally relevant stimuli, increase theta-band frequency, and increases low-frequency beta synchronization for attended stimuli. As we shall shortly see, top-down visual attention originating in Prefrontal Cortex (PFC) and PFC seems to modulate through low-frequency waves the activity in visual areas from V1 to V4, a modulation that results in increased synchronization at the gamma-band high-frequencies between either Front Eye Fields (FEF) in the case of spatial attention, or Inferior Frontal Junction (IFJ) in the case of object/feature-based attention) and the visual areas. These effects occur even during the delay period between test and target in VWM tasks, when firing rates in V4 cells are not affected (Fries et al., 2001).

Gregoriou et al. (2009, 2015) found that the distributions of the latencies of attentional effects in LFP gamma power in both FEF and V4 were significantly later than the distribution of the latencies for attentional effects on the firing rates in FEF, and significantly earlier than the distribution of latencies for attentional effects on V4 firing rates. These results indicate that significant attentional effects on LFP gamma power in either FEF or V4 occur later than the earliest attentional effects on firing rates in the FEF. Thus, rather than being caused by enhanced gamma oscillations, increases in firing rates in FEF with attention may initiate the coupled oscillations within and across areas. In contrast, firing rate changes in area V4 occur later and might result at least in part from enhanced gamma oscillations. Fries et al. (2001) found that the coupled oscillations remain even during the delay period where the firing activity in visual areas is subthreshold. More generally, increases in gamma synchrony are found among cells that decrease, or show no change in, their firing activity (Brunet et al., 2014).

During the delay period of VWM tasks, cells in PFC show increased activation rates, whereas it is likely that the activations in visual areas are subthreshold. According to the evidence discussed in the previous paragraphs, the increased firing rates activity in PFC, and the increased activity in visual areas, especially in FEF, due to top-down attention, induce coupled oscillations at the gamma band frequency both within PFC and across visual areas, starting from V4 where the attention effects are more pronounced and extending to other mid-level and high-level visual areas. The attentional effects on mid-level and high-level visual areas during the delay period, thus, manifest themselves in the LFP gamma power and not in the firing rates in the neurons in mid- or high-level visual areas. It should be noted that reports concerning the impact of attention on neuronal synchronization in early visual areas are conflicting (Gregoriou et al., 2015). For example, Chalk et al. (2010) found that attention reduced gamma-synchronization in V1, probably due to a decrease in the inhibitory drive that controls surround suppression.

Gazzaley et al. (2004) report increased top-down guided activity in FFA and PPA when the participants were asked to maintain during the delay period a mental image of a face or of a visual scene that they have been previously shown, a task that admittedly requires maintenance of attention. Gazzaley et al. (2004) also found significant functional connectivity between FFA and PPA and the parietal cortex during the delay period of the task. In this vein, Baldauf and Desimone (2014) show that the IFJ that is situated at PFC supplies top-down biasing signals for object/feature-based attention. It does so by driving the coupled oscillations at the gamma band frequency between PFC and FFA and PPA (which is the physical mechanism underlying the functional connectivity discussed by Gazzaley et al. (2004). IFJ includes areas that function as general executive modules, and, specifically, it is close to areas Ba45 and Ba46 that are thought to encode, in non-human primates, information about object categories (Baldauf & Desimone, 2014, 427)

We examined evidence showing the role of attention in increasing gamma-band high-frequencies synchronization between visual areas and between visual areas and higher cortical areas. Is this an attentional effect, or is this the way attention affects perceptual processing? In other words, how exactly does top-down attention affect the oscillations in the inter-communicating areas? The answer is that top-down attentional effects on visual areas are transmitted through low frequencies in alpha- and beta-bands and modulate the gamma-band activity in the modulated visual areas. Several studies in monkeys and humans show that spatial attention reduces local low-frequency alpha-band synchronization in visual areas and V4 (Fries et al., 2001; see also Gregoriou, 2015 for references), in contradistinction to the gamma band activity that is increased by attended stimuli in V4. Specifically, distractors increase alpha-band activity, whereas attended stimuli reduce alpha-band activity. Beta-band activity, on the other hand increases for attended stimuli. As Anderson et al. (2011) show, FEF that plays a predominant role in directing top-down spatial attention, excites inhibitory neurons in target areas on the visual cortex during attentional modulation. Similarly, Gregoriou et al. (2015) suggest that alpha band waves may play a role in suppressing irrelevant stimuli, enhancing, thus, the activations of neuronal assemblies representing the attended stimuli.

Research (Bastos et al., 2012; Fries, 2015; Fries et al., 2001; Michalareas et al., 2016) suggest that gamma band waves subserve feedforward signalling, whereas the alpha–beta band waves subserve top-down, feedback flow of information. When top-down attention affects visual processing, the predominantly bottom-up directed gamma-band (high frequency 30–90 Hz) influences are controlled by predominantly top-down directed alpha–beta band (8–20 Hz) influences. Plomp et al. (2014) and Van Kerkoerle et al. (2014) show that stimulation of V1 induces enhanced gamma-band activity in V4 (V1 to V4 feedforward projections), while stimulation of V4 under visual stimulation with a background stimulus induces enhanced alpha–beta activity in V1 (V4 to V1 feedback projections), which likely supresses the background stimulus. In general, attentional top-down influences carried by low-bad frequency waves are thought to modify gamma-band synchronization in the lower areas that receive the attentional feedback enhancing the feedforward signals emanating from these areas. Top-down signals increase both the synchronization strength, as measured by LPFs power, and the synchronization frequency of the gamma-band synchronization.

Fries et al. (2001) found that synchronization or coherence was modulated by attention very early in the response to the stimulus. The Spiked-Triggered Averages (STAs) for the 100-ms period after response onset (starting 50 ms after stimulus onset—it takes about 50 ms. for the brain to start responding to the incoming stimuli) contained large low-frequency modulations with superimposed gamma-frequency modulations. The low-frequency (10 Hz) synchronization was reduced by attention. Conversely, there was a smaller gamma-frequency peak at around 65 Hz that was enhanced by attention. Both the visual evoked potential (VEP) and the spike histogram contained strong stimulus-locked gamma-frequency oscillations in the first 100 ms of the response. This very early gamma-frequency synchronization was at least partially locked to stimulus onset. Since gamma-band synchronization is related to bottom-up processing, this finding shows the bottom-up, stimulus locked activity in visual areas. However, oscillatory synchronization during the later sustained visual response was not stimulus locked, which shows that effects of spatial attention that modulates in a top-down manner perceptual processing. This is shown by the fact that the low-frequency synchronization in the alpha-band was reduced by attention (alpha band synchronization is reduced for attended stimuli and increased for distractors). Since attention strengthens the representation of the attended stimulus in V4, it facilitates the bottom-up signals from V4 to IT and other cortical areas, which explains the increase in gamma-band frequency. Thus, attention increased gamma frequency and reduced low-frequency synchronization of V4 neurons representing the behaviorally relevant stimulus. This was observed even during the delay period and in the first few hundred milliseconds after response onset, when firing rates were not affected, because spatial attention owing to the fixation at a certain location modulates the preparatory activity of the relevant neurons before the presentation of the stimulus and this effect is carried on after stimulus onset before the attentional cue is being presented.

In contradistinction to alpha-band waves that suppress irrelevant stimuli, since distracting stimuli enhance alpha band oscillatory activity, beta-band wave activity may enhance the activation of attended stimuli by inducing stronger synchrony in lower frequency bands, which enhances the top-down signals. Bastos et al. (2012) argue that top-down signals that facilitate processing of attended stimuli are carried by beta-band (14–18 Hz) synchronization. In general, Bastos et al. (2012) suggest, cognitive tasks enhance top-down beta-band influences. Moreover, when attention selects a stimulus and enhanced top-down signals (carried by beta-band waves) reach the representations of the attended stimulus in visual cortical areas, this may lead to enhanced bottom-up signalling of that stimulus carried by gamma-band waves. Bosman et al. (2012) show that bottom-up causal influences from V1 to V4 are enhanced when they carry information about the attended stimulus, in accordance with Fries et al. (2001) finding that neurons activated by attended stimuli show increased gamma-frequency.

I claimed above that the top-down flow of information is subserved by alpha and beta-band synchronizations and the bottom-up flow of information by gamma (and theta)-band synchronizations. We also saw that there is substantial evidence that gamma-band causal influences between macaque FEF and V4 predominate in the FEF-to-V4 direction after an attentional cue (post cue condition), but subsequently predominate in the V4-to-FEF direction. In view of the fact that FEF are anatomically higher than V4 and, thus, the connections from FEF to V4 seem to be feedback/top-down projections, one might think that gamma-band synchronization subserves feedback flow of information, which contradicts the evidence that feedback is carried by low frequency waves. However, there is no discrepancy. FEF may be anatomically higher in the hierarchy than V4 but functional things are different.

FEF is situated in the PFC at a site heavily interconnected with the parietal cortex and is a part of the dorsal visual system. The mean activation latency of FEF neurons is 70 ms poststimulus. Signals arrive at FEF with a slight (if at all) time delay time with respect to the signals arriving at V1 (50–80 ms) and V2 (85 ms) and much earlier than they arrive at V4 despite the fact that FEF is anatomically higher than V4. Thus, functionally, FEF is at a lower level than V4 and, therefore, the projections from FEF to V4 are feedforward. This describes prestimulus conditions since it indicates the potential functional relations of FEF and V4. FEF contains visual and movement neurons (O'Shea et al., 2004).

Bastos et al. (2012) show that in the post-cue condition, that is, before spatial attention triggered by the cue intervenes, 8L area of FEF is lower than V4, which means that the projections from area 8L of FEF to V4 are feedforward. Thus, Gregoriou's et al. (2009) finding that gamma-band causal influences between macaque FEF and V4 predominate in the FEF-to-V4 direction after an attentional cue reinforces rather than contradicts the view that feedforward influence are subserved by gamma-band synchronizations. Notice that this description purports to explain the increase in gamma synchronization between FEF and V4 and the fact that the causal influence is from FEF to V4, in other words, that FEF projects feedforward signals to V4. It does not account for the top-down attentional effect of FEF, and especially of area 8M, on V4, because this is carried by low-frequency waves, as top-down signal propagation predominantly is, from FEF to V4.

Recall that Fries et al. (2001) found that spatial attention modulates synchronization in V4 at 150 ms. after stimulus onset, when large low-frequency modulations were superimposed by gamma-frequency synchronizations. The low frequency waves on V4 carry the top-down attentional modulation from area 8M of FEF to V4. Bastos et al. (2012) show that area 8M is slightly higher than V4. The gamma-band synchronization in the FEF to V4 direction before the cue probably shows the effect of the processing of the visual stimulus in area 8L of FEF in which the visual neurons of FEF are clustered and which process the stimulus and are independent of attention. Moreover, once the representation in V4 of an item at the attended location has been boosted by the top-down attentional effects carried by low-frequency waves, the feedforward projections from V4 to FEF dominate and enhanced bottom-up signalling of that stimulus carried by gamma-band waves prevails. This explains Gregoriou et al. (2009) finding that that gamma-band causal influences between macaque FEF and V4 predominate in the V4-to-FEF direction later in the post-cue period.

One might be puzzled by the fact that the same areas (for example, V4 and FEF) can have both feedforward and feedback projections between them, which leads to the functional hierarchy exhibiting dynamic changes. Fries (2015) argues that when two neuronal areas are bidirectionally connected, unidirectional entrainment (that is the causal influence of one on the other) occurs separately in both directions of the bidirectional link. Anatomical data show that for each direction of communication, the linked brain areas have specialized neuronal groups for sending and receiving signals; that is, a specific brain area has neurons receiving inputs and different neurons sending outputs (Fries, 2015; Markov et al., 2014).

Jensen et al. (2014) examine the role of alpha oscillations in relations to gamma oscillation in attentional effects on the visual areas. Irrespective of the exact nature of the signals that MVPA decodes, it is clear that VWM involves visual areas. When the test display appears, the perceptual areas also receive bottom-up, stimulus driven activation and neuronal assemblies in these areas encode the perceptual information in the test display. This information is compared to the perceptual information concerning the sample item that is stored in the connection weights of neuronal assemblies in visual areas that are reactivated by both the bottom-up signal from the test item owing to the distributed nature of representations, and by the top-down signals from PFC.

Work by Nakatani and Leeuwen (2006) elucidates the role of attention in perceiving ambiguous figures relating attentional activity to the synchronized activity in the right parietal areas that are responsible for perceptual awareness, and in the right frontal areas that correlate with perceptual flexibility and, hence, to perceptual switching between the two possible percepts of the ambiguous or bistable figure. Note that the same areas are also involved in top-down selective attention (Corbetta & Shulman, 2002). Nakatani and Leeuwen's (2006) research shows two cycles of synchrony in the gamma band; the first occurs 800–600 ms, and the second 400–200 ms before button pressing. The first period of synchronicity coincides with a drastic suppression of eye blinks that is related to attentional demands, as these demands make viewers to focus closely postponing saccades (Ito et al., 2003). The second period of synchronicity in the observed activity patterns in PFC coincides with the maximum saccade frequency that reaches its peak at about 250 ms before the switch response. Since saccade frequency is associated with shifts of attention (Leopold & Logothetis, 1999), the second period of synchronicity probably reflects the final focus of spatial attention after a series of attentional shifts, which, by determining the critical points on the image also determines which interpretation of the ambiguous figure will be perceived.

Nakatani and Leeuwen (2006) also explored the role of the activity in frontal and occipital cortex during switching episodes. They found that the theta activity in the frontal cortex is a general characteristic of the processing activity of viewers that perform frequent switches, but is not specifically related to perceptual switching. Increased theta band activity in the frontal cortex is related to the concentration of attention to a task and to the inhibition of eye blink (Yamada, 1998), as is the activity in the first period of synchrony in the gamma band. The alpha band activity observed in the occipital cortex is related to frequent perceptual switches. Increased alpha activity in the occipital cortex is related to attention to the stimulus by enhancing the efficiency of information processing (Yamagishi et al., 2003). Thus, the frontal and occipital cortex activity during perceptual switches signifies the crucial role of attentional modulation of the perception of ambiguous figures and its effects on the rate of perceptual switches.

The previous studies concern the role of synchronization in attention tasks. Attention is closely related to tasks involving VWM as well and, so, it would be interesting to see what role synchronization plays in VWM tasks. Examining the role of synchronic activity among brain regions in memory tasks, Roux and Uhlhaas (2014) proposed that gamma-band oscillations are specifically involved in the active maintenance of VWM information. Theta-band oscillations are specifically involved in the temporal organization of NWM items, a view that generalizes Nakatani and Leewen's (2006) proposal. Finally, alpha-band oscillations are involved in the inhibition of task-irrelevant information, which results in enhancing the efficiency of information processing, as Yamagishi et al. (2003) proposed. However, other factors may contribute to the enhancement of the efficiency of information processing in occipital area, and the increased alpha activity may reflect these factors as well. Such factors may a direct enhancement of task-relevant information (Miller & Cohen, 2001), or the sharpening of the representations of different object categories in the extrastriate cortex by an increase of the distinctiveness of their distributed neural representations (Fuster et al., 1985).

These results are based on studies demonstrating amplitude modulation of neural oscillations presumably emanating from particular brain regions involved in WM. During a delayed match-to-sample task while recording human EEG, Tallon-Baudry et al. (1999) observed that occipital gamma and frontal beta oscillations were sustained across the retention interval. Moreover, as the delay interval increased, these oscillations decreased in parallel with decreased performance on the task. Anderson et al. (2014) showed that the spatial distribution of power in the alpha frequency band (8–12 Hz) tracked both the content and the quality of the representations stored in visual working memory. Recall that in memory tasks there is increased power in the alpha frequency band in the occipital cortex related to the enhancement of the efficiency of information processing by attention to the stimulus (Yamagishi et al., 2003). These empirical findings together support both the view that neural oscillations are critical for VWM maintenance processes, and the view that in VWM tasks posterior visual processing areas play a critical role in sustaining the representations held in VWM. Finally, Lee et al. (2005) found evidence of enhanced local field potentials (4–10 Hz) in area V4 of the monkey during a visual working memory task.

Long-range synchronization of the oscillations between brain regions likely also plays an important role in VWM function (Crespo-Garcia et al., 2013) as well. In a human MEG study, synchronized oscillations in the alpha, beta, and gamma bands were observed between frontoparietal and visual areas during the retention interval of a delayed match-to-sample visual working memory task. These synchronized oscillations were sustained and stable throughout the delay period of the task, were memory load dependent, and were correlated with an individual's VWM capacity (Palva et al.,

2010). Monkey physiology data shed light on the mechanisms that underlie communication between brain regions during VWM. Liebe et al. (2012) investigated the neural interactions between V4 and lateral PFC during the performance of a visual delayed match-to-sample task. During the retention interval of the task the two areas exhibited synchronization of local field potentials (LFPs) in the *theta* frequencies. Furthermore, the neuronal spiking activity in these two regions was phase-locked to these observed *theta oscillations*, and the strength of this intercortical locking predicted the animal's performance, that is, it was higher for subsequently correctly remembered stimuli and session-to-session variability in memory performance.

These studies bring to the fore the crucial role of synchronous oscillations in alpha–beta–theta–and gamma-frequency-bands in top-down attention and in VWM tasks. Top-down attention effects are carried by low frequency oscillations that synchronize the LFPs oscillations between the affecting and affected brain areas, while bottom-up signals are carried by high-frequency gamma-band oscillations.

The foregoing discussion shows the close collaboration between cognitive centers and visual areas in the brain in VWM, since the higher-level cognitive centers guide attention through which they sustain the perceptual representations in visual areas during memory tasks. This suggests that the perceptual information used in memory tasks is most likely represented in visual areas and, in this sense, memory recruits the representations in these areas to achieve its goals. This is the basic premise of sensorimotor recruitment models of VWM, a class of models that hold that the systems and representations engaged to perceive information can also contribute to the short-term retention of that information in VWM (D'Esposito & Postle, 2015).

We can bring in, now, the upshot of the discussion about Mereology and Topology, which is that to solve the problem of how to constrain the fusion of parts so that only natural objects be accepted as proper wholes, one must consider both "parthood" and "connectedness". Since the visual system has solved this problem, it is plausible to assume that its computations directly implement, in one form or another, the Mereotopological principles of, say, "parthood" and "connectedness" and combine them in such a way that only natural wholes are represented in visual perception under normal conditions. The reader has perhaps made this connection as a result of our discussion of the neural mechanisms of synchronicity that underlie the compositionality of visual representations. If iconic compositionality is realized by neural mechanisms and the ways parts are composed are hardwired in the visual system, and if these compositions are expressed by the principles of Mereotopology, it follows that the principles of Mereotopology express the functioning of the relevant neural mechanisms.

At the same time, the operation of the visual system, as we have seen, is characterized by a set of operational constraints. Thus, these constraints must be realized by the neural circuits in our visual system. This is what Han et al. (2002) studies show. Han et al. (2002) studied the neural mechanisms underpinning the operational constraints of "local proximity" and "similarity". The findings of their study suggest that proximity grouping resulted in short latency modulations of medial occipital activity that was followed by longer latency modulations in the occipito-parietal cortex. Proximity grouping that relies on "local proximity" induced similar medial occipital modulations at 110 ms, which suggests that it depends mainly on representations of spatial relationships between local elements, and is independent of visual features. Grouping by color similarity that relies on feature similarity produced only long-latency occipito-temporal modulations.

In view of these considerations, it follows that the operational constraints in vision should reflect the principles or Mereotopology. I shall argue now that they do so, starting with a brief account of the operational constraints in vision. As I have repeatedly said (Author), perceptual computations are constrained by a set of what I have called operational constraints. Burge (2010) calls them “formation principles”, Echeverri (2017) calls them “object constraints”, and some among them are also known as Spelke (1990) principles of “object perception”. These principles also figure in Haugeland’s account of perception, where he claims (Haugeland, 1998, 261) that non-concept possessing creatures and we share various innate object-constancy and object-tracking mechanisms that automatically ‘lock onto’ medium sized lumps of matter. These mechanisms implement the operational principles. The operational constraints reflect general or higher-order physical regularities that govern the behavior of objects in our world and the geometry of our environment, and which have been ingrained in the perceptual system through causal interaction with the environment over the evolution of our species. This means that they reflect generalities in the world given our physical constitution and needs for survival.

Empirical studies by Spelke (1990), Spelke et al. (1995), and Karmiloff-Smith (1992) strongly support the assumption that the infant, from the beginning of life, is constrained by a number of domain-specific principles about material objects and some of their properties. These constraints involve ‘attention biases toward particular inputs and a certain number of principled predispositions constraining the computation of those inputs.’ (Karmiloff-Smith, 1992, 15) Among these predispositions are the conception of object persistence, and four basic principles: boundness, cohesion, rigidity, and no action at a distance.

The operational constraints function at almost all levels of visual processing, are hardwired in the brain, and do not entail that perception is cognitively penetrated since they do not constitute cognitive states that affect perceptual processing (Raftopoulos, 2009, 2019, Chap. 3). One of these *principles is cohesion* (Bloom, 2000). ‘Objects are connected and bounded bodies that maintain both their connectedness and their boundaries as they move freely’ (Spelke et al., 1995, 45). That is, the *cohesion principle* dictates that two surface points lie on the same object only if the points are linked by a path of connected surface points. This entails that if some relative motion alters the adjacency relations among points at their borders, the surfaces lie on distinct objects, and that all points on an object move on connected paths over space and time. When surface points appear at different places and times such that no connected path could unite their appearances, the surface points do not lie on the same object. This constraint reflects the principle of “connectedness” of Topology that, as we have seen, supplants the rules of Mereology to restrict the definition of “sum” so that only compiling parts that yield natural whole objects is an acceptable notion of “sum”. This is so because if two surface points lie on the same object only if the points are linked by a path of connected surface points, then the parts on which the two points lie are connected. The importance of the cohesion principle is manifested by the finding that some violations of cohesion seem to destroy infants' representations of enduring object (Chiang & Wynn, 2000; Huntley-Fenner et al., 2002). Mitroff et al. (2004) and vanMarle and Scholl (2003) show that even adults' visual processing is critically affected by cohesion violations.

Another principle, closely related to "cohesion" is the principle of solidity; 'objects move only on unobstructed paths: no parts of two distinct objects coincide in space and time' (Spelke et al., 1992, 606). This is clear expression of Theorem (P4c) of Mereology presented above, which states that no two distinct objects can share parts and shows how the perceptual system has hardwired basic regularities of the environment, in this case, a basic mereological property of solid objects.

Another constraint is the *boundness principle*, according to which two surface points lie on distinct objects only if no path of connected surface points links them. This principle determines the set of those points that define an object boundary, and entails that two distinct objects cannot interpenetrate, because two distinct bodies cannot occupy the same place at the same time. Finally, the *rigidity* and no *action at a distance* principles specify that bodies move rigidly (unless the other mechanisms show that a seemingly unique body is, in fact, a set of two distinct bodies) and that they move independently of one another (unless the mechanisms show that two seemingly separate objects are in fact connected). These constraints guide the perception of the motions of objects, of the layout of adjacent objects, of object boundaries, and of object segmentation by both adults and infants, and play a crucial role in the segmentation processes that take place in the visual system upon viewing a scene. These constraints relate both to "connectedness" and to theorem P4a of Mereology. The former because if there is no path of connected surface points lining the two surface points, they belong to unconnected parts that form distinct objects since they cannot be combined to form a natural whole. Theorem P4a applies because it rules out the possibility that two object parts that combine to form a whole object can have any common parts (they cannot overlap), because if that were the case, two distinct object wholes formed of parts that can overlap could penetrate one the other over the points of overlapping.

There are more constraints at work in perception than those mentioned above. The formation of the *full primal sketch* in Marr's (1982) theory, for instance, which involves the grouping of the edge fragments formed in the *raw primal sketch*, relies on the principles of "local proximity" (adjacent elements are combined), which shows the Topological principle of "connectivity" at work, and of "similarity" (elements with similar features are combined). All these principles are parts of "perceptual grouping", which refers to the function of the human visual system to organize discrete entities in the visual field into chunks or perceptual objects. The principle of local proximity states that spatially close objects or object parts tend to be grouped together, constraining thus which parts can be combined to form natural wholes. The principle of similarity claims that element s with similar features in the field tend to be grouped together. Grouping processes have been assumed to take place at an early stage in the visual processing stream. Perceptual grouping also relies on the more general principle of "closure" (two edge segments could be joined even though their contrasts differ because of illumination effects) (Bruce & Green, 1990, 131–132). Other assumptions that are brought to bear upon the early visual processing to solve the problem of the underdetermination of perception by the retinal image are those of "continuity" (the shapes of natural objects tend to vary smoothly and usually do not have abrupt discontinuities), "proximity" (since matter is cohesive, adjacent regions usually belong together and remain so even when the object moves), and "similarity" (since the same kind of surface absorbs and reflects light in the same way, the different subregions of an object are likely to look similar).

The formation of the 21/2D sketch is similarly underdetermined, in that there is a great deal of ambiguity in matching features between the two images form in the retinas of the two eyes, since there are usually more than one possible matches. Stereopsis requires a unique matching, which means that the matching processing must be constrained. The formation of the 21/2D sketch, therefore, relies upon a different set of operational constraints that guide stereopsis. 'A given point on a physical surface has a unique position in space at some time' (Marr, 1982, 112), and matter is cohesive and surfaces are generally smooth. These operational constraints give rise to the general constraints of "compatibility" (a pair of image elements are matched together if they are physically similar, since they originate from the same point of the surface of an object), "uniqueness" (an item from one image matches with only one item from the other image), and "continuity" (disparities must vary smoothly).

I think that it is not hard to see, and I have given some examples, that almost all of these operational constraints are directly or indirectly related to the connectedness and parthood Mereotopological relations among elements in perceptual computations and that their role is exactly to ensure the formation, and thereby visual representation, of natural wholes. All these suggest that the *Picture Principle* as it applies to perception is severely constrained and that it is not the case that any way you cut a visual representation of a visual scene you get a representation of some part of this scene.

We saw that adjacent regions in a visual scene are registered by adjacent regions in the retina, and, through retinotopic projections, are represented by adjacent regions in most of the brain. Thus, neighborhood relations, which is a topological notion strongly associated with 'connectedness,' are retained in the brain. Moreover, parthood relations in a visual scene are also retained in the brain because a neural representation of, say, a whole object or scene consists, in the sense of containment, of the neural representations of its parts, the same way a picture of a scene contains as parts pictures of the constituents of the scene. Thus, mereological relations are also retained in the brain. It follows that even though mental states are not spatially arrayed in space as pictures are, mereotopology is still applicable and can be put into use to show how the brain distinguishes natural wholes from mere collections of parts (recall that mereology by itself cannot solve this problem but requires the topological notion of 'connectedness'). This means that the mathematical tools of mereotopology can be used not only to describe the compositionality of visual perceptual contents, but also the compositionality of their neural vehicles.

I have argued that the functioning of our perceptual systems function restricts the way parts of the image can form composites that lead eventually to the formation of the percept relying on the various operational constraints, and I built these arguments partly on the idea that there is a tendency in Mereology to find ways to restrict the operation of summation over parts to yield a composite sum; restricting summation so that the sum of some parts outputs a natural, whole object would fit my general idea. Fine (2010, 567) provides a means to incorporate this idea in Mereology, by proposing that compositional operations and not the notion of parthood should be given priority. In such a scenario, one could define the general compositional operation Σ, such that Σ(x1, x2, ...) = K, where K depends on the type of operation Σ. If, Σ, for example, is a set builder, then Σ(x1, x2, ...) = {x1, x2, ...}. If Σ is a simple summation mechanism, then Σ(x1, x2, ...) = (x1 + x2 + ···). One could extend this idea to perceptual processing and the gradual construction of composites that lead to the percept by assuming that the operation Σ instantiates all the restrictions on perceptual processing imposed by the operational constraints.

To see how this bears on the discussion concerning PP consider the following argument. If the whole is a mere summation of its parts without any restrictions whatsoever, as envisaged by the definition of sum or fusion in classical Mereology, then each component is part of the sum. In iconic representations like a percept or any image, this would mean that any way one cuts the image, one gets a part of the image. It follows that any way an image is cut, one gets a part of the represented entity, that is, if P is a picture of X, then parts of P are pictures of parts of X, which is PP. However, not all compositional operations behave like mere summation. If, as I argued above, one has to restrict the compositional operations at work in perception so that the operational constraints be satisfied, then even though two items in a scene, say an object O and a piece of the background B, have parts O1 and O2 and B1 and B2 respectively and O is the sum of O1 and O2 and B is the sum of B1 and B2, it does not follow that, say, O1 and B2, are necessarily proper parts of the scene and, thus, have any representational import, simply because, unlike mere summation that is transitive and components of component of a sum are components of the sum, the compositional operation in perception is not transitive and, thus, components of components and their sums may not be parts of the iconic representation. Here is a simple example from set theory (which I borrow from Fine (2010)) to make the point clear. In mere summation, σ((x1 + x2) + (x3 + x4)) = σ(x1 + x2 + x3 + x4), and, thus, x, x2, x3, and x4, that are parts of the (x1 + x2) and (x3 + x4) are also parts of (x1 + x2 + x3 + x4). In set theory, however, even though the x, y, z are parts of the sets {x, y} and {y, z, and {x, y} and {y, z) are parts of the set {{x, y}, {y, z}}, x, y, z are not parts of the set {{x, y}, {y, z}}. The reason that O1 and B2 are not parts of the scene is that they do not form a natural whole that satisfies the relevant principles that, as we saw, are operationalized in perception by the operational constraints. As Fine (2010, 576), puts it summation is blind to all aspects of the whole that the parts compose other than the parts from which it is formed. I have argued that the same holds for PP as it bears on perception.

If the theses defended above are correct, it follows that the structure of the information-carrying vehicles and events in the environment structures the make-up of the representational perceptual vehicles of organisms (by hardwiring, as it were, in them, the basic principles that govern the ways objects in the environment compose and interact) in order to enable them to extract successfully the information in the environment and organize the information flow. The reason for this is that way objects in the environment compose must be reflected in the ways the respective perceptual representational vehicles compose for the perceptual system to be able to extract successfully the information carried by the environmental structures. This make-up, in turn, determines the way the organism manipulates the incoming information and, hence, interacts with the environment.

This conclusion rests on the assumption that material objects and physical events, as well as social constructions, constitute informational domains and, thus, can be construed as information-carrying vehicles; the information they carry can be construed as exogenous information since it is deposited in material structures that lie outside the organisms that exploit this information. These informational domains have a structure expressed by the organized relations among the information-carrying vehicles. '[D]evelopmental environments contain heterogeneous vehicles of informational content (external artefacts, linguistic labels, the behavior of other agents) organized in systems.' (Buskell, 2022, 8540).

This process is nicely summarized by Buskell (2022, 8535):

> the organization and flow of information is important for understanding both the development of cognition and its occurrent functioning. Structured informational domains like material artefacts, linguistic systems, or workshop apprenticeships can facilitate the acquisition of abilities and regulate their deployment—and here I extend this point to help explain cognitive novelties.

And

> the structural relationships between these vehicles can make a difference to the development of cognitive skills, even if they are only present for short periods of time. (Buskell, 2022, 8540)

Buskell's point is important because it shows how cognitive capacities can be created by pressures to agents to exploit effectively environmental information, irrespective of whether these cognitive novelties are genetic, epigenetic, or the result of the restructuring of the brain. These pressures cause the brain to reorganize its structure, or get new structures, which, in turn, render cognitive innovations possible. It is interesting to note that, if this view is correct, the external world imposes, as it were, its structure on the mind so that the transformations in the worldly states of affairs map in a natural way and not by some convention onto the transformations in the states of the mind. As the reader recalls, this is a characteristic trait of iconic representations, which may suggest that at the level of the interactions of the mind and the environment at the level of actions, the mind represents iconically the world, transferring the structure and processes of the world and of the actions on the world into the mind.

Cognitive capacities can also emerge, accompanied by the necessary brain rewiring or development, as a result of exercises on particular actions that may be beneficial for survival, which is another example of the material Engagement Theory. Lombard (2024) explores the relationship between attention development in modern archers and attention as a cognitive requirement for ancient bow hunting, a hunting behaviour that may have originated sometime between 80 and 60 thousand years ago in sub-Saharan Africa. He argues that practicing to use bimanual technologies shapes aspects of our cognition, including attention. In her study, Lombard uses cognitive-motor neuroscience to demonstrate the role of attention in modern archery and highlight the brain regions that are activated during aiming. She found evidence that the precuneus and the default mode network serve as neurological hub for accurate bimanual material engagement practiced over a distance. She notes (Lombard, 2024, 1)

> [t]he likely development of the precuneus, in tandem with the unique globularisation of the human skull, can be traced in the *Homo sapiens* fossil record since about 160 thousand years ago, reaching the modern range by around 100 thousand years ago within a continuum of brain modification. Variation in human neuro-genetic adaptations since our split from the Denisovan and Neanderthal groups further suggest differences in attention as a cognitive trait between recent big-brained humans. I suggest that these observations may serve as bridging theory for understanding how some aspects of the sapient ability to pay attention was developed.

In the preceding paragraphs, I argued that the hominin visual perceptual system has been structured in part by pressures put upon it by the organization and relations of the objects in the environment, by means of hardwiring the basic principles, expressed mathematically by Mereotopology, that govern the way objects and events in the environment compose to form more complex objects; in this sense, these principles are internalized. This organization of the visual perceptual system is not, strictly speaking, a cognitive novelty, but the account I provided lays the groundwork for finding similar explanations of how other environmental, including the social domain, factors enabled cognitive innovation through brain structuring and restructuring that resulting in the internalization of the structure of the vehicles bearing the exogenous information. The view that the structure of the information-bearing vehicles (that is, their interrelations and organization) of exogenous information can be internalized into cognitive routines has also been defended by Sterelny (2012). I shall explore this issue in the next chapters, where I discuss, first, the cognitive capacities required for iconic representations to emerge from non-representational indices and, second, the cognitive capacities the enabled the emergence of symbolic representations on the basis of iconic representations on which symbolic representations supervene.

Not only can the structure of information-bearing vehicles of exogenous information be internalized, but, also, the vehicles themselves can be internalized in iconic representations. The reason is that, as we have seen while discussing the nature of iconicity, in iconic representations, the physical transformations of the information-bearing vehicles in the environment must be reflected and matched by the physical transformations of the representational vehicles, whether it be external representations or internal, mental representations.

References

Abell, C. (2009). Canny resemblance. *The Philosophical Review, 118*(2), 183–223.
Anderson, J. R., & Gallup, G. G. (2011). Which primates recognize themselves in mirrors? *PLoS Biology, 9*(3), e1001024. https://doi.org/10.1371/journal.pbio.1001024
Anderson, J. C., Kennedy, H., & Martin, K. A. (2014). Pathways of attention. *Synaptic relationships of frontal eye fields to V4, lateral intraparietal cortex, and area 46 in macaque monkey* (30), 10872–10881.
Ayers, M. (2019). *Knowing and seeing: Groundwork for a new empiricism.* Oxford University Press.
Baldauf, D., & Desimone, R. (2014). Neural mechanisms of object-based attention. *Science, 344*, 424–427.
Bastos, A. M., Usrey, W. M., Adams, R. A., Mangun, G. R., Fries, P., & Friston, K. J. (2012). Canonical microcircuits for predictive coding. *Neuron, 76*(4), 695–711.
Beck, J. (2017). Marking the perception-cognition boundary: The criterion of stimulus-dependence. *Australasian Journal of Philosophy, 96*(2), 319–334.
Beck, J. (2018). Analog mental representation. *Wires Cognitive Science.* https://doi.org/10.1002/wcs.1479
Beck, J. (2019). Perception is analog: The argument from Weber's law. *Journal of Philosophy, 116*(6), 319–349.
Bermudez, J. L. (1995). Nonconceptual content: from perceptual experience to subpersonal computational states. *Mind and Language, 10*(4), 333–369.
Biederman, I. (1985). Human image understanding. *Recent research and a theory. Computer Vision, Graphics, and Image Processing, 32*, 29–73.
Blachowicz, J. (1997). Analog representation beyond mental imagery. *Journal of Philosophy, 94*(2), 55–84.
Block, N. (2014). Seeing-as in the light of vision science. *Philosophy and Phenomenological Research, 89*(3), 560–572.
Bloom, P. (2000). *How children learn the meanings of words.* MIT Press.
Bosman, C. A., Schoffelen, J. M., Brunet, N., Oostenveld, R., Bastos, A. M., Womelsdorf, T., Rubehn, B., Stieglitz, T., De Weerd, P., & Fries, P. (2012). Attentional stimulus selection through selective synchronization between monkey visual areas. *Neuron, 75*(5), 875–888.
Bruce, V., & Green, P. R. (1990). *Visual perception* (2nd ed.). Sussex: Lawrence Erlbaum Associates.
Brunet, N. M., Bosman, C. A., Vinck, M., & Fries, P. (2014). Stimulus repetition modulates gamma-band synchronization in primate visual cortex. *PNAS, 111*(9), 3626–3631.
Burge, T. (2010). *Origins of objectivity.* Oxford University Press.
Burge, T. (2014). Reply to block: Adaptation and the upper border of perception. *Philosophy and Phenomenological Research, 89*(3), 573–583.
Burge, T. (2018). Iconic representation: Maps, pictures, and perception. In W. Shyam & F. A. Dorio (Eds.), *The map and the territory* (pp. 79–100). Springer.
Burge, T. (2022). *Perception: The first form of mind.* Oxford University Press.
Burns, B. (1987). Is stimulus structure in the mind's eye? An examination of dimensional structure in iconic memory. *The Quarterly Journal of Experimental Psychology, 39*(A), 385–408.
Burnston, D. (2017). Cognitive penetration and the cognition-perception interface. *Synthese, 194*, 3645–3668.
Buskell, A. (2022). Cumulative culture and complex cultural traditions. *Mind and Language, 37*(3), 284–303.
Camp, E. (2018). Why maps are not propositional. In A. Grzankowski & M. Montague (Eds.), *Non-propositional Intentionality* (pp. 19–45). Oxford University Press.
Carey, S. (2009). *The origin of concepts.* Oxford University Press.
Chalk, M., Herrero, J. L., Gieselmann, M. A., Delicato, L. S., Gothardt, S. & Thiele, A. (2010). Attention reduces stimulus driven gamma frequency oscillations and spike field coherence in V1. *Neuron, 66*, 114–125

Chiang, W.-C., & Wynn, K. (2000). Infants' representation and tracking of multiple objects. *Cognition, 77*, 169–195.
Corbetta, M., & Shulman, G., L. (2002). Control of goal-directed and stimulus-driven attention in the brain. *National Review of Neuroscience, 3*, 201–215.
Crane, T. (1992). The nonconceptual content of experience. In T. Crane (Ed.), *The contents of experience: Essays on perception* (pp. 136–157). Cambridge University Press.
Crane, T. (2003). *The mechanical mind* (2nd ed.). Routledge.
Crane, T. (2009). Is perception a propositional attitude? *The Philosophical Quarterly, 59*, 452–469.
Crespo-Garcia, M., Pinal, D., Cantero, J. L., Dıaz, F., Zurron, M., & Atienza, M. (2013). Working memory processes are mediated by local and long-range synchronization of alpha oscillations. *Journal of Cognitive Neuroscience, 25*, 1343–1357.
Cummins, R., & Roth, M. (2012). Meaning and content in cognitive science. In R. Schantz (Ed.), *Prospects for meaning* (pp. 365–382). De Gruyter.
D'Esposito, M. D., & Postle, B. R. (2015). The cognitive neuroscience of working memory. *Annual Review of Psychology, 6*, 115–142.
Deutscher, G. (2005). *The unfolding of language: An evolutionary tour of mankind's greatest invention.* Picador.
Dretske, F. (1981). *Knowledge and the flow of information.* MIT Press.
Echeverri, S. (2017). Visual reference and iconic content. *Philosophy of Science, 84*, 761–781.
Evans, G. (1982). *The varieties of reference.* New York, N.Y: Oxford University Press.
Everett, D. L. (2017). *How language began.* Liverlight.
Fine, K. (2010). Towards a theory of part. *Journal of Philosophy, 107*(11), 559–589.
Fletcher, S. C., & Lackey, N. (2022). The introduction of topology into analytic philosophy: Two movements and a coda. *Synthese, 200*, 197. https://doi.org/10.1007/s11229-022-03689-9
Fodor, J. (2007). The revenge of the given. In B. P. McLaughlin & J. Cohen (Eds.), *Contemporary debates in the philosophy of mind* (pp. 105–117). Blackwell.
Fodor, J. A. (2008). *LOT 2: The language of thought revisited.* Cambridge, MA: MIT Press.
Fodor, J., & Pylyshyn, Z. (1981). How direct is visual perception? *Cognition, 9*, 139–196.
Fries, P. (2015). Rhythms for cognition: communication through coherence. *Neuron, 88*(1), 220–235.
Fries, P., Reynolds, J. H., Rorie, A. E., & Desimone, R. (2001). Modulation of oscillatory neuronal synchronization by selective visual attention. *Science, 291*, 1560–1563.
Frege, G. (1962), (1892). Begriff und gegenstand. In G. Patzig (Ed.), *Funktion, Begriff, Bedeutung.* Vandenhoek and Ruprecht.
French, S. (2003). A model-theoretic account of representation (or, I don't know much about art but I know it involves isomorphism) *Philosophy of Science, 70*(5), 1472–1483.
Fuster, J. M., Bauer, R. H., & Jervey, J. P. (1985). Functional interactions between inferotemporal and prefrontal cortex in a cognitive task. *Brain Research, 330*, 299–307.
Gallistel, C. R., & Gelman, R. (1992). Preverbal and verbal counting and computation. *Cognition, 44*(1), 43–74.
Gazzaley, A., Rissman, J., & D'Esposito, M. (2004). Functional connectivity during working memory maintenance. *Cognitive and Affective Behavioral Neuroscience, 4*, 580–599.
Giardino, V., & Greenberg, G. (2015). Introduction: Varieties of iconicity. *Review of Philosophy and Psychology, 6*, 1–25.
Gombrich, H. E. (1951). *The story of art.* Oxford University Press.
Goodman, N. (1972). Seven strictures on similarity. In *Problems and projects* (pp. 437–446). Bobbs-Merrell.
Goodman, N. (1976). *Languages of art* (2nd ed.). Hackett.
Greenberg, G. (2013). Beyond resemblance. *Philosophical Review, 122*(2), 215–287.
Gregoriou, G., Gotts, S. J., Zhou, H., & Desimone, R. (2009). High-frequency, long-range coupling between prefrontal and visual cortex during attention. *Science, 324*, 1207–1210.
Gregoriou, G., Paneri. S., & Sapounzis, P. (2015). Oscillatory synchrony as a mechanism of attentional processing. *Brain Research, 1626*, 165–182.

Grzankowski, A. (2015). Pictures have propositional contents. *Review of Philosophy and Psychology, 6*, 151–163.

Han, S., Ding, Y., & Song, Y. (2002). Neural mechanisms of perceptual grouping in humans as revealed by high density event related potentials. *Neuroscience Letters, 31*(9), 29–32.

Haugeland, J. (1987). An overview of the frame problem. In Z. Pylyshyn (Ed.), *The robot's dilemma: The frame problem and artificial intelligence*. Ablex Publishing Company.

Haugeland, J. (1998). Representational genera. In J. Haugeland (Ed.), *Having thought* (pp. 171–207). Harvard University Press.

Hausdorff, F. (1914). *Grundzuge Der Mengenlehre*. Veit.

Heck, R. G., Jr. (2007). Are there different kinds of content? In B. P. Mclaughlin & J. Cohen (Eds.), *Contemporary debates in the philosophy of mind* (pp. 117–139). Blackwell.

Hopkins, R. (1995). Explaining depiction. *The Philosophical Review, 104*(3), 425–455.

Hopkins, R. (1998). *Picture, image, and experience*. Cambridge University Press.

Hubel, D. H., & Wiesel, T. N. (1968). Receptive fields and functional architecture of monkey striate cortex. *Journal of Physiology, 195*, 215–243

Huntley-Fenner, G., Carey, S., & Solimando, A. (2002). Objects are individuals but stuff doesn't count: Perceived rigidity and cohesiveness influence infants' representations of small groups of distinct entities. *Cognition, 85*(3), 203–221.

Hyman, J. (2006). *The objective eye*. Chicago University Press.

Ito, J., Nikolaev, A. R., Luman, M., Aukes, M. F., Nakatani, C., & Leeuwen, Van C. (2003). Perceptual switching, eye movements, and the bus paradox. *Perception, 32*(6), 681–698.

Jackendoff, R. (1987). *Consciousness and the computational mind*. MIT Press.

Jensen, O., Gips, B., Bergmann, T. O., & Bonnefond, M. (2014). Temporal coding organized by coupled alpha nd gamma oscillations prioritize visual processing. *Trends in Neuroscience, 37*(7), 357–369.

Karmiloff-Smith, A. (1992). *Beyond modularity: A developmental perspective on cognitive science*. Cambridge University Press.

Kjørup, S. (2004). Pictograms. In R. Posner, K. Robering & T. Sebeok (Ed.), *4. Teilband: Ein Handbuch zu den zeichentheoretischen Grundlagen von Natur und Kultur* (pp. 3504–3510). De Gruyter Mouton.

Kosslyn, S. M. (1980). *Image and mind*. Cambridge, MA: Harvard University Press.

Kosslyn, S. M. (1984). Mental representations. In J. Anderson & S. Kosslyn (Eds.), Tutorials in learning and memory: Essays in honor of gordon nower. New York, NY: Freeman.

Kosslyn, S. M. (1994). *Image and brain*. MIT Press.

Kulvicki, J. V. (2006). *On images: Their structure and content*. Clarendon Press.

Kulvicki, J. V. (2014). *Images*. Routledge.

Kulvicki, J. (2015). Analog representation and the parts principle. *Review of Philosophy and Psychology, 6*, 165–180.

Lee, S. H., Simpson, G. V., Logothetis, N. K., & Rainer, G. (2005). Phase locking of single neuron activity to theta oscillations during working memory in monkey extrastriate visual cortex. *Neuron, 45*, 147–156.

Leopold, D. A., & Logothetis, N. K. (1999). Multistable phenomena: changing views in perception. *Trends in Cognitive Science, 3*(7), 254–264.

Leyton, M. (1992). Symmetry, causality, mind. Cambridge: Cambridge University Press

Liebe, S., Hoerzer, G. M., Logothetis, N. K., & Rainer, G. (2012). Theta coupling between V4 and prefrontal cortex predicts visual short-term memory performance. *Nature Neuroscience, 15*, 456–462, S1–2.

Lombard, M. (2024). Paying attention: The neurocognition of archery, middle stone age bow hunting, and the shaping of the sapient mind. *Phenomenology and the Cognitive Sciences*. Advance online publication. https://doi.org/10.1007/s11097-024-09980-z

Lyons, J. C. (2022). Three grades of iconicity in perception. *Asian Journal of Philosophy, 1*(50), https://doi.org/10.1007/s44204-022-00039-x

Maley, C. (2011). Analog and digital, continuous and discrete. *Philosophical Studies, 155*, 117–131.

Markov, N. T., Vezoli, J., Chameau, P., Falchier, A., Quilodran, R., Huissoud, C., Lamy, C., Misery, P., Giroud, P., Ullman, S., Barone, P., Dehay, C., Knoblauch, K., & Kennedy, H. (2014). Anatomy of hierarchy: feedforward and feedback pathways in macaque visual cortex. *The Journal of Comparative Neurology, 522*(1), 225–259. https://doi.org/10.1002/cne.23458

Marr, D. (1982). *Vision*. Freeman.

Michalareas, G., Vezolli, J., Pelt, S. V., Schoffelen, J-M., Kennedy, H., & Fries, P. (2016). Alpha-beta and gamma rythms subserve feedback and feedforward influences among human visual cortical areas. *Neuron, 89*(2), 384–397.

Miller, E. K., & Cohen, J. D. (2001). An integrative theory of prefrontal cortex function. *Annual Review of Neuroscience, 24*, 167–202.

Mitroff, S. R., Simons, D. J., & Levin, D. T. (2004). Nothing compares 2 views: Change blindness can occur despite preserved access to the changed information. *Perceptual Psychophysics, 66*(8), 1268–1281.

Nakatani, H., & van Leeuwen, C. (2006). Transient synchrony of distant brain areas and perceptual switching in ambiguous figures. *Biological Cybernetics, 94*, 445–457.

Newall, M. (2011). *What is a picture?* Palgrave Macmillan.

Norton, J. D. (2011). Challenges to Bayesian confirmation theory. In P. S. Bandyopadhyay & M. R. Forster (Eds.), *Philosophy of statistics* (pp. 391–439). Handbook of the Philosophy of Science 7. North-Holland.

O'Shea, J., Muggleton, N.G., Cowey, A., & Walsh, V. (2004). Timing of target discrimination in human front eye fields. *Journal of Cognitive Neuroscience, 16*(6), 1060–1067.

Palmer, S. (1978). Fundamental aspects of cognitive representation. In E. Rosch & B. Lloyd (Eds.), *Cognition and categorization* (pp. 259–303). Erlbaum.

Palmer, S. (1999). *Vision science: Photons to phenomenology*. MIT Press.

Palva, J. M., Monto, S., Kulashekhar, S., & Palva, S. (2010). Neuronal synchrony reveals working memory networks and predicts individual memory capacity. *Proceedings of the National Academy of Science USA, 107*, 1580–1585.

Peacocke, Cr. (1986). Analogue content. *Aristotelian Society Supplementary Volume, 60*(1), 1–18.

Peacocke, Cr. (2015). Magnitudes: Metaphysics, perception, and explanation. In D. Moyal-Sharrok, V. Munz, & A. Coliva (Eds.), *Mind, Language and Action: Proceedings of the 36th International Wittgenstein Symposium* (pp. 357–388). De Gruyter.

Peacocke, Cr. (2019). *The primacy of metaphysics*. Oxford University Press.

Peacocke, C. (2001). Does perception have a nonconceptual content? *Journal of Philosophy, 98*(5), 239–269.

Peirce, C. S. (1977, 1903). Logic as semiotics: the theory of signs. In J. Buchler (Ed.), *The philosophical writings of Peirce* (pp. 98–119) (1955). Dover.

Pianesi, F., & Varzi, A. C. (1996). Events, topology and temporal relations. *The Monist, 79*(1), 89–116.

Planer, R. J., & Kalkman, D. (2019). Arbitrary signals and cognitive complexity. *British Journal for the Philosophy of Science, 72*(2), https://doi.org/10.1093/bjps/axz018

Plomp, G., Quairiaux, C., Kiss J. Z., Astilfi, L., & Michel, C. M. (2014). Dynamic connectivity among cortical layers in local and large-scale sensory processing. *European Journal of Neuroscience, 40*, 3215–3223.

Pylyshyn, Z. (1984). *Computation and cognition*. MIT Press.

Quilty-Dunn, J. (2019). Perceptual pluralism. *Nous*, 1–32. https://doi.org/10.1111/nous.12285

Quilty-Dunn, J. (2020). Is iconic memory iconic? *Forthcoming in Philosophy and Phenomenological Research, 101*(3), 660–682.

Quilty-Dunn, J. (2016). Iconicity and the format of perception. *Journal of Consciousness Studies, 23*(3–4), 255–263.

Quine, W. V. O. (1960). *Word and object*. MIT University Press.

Raftopoulos, A. (2008). Perceptual systems and realism. *Synthese, 164*(1), 61–91.

Raftopoulos, A. (2009). *Perception and cognition: How do psychology and neural sciences inform philosophy*. Cambridge, MA: The MIT University Press, a Bradford Book.

Raftopoulos, A. (2019). *Cognitive penetrability and the epistemic role of perception*. Palgrave, McMillan.

Raftopoulos, A., & Muller, V. (2006). Nonconceptual demonstrative reference. *Philosophy and Phenomenological Research, 72*(2), 251–285.

Recanati, F. (1997). *Direct reference: from language to thought*. Blackwell.

Rescorla, M. (2009). Predication and cartographic representation. *Synthese, 169*, 175–200.

Roux, F., & Uhlhaas, P. J. (2014). Working memory and neural oscillations: alpha–gamma versus theta–gamma codes for distinct WM information? *Trends in Cognitive Science, 18*, 16–25.

Russell, B. (1927). *The analysis of matter*. Keagan Paul.

Sainsbury, R. M. (2005). *Reference without referents*. Oxford University Press.

Schonbein, W. (2014). Varieties of analog and digital representation. *Minds & Machines, 24*, 415–438.

Shepard, R. N. (1978). The mental image. *American Psychologist, 33*(2), 125–137.

Spelke, E. S. (1990). Principles of object perception. *Cognitive Science, 14*, 29–56.

Spelke, E. S., Breinlinger, K., Macomber, J., & Jacobson, K. (1992). Origins of knowledge. *Psychological Review, 99*, 605–632.

Spelke, E. S., Phillips, A., & Woodward, A. L. (1995). Infants' knowledge of object motion and human action. In D. Sperber & D. Premack (Eds.), *Causal cognition: A multidisciplinary debate* (pp. 44–78). Clarendon Press.

Sterelny, K. (2012). *The evolved apprentice: How evolution made humans unique*. MIT Press.

Sterelny, K. (2020). Afterword: Tough questions; hard problems; incremental progress. *Topics in Cognitive Science, 12*(3), 766–783.

Strawson, P. F. (1959). *Individuals*. Methuen.

Tallon-Baudry, C., Kreiter, A., & Bertrand, O. (1999). Sustained and transient oscillatory responses in the gamma and beta bands in a visual short-term memory task in humans. *Visual Neuroscience, 16*, 449–459.

Tye, M. (1991). *The imagery debate*. MIT Press.

Van Kerkoerle, T., Self, M. W., Dagnino, B. & Roelfsema, P. R. (2014). Alpha and gamma oscillations characterize feedback and feedforward processing in monkey visual cortex. *PNAS, 111*(40), 14332–14341.

VanMarle, K., & Scholl, B. J. (2003). Attentive tracking of objects versus substances. *Psychological Science, 14*, 498–504.

Varzi, A. C. (2016). Mereology. In Zalta, E. N. (Ed.), *The Stanford Encyclopedia of Philosophy*. Winter 2016 edition.

Werning, M. (2012). Non-symbolic compositional representation and its neuronal foundation. In M. Werning, W. Hinzen, & E. Machery (Eds.), *The Oxford handbook of compositionality* (pp. 633–724). Oxford University Press.

Yamada, F. (1998). Frontal midline theta rhythm and eye linking activity during a VDT task and a video game. *Ergonomics, 41*, 678–688.

Yamagishi, N., Callan, D. E., Goda, N., Anderson, S. J., Yoshida, Y., & Kawato, M. (2003). Attentional modulation of oscillatory activity in human visual cortex. *Neuroimage, 21*, 98–113

Chapter 5
From Non-representational Indices to Representational Icons: What Is Cognitively Required for Representations and Icons?

In this chapter, I examine various cognitive changes or innovations realized by specific cognitive mechanisms that could explain both what made possible the emergence of representations from non-representational indices, and, also, what allowed the brain to be able to support icons and symbols. It is understood that these mechanisms survived because they played an adaptive role in the evolution of our species. To play such a role, these mechanisms must have enabled functional capacities that were advantageous for the survival of our species. This means that not only were these new functional capacities a successful answer to some environmental and cultural pressures, but also that the mechanisms that enabled them were best suited to fulfill these functions compared to other cognitive mechanisms that could have performed these same functions. Hence, I discuss the cognitive prerequisites that allowed our ancestors to move from indices (which is the first form of non-representational signs used) to representations the first form of which was that of iconic representations. I also briefly address the problem of what necessitated the emergence of representational capacities, that is, why having representations was in evolutionary terms beneficial to a species.

I have argued, following Peirce's views, that "indices" are involuntary, albeit with the proviso discussed above, non-representational signs whose (physical) form bears a physical, factual, or causal relationship with what they indicate. They have no representational content and just indicate to the perceiving animal potentialities of interaction with the indicated animal (it is a possible source of food, it is a predator, a possible mate, etc.)

Icons, the first representational sign that appeared after indices, are, according to Peirce, non-arbitrary representational signs that resemble (are similar to) what they stand for or represent. Icons physically evoke through their resemblance what they refer to and, hence, there is a non-arbitrary, natural linkage between the form and the content, that is, between the iconic sign and the way the referent is presented in the icon; this last remark is necessary in view of the distinction, which we discussed

A. Raftopoulos, *The Origins of Representations: Cognitive and Brain Development*, Studies in Applied Philosophy, Epistemology and Rational Ethics 73,
https://doi.org/10.1007/978-3-031-91213-9_5

extensively in Chapter Three, between the intentional content of an iconic representation and the object of intention, that is, the object in the represented world that the icon represents. This natural linkage extends to the relation between the icon and the referent because the representational content of the icon must resemble the physical characteristics of the referent. Thus, semantics restricts syntax because the referent determines, through the representational content of the sign, the physical form of the sign. It is easy to see that icons satisfy all three criteria for being representations, that is, for having representational contents. Our discussion in the preceding chapter suggests that to encompass all sorts of iconic representations, "resemblance" should be understood to include second-order structural resemblance that is strong enough to allow an isomorphism or homomorphism between the icon and what it represents that preserves the semantic or spatial structure in the represented domain.

What made possible the emergence of iconic representations, first in the mind of agents and then, in the world when hominins started constructing iconic representations or started understanding physically-made icons (as when the painters of the icons in the caves recognized in rock formations characteristics of animals and used them to depict the animals)? That is, what made possible the appearance of both internal and external icons? However, before one attempts to answer this question, one should first attack another problem, namely, what made representations *simpliciter* possible, that is, what allowed to passage from non-representational indices to representational signs? The examination of this problem must cover two domains, first the cognitive level, that is, what cognitive capacities were required in order for representations to become possible, and second, what neural underpinnings were required to subserve these capacities. In this chapter, I discuss the "cognitive revolution" that enabled, first the emergence of representations and then, the emergence of iconic representations.

I also address the problem of which among the hominin species was the first to have representations and be aware they had these representations, that is, the first species to have self-awareness. This is important because all species "take in" information from the environment and use it to interact with it, and some among these species have neural (although some of these are quite simple) states that present the words as being some way and are, thus, semantically evaluable. In addition, many species have states that satisfy Bermudez's criteria for something to be a representation, but these species use these representations only in the context of a motor-action schemes that are activated by these representations. In addition, there are representations in our perceptual systems that do not require any awareness of having them, and thus do not require self-consciousness. However, as I have said in the introduction, I am interested in species that have self-awareness (and, therefore, also some theory of mind), with respect to their representational states, and can use them independently of any motor-action schemes. The reason is that only these species would be capable for exhibiting complex intentional behaviors, such as planning, extended social-cooperation, relatively complex social relations, and, perhaps, eventually some form of language; in other words, I am interested in examining which hominins were the first to employ and understand such signs that I have designated as representations. Thus, the notion of representation I have in mind for the rest of the book concerns the signs that such creatures may intentionally employ for various purposes in their effort to interact with their environment.

Before one engages in this discussion, one has to ask why representations became necessary and, thus, were eventually selected for by evolution. As we saw in the introduction, representations are detached from their referents in space and time. They have "displaced reference" that allows signs to *refer* to entities and events that are removed in space or time from the immediate communicative context. Animals that live in the here-and-now do not need representations since they can share information by using indices, such as pointing. Even when these animals engage in some activity that needs coordination and, thus, sharing of information, indices suffice to carry the useful information because all the participants are in the same immediate context and, thus, most of the relevant information is perceptually available. It follows that these animals do not have to share information about items that are not in the here-and-now and outside the immediate context. In other words, they need no representations.

However, when the collaborative activity extends to actions that have to take place in another context and in the future, the agents in order to convey information beneficial for the common enterprise (Tomasello, 2008) must be able to think and "talk" about items that are not present and cannot be perceived, and about situations that could occur in the future. To accomplish this, they need to use signs that have displaced reference, that is, representations. The benefits of using representations for exchange of information and collaborative actions are such that selective pressures would lead to the acquisition of representations. Of course, the ability to use signs with displaced reference is conditioned on certain cognitive capacities, such as a TOM, an increased WM capacity, causal reasoning and the use of mental models it allows. As I have said, the great apes most likely have them, and, thus, they are part of our genetic inheritance. A genetic change that made the requisite cognitive capacities available need not be the cause for representations to emerge, however. Since representations are useful for exchange of information in collaborative activities that require effective communication, once such activities and communication were a significant part of the hominis lifestyle, they were the cause of the selection for such a genetic change.

The comparable or homologue brain structures in humans and great apes are taken by researchers to suggest an evolutionary continuity from great apes to hominins and to modern humans, a view that is compatible with gradualism according to which qualitative differences in cognitive capacities appeared gradually in the course of evolution. As I said in the introduction, to examine the cognitive capacities of the H.E and other hominis one should start from a baseline that consists in the cognitive capacities of the great apes because if gradualism is correct, hominins started with the capacities that continuously and incrementally evolved. In the course of the discussion, we shall discuss in detail many of these capacities (such as self-consciousness, WM, attention, executive control, theories of mind, and social intelligence). However, it would be useful to synopsize here what recent research shows or suggests concerning the cognitive capabilities of the great apes. Here is what Read et al. (2022, 1) report:

> The last decade has seen a dramatic change in the way we think about commonalities between us and the pongids. This change in thinking has been driven by experimental discoveries that attribute to them many cognitive traits traditionally reserved for humans. It has been reported that pongids can save and make tools for future use (Brauer & Call, 2015; Mulcahy & Call, 2006), are able to adopt another individual's visual perspective (Hare et al., 2000; Karg et al., 2015), attribute knowledge of the world to conspecifics (Kaminski et al., 2008), and perhaps attribute beliefs to humans (Krupenye et al., 2016). They also have been observed to cooperate to solve common feeding problems, both in the wild (Boesch, 1994, 2002) and in captivity (Melis et al., 2006, 2011). Moreover, complex planning and the capacity to remember past personal events are claimed to be among their cognitive repertoire (Lewis et al., 2019; Martin-Ordas et al., 2010; Mulcahy & Call, 2006). There are even studies reporting that they may incur personal costs to witnessing how justice is dispensed (Mendes et al., 2018).

Thus, some form of executive control, working memory, semantic/episodic/autobiographical memory, mental models, social learning, and theory of mind form the baseline from which our evolutionary started. The same conclusion can be reached when one considers the evidence concerning the ability of encultured apes to learn to use composite signs (Schoenemann, 2022). The fact that in an appropriate context great apes can enhance their communicative abilities by employing composite signs means that the great apes have TOM, WM, and the other capacities required form this sort of learning, in such an extent that suffices for composite signal acquisition and use. These cognitive capacities have without a doubt evolved and are found in our species in their most advanced form. For the same reason, in the 5–8 million years that separate H.E from when our line diverged chimpanzees, one would expect these the homologue capacities of H.E to be more sophisticated than those of the great apes and, of course, much less advanced than the same capacities of modern humans.

In fact, there is some indirect evidence of such capacities in earlier Pliocene hominis that used stone tools 3.3 Ma (Harmand et al., 2012). The tools were most likely used to extract meet from bones of large animals, meet that had to be carried away from the finding site to a safer place to avoid unpleasant meetings with the carnivore animals that did the killing. These seemingly simple tasks require some degree of intragroup cooperation and social tolerance because they are collective actions that require teamwork. Intragroup cooperation and social tolerance, in turn, require some sort of communicative capacity that exceeds the simple gesturing repertory of the great apes (who eat at the site and do not carry their food away). It follows that simple forms of theory of mind and social intelligence in general are already in place, as is an advanced form of WM involved in remembering where the home-base is and how to return to it. In the next sections, I expand on and examine each of these capacities as they relate, mostly, to H.E.

5.1 The Emergence of Representations

To be able to represent means to be able to re-enact, as it were, an object or event in the mind when that object is no longer present. In other words, the ability to use spatio-temporally detached signs is inextricably tied to the capacity to represent. One might be tempted to argue that when an animal perceives a footprint on the ground and assesses it to be the footprint of a possible pray and follows it in an attempt to catch the prey, the footprint creates in the mind of the animal a representation of the prey since the prey is not present. In other words, the index is a representational state with a certain content, namely the pray. As I have argued elsewhere (Raftopoulos, 2008), however, what plausibly happens is that the footprint activates some affordances in the perceiving animal (that is, potentialities for action on objects on the part of the animal afforded by these objects) that are related to some basic sensory-motor causal schemata of interaction that are either hardwired or are developed fast after birth owing to some genetic predisposition. Moreover, a representation is not simply the re-enactment of an absent object; representations are true or false, have a meaning, and must have structured contents. The content of the state that the animal causally forms when perceiving the footprint does not have these properties.

Let me explain this starting from a simple organism-environment interaction, such as a frog's interaction with a fly when a frog visually perceives a fly. What is going on in the frog's "mind"? The fly produces a neural state in the frog's brain; the content of the neural activity induced by the fly in the frog's brain consists in the recognition of the perceived object as a possible source of food, and in the possibility of tongue flicking and eating on the basis of indications about potentialities that are afforded by specific objects in the environment. For an object to afford eating possibility, it must be within range, that is, at a certain distance, and it should be moving with a certain velocity. These parameters co-determine the relevant affordances and, thus, their values must be taken into account by the frog's brain. This does not mean that the brain performs complicated computations or any computations at all. It must just solve an equation whose solution is hardwired in the brain, the same way we do not have to perform any computations to compute the aperture of our step to climb a pair of stairs; our brain just solves an equation.

It follows that the content of the perceptual state of the frog is in the indication in the frog that it is in a tongue flicking and eating situation. This content is about the potentialities, or possibilities of further interactions that are afforded by the environment for the system's interactions with it, not about the fly itself; the relevant states of the frog are intimately connected with action (the sensory-motor scheme that I alluded to in the previous paragraph). The potentialities implicitly predicate those interactive properties of the environment that could support the indicated interactions of the frog with it. This is a kind of functional predication. Since this kind of content implicitly predicts the properties of the environment that afford actions on the part of the cognizer, it is about these propertiese even though the frog does not represent them. Furthermore, these sensory-motor causal schemes are tied to the here-and-now; they are activated by the presence of the fly and cannot function when a fly is not present; no displaced referential capabilities in these schemes.

In this sense, the sensory-action causal schemes act like other animal signaling activities, such as the vervet danger calls. In the action scheme there is no reference to a fly or to its attributes; only to the possibility of tongue-flicking and eating. In the same way, in a vervet call that signals leopard-here-now, which is a situation, its elements do not correspond to the elements in the situation; no reference to a leopard or its attributes but only to an expected, hardwired or learnt, reaction. And, of course, in both the frog and the vervet scenarios there is no displaced reference; a vervet monkey does not issue a warning signal thinking that a leopard will approach in a few hours. Recall that one of the four characteristics of representations is that they must have structure. This means that its constituent elements could be constituent parts of other contents, and the content could be decomposed into its elements that could be recombined with the elements of other states having representational content. Sensory-action causal schemes and the vervet calls have no such structure, which is why they are not representational.

Having structure also means that a (complex) representation consists of parts that purport to represent a state of affairs. As the reader recall, a representation acquires its meaning by combining following syntax (in symbolic systems) or Mereotopology (in icons) the meanings of the individual constituent parts of the representation. Since these meanings are partly determined by the referents of the constituent parts, it follows that each representational constituent maps on an element in the represented state of affairs. This is a property of representations that causal sensory-action schemes and vervet call systems do not have.

To recapitulate, to catch the fly the frog does not have to calculate the courses of action that are available to it, the chances of success, the outcome of each course of action, and estimate utilities. The possibility of eating that the fly affords to the frog is indirectly perceived by the frog; when it sees the fly it sees the affordance of food. Furthermore, the instrumentality of a particular course of action reveals itself in the perceptual content; there is no need for a decision-making. All the information needed to initiate action is directly present in the perceptual content. However, and this is important, the frog has a perceptual state with some content, namely a black dot moving to a certain direction at some distance, even though it need not make any calculations in order to act on this information and, more importantly, it need not represent any of these facts; the perceptual system takes care of this. This content is not a representational content because it fails to satisfy the criteria for being a representation. The sighting of a fly merely causes a sensory-motor schema to get activated and this does not require a representation. The activation of this scheme brings into the picture the aforementioned affordances; such causal schemata regulate interactions with the world rather than model or represent the world.

If my argument above is sound, it is likely that the brain started phylogenetically and starts ontogenetically as a purely causal mechanism and in the course of development acquires representational powers. The question is how? How is it that schemes that started as simple causal functions that work on the basis of affordances acquire representational properties and are transformed into states with representational contents? To answer this question, one should start with seeking the basic cognitive abilities that an organism must possess in order to be able to form representations.

Representations, unlike hardwired causal sensory-motor schemes that are activated in the presence of the object or class of objects with which they are causally associated, presuppose the capacity to hold in mind an object in its absence. This means that the representation of an object may be activated, and agents may use this proxy of the object in their minds for a wide variety of functions, even though a sensory-motor scheme is not active. Thus, the representation is detached from the sensory-motor scheme.

The perceptual system already has this ability, first, in the form of iconic memory or sensory persistence that has an almost unlimited storage capacity. Iconic memory is divided into two subsystems (Coltheart, 1980); *visible persistence* that lasts for only 80–100 ms. after the onset of a stimulus event (Di Lollo, 1980) and produces the phenomenology of a persisting visible image, and *informational persistence* that is not visible but supports the retention of sensory information for 300–500 ms. after stimulus offset (Irwin & Yeomans, 1986). Second, in the form of fragile-visual-short-term memory (FVSTM) (Sligte et al., 2011; Vandenbroucke et al., 2011). Both systems make an image available to perceivers for a few seconds to allow them to finish processing a visual scene even when the scene is no longer present. The perceptual content held in mind is conscious, personal-level content, even though it may be outside of cognitive access consciousness, that is, even though perceivers may not be aware that they have that content (Burge, 2022; Raftopoulos, 2015). This is Dretske's (2000) distinction between "thing awareness" and "fact awareness", or Block's (2007) distinction between "phenomenal content" and "cognitive access content". Holding an item in perception does not entail a conscious awareness of, or thinking about, this item even though it is experienced. However, this holding in mind eventually gives rise to the emergence of a representation that is accessible by cognition. This leads to object recognition, and to beliefs, thoughts, or other propositional attitudes. This is the level at which symbols become available.

Thus, the first step in the emergence of representations is that a sensory-motor purely causal schema is decoupled from its normal causal sensory-motor functions and serves as a representation of the object or event that initially called it into play. Sensory-motor schemes are tied to specific contexts of the organism-environment interactions and are activated only within these contexts. For a representation of an object to be possible, this scheme must be de-contextualized so that it would be possible for the same object to be represented in different contexts, as, for example, the perceptual content related to the occurrence of an object is retained in iconic memory and processed even when the object disappears from the scene. Notice that the argument is that an object may be represented across different contexts, not that it is necessarily represented outside any context. Iconic perceptual representations, as we saw in Chap. 4, require some context to be formed and used, while symbols may be used independent of any context. However, even with symbols, an environmental context was necessary for the creation of many among them—those, for example, referring to natural kinds, or names (see that discussion of Kripke's (1980) and Putnam's (1983) view, in Chapter Two). The representation becomes a copy of the object and has the properties usually associated with representational states, namely, possibility of misrepresentation through failure of reference, and having a truth value. In other words, a decoupled schema loses its direct ties to reality and becomes a mental content properly speaking. To return to the fly example, a humanly developed frog would have been able to detach the perceptual content associated with a fly from the affordances associated with perceiving a fly and use it as an internal copy of a fly that may be evoked even in the fly's absence.

It is understood that this decoupling was gradual and lasted, probably for millennia. A characteristic example of this gradual process that culminated in the fully fledged symbolic capacities of modern humans is the case of H.E and Neanderthals. H.E first used the Oldowan technology and gradually, by 1.76 Ma, developed the much more advanced Acheulean technology that allowed the construction of hand-axes. Neanderthals used and produced blades, scrapers, and single-edged hand-axes, a technology called Mousterian that was markedly more sophisticated than the Acheulean industry of H.E. Late Neanderthal sites in Portugal and France, dated to after 40,000 years ago (the beginning of the Upper Paleolithic), preserve a blade tool technology that was similar to that of early modern humans from the same time period. This Upper Paleolithic technology, known as the Châtelperronian industry required much planning and precision to produce and suggests that early Homo sapiens were not the sole makers of this sophisticated technology.

Furthermore, as we shall see, many hand-axes constructed by H.E were found intact without any trace of use. They may have been left unused for several reasons, but the copious quantities found suggest that they had some social or ceremonial significance, and they were not designed to be used for cutting meat or other menial tasks (Kohn & Mithen, 1999). The same non-utilitarian practice has been attested in Neanderthals. Nonutilitarian artifacts have been recovered with Neanderthal remains and Mousterian tools. These artifacts indicate cognitive function beyond that needed for basic survival. Some of these artifacts seem to have been collected because of their intrinsic value or beauty rather than for some utilitarian function. Studies have indicated the use of ornaments, colorants, bone tools decorated with designs, and natural, unmodified objects that have been moved by people, which were carried away from their original context and sometimes altered by using stone tools. The occurrence of non-utilitarian lithic objects in Middle Stone Age or Upper Paleolithic sites is accepted as evidence of symbolic behavior associated with the appearance of Homo sapiens. However the occurrence of non-utilitarian and unusual objects in far earlier sites is a puzzle. In fact, from the earliest beginnings of mankind, various objects have been found in the occupation sites that have no apparent functional link with any technical activity or food procurement. This type of object is more obvious from the Acheulian times (*Homo erectus* time) onwards and then becomes common in the Late Pleistocene. Perhaps they were intended to convey some, unknown to us, symbolic expression, but at least their occurrence suggests some non-utilitarian concerns among the people who constructed or used them (Edwards, 1978; Mellars, 1996; Moncel et al., 2012). Finally, H.E has been credited with construction of iconic representations, and there is also evidence for Neanderthal symbolic engravings 75,000 years ago (Marquet et al., 2023).

The development of the Acheulean and Mousterian technologies presupposes a multitude of cognitive abilities associated with finding the appropriate materials, then creating a mental image of the tool in mind, overcoming the various difficulties to construct the tool etc. By 1.7 Ma, the toolmakers were imposing the final form of the tool to be produced that was held in their mind. They also, probably, had to create in mind the series of action-steps that should be followed to achieve the goal and put these sequences into effect to construct the actual tool. This, of course, requires, first, the ability to entertain structured representations in mind and, second, a WM with enough holding capacity that would allow keeping in mind both the final product whose production guides the series of actions and the sequence of the actions that should be performed in order to achieve the goal. Homo Sapiens have this capacity with a WM capacity of 7 ± 2. Chimpanzees with a capacity of 2 ± 1 cannot construct this kind of tools. H.E with an estimated WM holding capacity of about 5 (Read et al., 2022, 22) developed and expanded the Acheulian technology, which suggests that their WM was strong enough for the task at hand. The construction technique must also be taught. All these call for planning, communication, and social roles within the group. All these require not only representations but, in fact, advanced representations capable to support the relevant cognitive abilities.

Consider, to take another example, planning for group hunting. The aim is to organize a successful hunt so that enough meet be secured. To achieve this aim, a series of actions must be planned ahead. The success or fail of actions depends on their causal consequences and, when in planning, different courses of actions are considered so that the most promising be selected, the hunters that consider these possible actions must also think in advance of their causal consequences, because only then could they make an informed choice. To achieve all these, these hunters must activate in their mind mental models of the whole hunting situation, since mental models of situations are representations that allow making predictions about the behavior of the objects involved in the situation in novel situations that may be familiar to the agents using the mental models, or they may be not. Making predictions requires if X then Y modes of thinking and choosing among the possible courses of actions requires comparing the possible outcomes of the different actions and these comparisons involve possible in the future outcomes. These require quite advanced representational abilities indeed and a WM with sufficient holding capacity.

Moreover, the advanced sophistication of the Neanderthal tool-making compared to the H.E tool-making, and the plausible use of iconic representations constructed by the former strongly suggest that there was a development of the representational capacities, enabled by, and also causing in a bootstrapping relationship, a bigger brain and a re-organization of its neural circuits. Relatedly, mental images (which are iconic representations) of tools used by the hominins in Upper Paleolithic were increasingly more faithful copies of the tools that were designed on the basis of these same mental images because the form of the tools became as important as their function, and the construction of tools required that the craftsmen had a clear, precise mental picture of the tool that they had to create (Lewis-Williams, 2004, 75–76).

Still, Lewis-Williams (2004, 93) argues, even though Neanderthals, who appeared about 400 Ka or perhaps even earlier (Gómez-Robles, 2019), undoubtedly had mental images of what their tools looked like, a prerequisite to constructing them and transmitting this knowledge, this mental image was very different from the mental images of animals that could stand for representations of these animals. 'Neanderthal mental imagery was, I suggest, intricately linked to motor skills-such as a series of manual actions that produced a flint tool of the desired shape. Physical action activated mental imagery, and, to a limited extent, vice versa.' As it has been done several times in this book, using the "principle" that ontogenesis may shed light on phylogenesis, *and* vice versa, proves useful. This time it cuts deeper though, because what follows does not concern the ontogenetic development of the individual, but, rather, the vestiges of the ancient parts of our brains in our modern brains, as they shape the interaction between vision and action. This should not surprise us. Successful vision-action interaction is a prerequisite for survival and, thus, it is hardly surprising that the basic mechanisms underlying such interactions are well entrenched to still be with us and characterize the first stages of our visual system. Let me dwell on this.

The importance of the interaction between visual perception and action hardly needs explaining. Successful interaction requires some form of action memory so that past reliable actions be learned and repeated when needed (this is why training in the early years, or months, is usually important in the animal kingdom). Indeed there is evidence that there is a motor element in memory for enacted events stored in the brain that facilitates the vision-action coordination. The object files (the repositories of information pertaining to specific objects) in the brain do not store these actions memories. The perceptual and mnemonic circuits use the objects files in their capacity as reference fixers (Raftopoulos, 2009) to objects and their role is to parse and track objects in a scene. Their function precedes the storage of action memories since the latter are actions on objects that must have been picked out in a scene first. Otherwise, one would have to assume that when a first encounter with an object occurs and since there is no action attached to it, one does not detect the new object, a quite counter intuitive and false assumption.

However, as a result of repeated interactions with the environment, the object files become attached to specific actions, that is, they are linked with the feature codes of responses (which are the representational contents of the representational states in the premotor cortex associated with the specific action). Notice that as the object files bind together features of objects that are useful for the task at hand, action planning is viewed as a binding process that integrates codes of the features that the intended action ought to have (planning a movement with the left hand, for example, is such an action feature); these are the action files (Hommel, 2004). The linking of object feature codes with action feature codes, that is, the linking of object files with action files creates the event files. This link is a form of binding too, where the binding is across representational domains, namely perception (the dorsal system) and action (pre-motor and motor system). Event files for repeated actions can be stored in memory and the activation of the object file or of the action file may through the event file activate the action and the object file respectively. Thus, during the perception of items in a visual scene, action-based representations are automatically activated and can feed back and influence the perception of people and objects. For example, when one is merely viewing an object such as a coffee cup, the action it affords, such as a reach to grasp, is activated even though the viewer may have no intention to act on the cup. Similarly, when one is observing a person's behavior, their actions are automatically simulated, and such action simulation can influence our perception of the person and the object with which they interacted (Tipper, 2010).

The close link between vision and action is not only due to the binding of object-files with action-files that binds visual and motor representations. There is evidence that the final stages of perception and the initial stages of motor control share common representations (and, thus, neural populations), in which planned actions are represented in the same format as perceived events (Hommel et al., 2001). It should be noted, however, that only percept-dependent actions affect perception (Beets et al. 2010). To make the link even tighter, and to encompass cases of actions involving tools as well, the brain has seen to it not only that there be bimodal visual-haptic cells but also that these cells can expand their receptive fields to encompass the tool as well as the hand. This way the brain solves the problem of the displacement of perception from action. When a tool is held by the hand to manipulate an object, visual focus is on the end of the tool, which is the source of feedback about the contact with the manipulated object, and away from the movements of the hand itself that does the manipulatory movements. This is a case of perception/action decoupling since the eye is separated from the hand (Klatzky et al. 2010).

The inextricable relation between perceptual object-files and action-files can explain the interference effects of action on perception or of perception on action that are widely observed and studies in psychology (Beets et al. 2010; Klatzky et al. 2010; Tipper 2010). Planning a manual hand movement and maintaining that plan in memory impairs the detection and perception (that is, conscious awareness) of spatially corresponding stimuli that are either briefly presented or masked (Müsseler & Hommel, 1997). Or, consider the Simon effect that demonstrates the automatic link between perception and action. Subjects are asked to report the color of a stimulus by pressing a key. For example, they are asked to press the right key with the right hand if the color is red and to press the left key with the left hand if the color is green. There is no correlation between the color of the stimulus and its spatial position on the screen so that spatial location be irrelevant to the task. However, reaction times are consistently faster when the red object appears on the right side of the screen and the green object appears on the left side of the screen. The reason is that the right-hand stimulus is closer to the responding hand and this reveals an automatic link between a stimulus and goal-directed reaching actions (Hommel, 2004, Tucker & Ellis, 1998). Finally, there is evidence that previously learned movements improve visual discrimination of the same movement and induce increased cortical activity of the motor areas when observing the movement even when no action is planned on the part of the viewer. This is evidence for reciprocal effects between action and perception (Beets et al., 2010; Casile & Giese, 2006).

It also explains why sometimes the potentiality for action afforded by an object helps the visual system to construct an object when the sensory evidence is fragmentary. Consider patients with visual extinction. (Visual extinction is the phenomenon in which damage to the parietal lobe impairs spatial attention; patients cannot focus attention on two objects that are presented concurrently on the two visual fields and, thus, they cannot report seeing an object in their bad field even though if the same object were presented alone in the bad field they would have reported it.) When the patients are presented with pairs of objects which are either positioned correctly for action (a cork-screw attached to the cork of a bottle), or which were positioned incorrectly for action (a cork-screw attached to the bottom of the bottle), they show extinction when the object in the bad field is positioned incorrectly but the extinction is reliably reduced when the object is correctly positioned for action. This suggests 'that co-locationing objects for action enables them to be coded as a single perceptual unit, overcoming the competition for selection in patients with extinction which the ipsi- and contralesional stimuli are coded as separate objects' (Humphreys & Riddoch, 2007, 538).

Finally, the vision action link can explain the consequences of displacing perception from action (Klatzky et al. 2010). When eyes are disembodied in this way and perception is displaced from the space of action (as when we observe the results of our actions on a screen away from the locus of action), the ability to localize targets by pointing is impaired, as it is the ability to construct a unified mental representation of a 3D object by exploring it with 2D slices. The reason is that when eyes follow actions, the actions produce motor efference (from the movements of the hand) and kinesthetic feedback (from holding the tool) and, since the actions themselves are viewed, they also produce visual feedback. When the eyes are displaced from action the visual feedback is lost and when the task is, say, to combine temporarily linked 2D images that represent spatially contiguous source data the perception/action dislocation precludes the formation of a common spatial frame of reference that would allow the integration of the images so that they form a 3D representation. The integration of information becomes a difficult cognitive task involving mental rotations and the use of knowledge about the tools and the targets rather than a purely perceptual task.

This evidence suggests that there are intricate vision-action links embedded in the visual system. It follows that, during the evolution of our species, the first steps towards the acquisition of, first, representational, and then, later on, symbolic abilities included a phase in which representations were not fully detached from their initial visuo-motor connections. Still, even though fully abstract representations were not possible, the distinction between object files and motor scheme points to the first signs of a dissociation between these, a dissociation which, as we have seen, is necessary for representations to appear.

The ability to decouple sensory-motor purely causal schemes from their causal sensory-motor functions has some striking consequences. First, when agents represent an object or event in their sensory memory or in their mind in general, at the same time their perceptual systems or minds are free to process a second object or event. This allows the formation of a relation, which may be a comparison, between the object "held in mind" and the current object of perception. When 10-months old children are surprised to see one object reappearing from a screen into which two objects entered (Carey, 1995; Carey & Xu, 2001; Xu & Carey, 1996), this is possible because they hold in mind the representations of two objects entering and compare this with the representation of one object exiting; since infants "know" that material objects do not vanish in thin air, they get surprised. Without these representations, in the form of object files (Carey, 1995; Carey & Xu, 2001; Xu & Carey, 1996), the infants would not be able to be surprised.

Another consequence of this ability is that the relation between the object held in mind and the current object of perception can be a relation of predication. We have discussed "predication" in symbolic and iconic representations. Let us revisit some of the main themes. Mental representational states have representational contents. Those mental states that are sentences or sentence-like entities, that is, they are symbols, have propositional contents that form propositional structures. Propositional structures are complexes of representational contents that are formed by means of various combinations of atomic propositional units that are performed by means of specific compositional mechanisms. The simplest and most fundamental mechanism is predication, where an unsaturated, predicative part of thought is completed by a term that denotes some entity in the world (such as a proper name, for example) and becomes a complete part of thought, that is, a proposition that has some truth value. For example, the proper names "John" and "Mary" are denoting terms that when they fill the two argument-places of the 2nd order (relational) predicate "Love" form the proposition 'John loves Mary', which is true or false depending on whether John loves Mary or not.

To move to perception, suppose someone perceives a visual scene in which includes an object O that is X and comes to think that 'O is X.' They can do that because they hold in mind O and, based on their perceptual evidence, form in working memory the hypothesis that X is O. This presupposes that viewers form first a representation of an object in a visual scene and then assign to it its properties, a thesis that is amply supported by psychological and neuropsychological evidence (Raftopoulos, 2009). This modulates perceptual processing, and the hypothesis is tested against the perceptual evidence by searching the perceptual content to find out if indeed O is X; this is done by making predictions about the traits the retinal input should have if the hypothesis were correct and comparing these predictions with the actual retinal input. If it is confirmed, the hypothesis becomes a belief. Things could have turned otherwise; testing may have revealed that though O seemed to be X, in reality it was Y. A belief, thus, is the product of holding O in mind, perceiving X, and eventually attributing X to O after searching the perceptual evidence. This procedure turns denoting parts of sentences (such as names that designate objects) and their properties into subjects and predicates of propositions. Before the formation of the belief, however, the viewers had formed an icon of O having the property X, that is, an icon is which X is attached or predicated to O. This presupposes that an independent representation of O be formed and held in perceptual circuits and to which perceptual properties are eventually attached. This means that O is first individuated from other objects in the visual scene on the basis of its spatio-temporal and featural properties but these properties are not at first predicated, even indirectly, to O (see Raftopoulos, 2009), O, thus, is detached from any causal sensory-motor schemes because these schemes are activated from some properties of objects (the fly in the frog case, for example, must be at a certain distance from the frog to activate the affordance of eating. Moreover, the frog when seeing a fly does not perceive an object and then assigns or predicates to it its properties, which is what our perceptual system does after O has been individuated from other objects in the visual scene, but perceives the fly-properties-affordances causal scheme in which the ingredients are undifferentiated.

We started from purely causal sensory-motor schemes, where the problem of truth and falsity does not arise, but issues of success and adaptation do, and ended up with propositional attitudes that are true or false, or perceptual states whose content are correct or not. What develops is the ability to hold in perception first and then, in the mind an object, and later a relation between the object and a property by means of a symbolic or an iconic representation

Suppose that the account I offered concerning the transition from non-representational indices to representations is on the right track. What are the cognitive abilities required to make this transition possible? What allows, cognitively speaking, the decoupling of sensory-motor schemes from the purely causal context of their activation so that mental representations of objects or events could be formed?

First, it is required that agents have a sense of self-awareness, that is, the ability of agents to recognize themselves as beings distinct from other beings. Self-awareness is knowing that one exists, that one stands in relations to other things that exist but are separated from one, and that these things, including one-self, can become the object of one's attention and thought. This also means that one can recognize one-self when seeing a picture of oneself or looks one-self in the mirror. One might think that anyone, let alone a human being, should have this capacity from the beginning. However, psychological experiments suggest that this is not true. In human children, this capacity gradually develops from birth (where there is a self-world confusion) up to the 4–5 years of age, where full self-awareness arises (Rochat, 2003). Children younger than 18 months, for example, do not have this capacity as they do not recognize themselves upon looking at a mirror, a capacity which is acquired between twenty and twenty-four months by two-thirds of the children of that age (Amsterdam, 1972).

Let us see what studies on this theme suggest. It is thought that the development of self-awareness includes five main levels. In the first level, babies can from birth differentiate themselves from their surroundings. They quickly understand that their own body is different from their environment. This is suggested by the fact that babies reach for objects around them. At this stage, they still have no sense that the reflection they see when looking at a mirror has anything to do with them.

The second level is that of the situationally aware self. Around 2 months old, babies develop a situational awareness of themselves and their environment. This means they not only recognise the difference between themselves and the environment, but, in addition, they have a sense of how their body is situated relative to that environment. At this age babies begin to reciprocate facial expressions. This social back and forth shows their understanding of their presence in the world *alongside* others.

Between 18 months and 2 years, children begin to develop the understanding that the baby looking back at them is them. This is the level of identification, Rochat (2003) tested this theory by surreptitiously placing a Post-it note on the children's forehead before asking them to look in a mirror. On seeing their reflections, children who had reached this stage, reached for their heads to remove the sticky note, whereas younger children reached for the mirror or ignored it completely.

The fourth level is that of the development of permanence. The time between 2 and 4 is developmentally interesting in that children often appear to have grasped that the person they see in the mirror is them while also sometimes referring to the person they see by name rather than saying "me". As Rochat observes, it is not unusual for children of this age to oscillate between an awareness of the self and an awareness of seeing someone else facing them.

The final level is that of meta-self-awareness (that is, self-consciousness). From around 4 years old, children have a well-entrenched understanding that the person they see in the mirror is them. They also become aware that what they see in front of them is also seen by others and more than that, they recognise that what they think of themselves may not be the same as what others think of them. It is obvious that this development is facilitated by the concurrent development of an advanced theory of mind; Lage et al. (2022) argue that theory of mind is a component of self-awareness. This realisation can explain why some children are shy or hesitant in social situations.

It seems, thus, that there is a gradual development from understanding a differentiated self to acquiring full self-consciousness. These stages have in common that they are processes in which agents become the object of their own attention. Since it is known that phylogeny may shed some light on ontogeny, that is, that the sequence of acquisition of cognitive capacities as humans grow roughly parallels the sequence in which these capacities developed in human evolution, the fact that children need to reach a certain age to acquire self-consciousness entails that it took some time for hominins to develop fully this capacity, and, hence, that this capacity developed gradually in the course of evolution. This is the idea underlying "gradualism" in evolution, that is, the view that evolution occurs gradually, and incremental changes are accumulated over a period of time (Laland et al., 2015). Hominins, however, are not the only species to have self-awareness. The view that self-awareness is a multifaceted capacity consisting of various components is useful in this context because it suggests that as these components develop so does self-awareness and that, in consequence, if two agents share one component but are developmentally speaking at different evolutionary or developmental stages, they will be at different levels with respect to self-awareness. If, for, example, as numerous studies suggest a theory of mind is a critical component of self-awareness, agents who possess a developed theory of mind will be developmentally or evolutionary in higher stages of self-awareness. In the developmental literature, as we saw, the fifth and last stage, that of meta-awareness in which self-consciousness is achieved requires an advanced theory of mind because the child must have mastered the ability to become aware that what they see in front of them is also seen by others and more than that, they recognise that what they think of themselves may not be the same as what others think of them; this is an advanced theory of mind, indeed.

If gradualism is correct and self-awareness is not an all or nothing concept but comes in degrees of complexity, it comes as no surprise that it is thought that fish, cats and dogs display a lower level of awareness that elephants, dolphins, apes, and hominins, since fish and dogs do not recognize themselves in a mirror, while elephants dolphins and some apes do. Great apes, dolphins, Asian elephants, and Eurasian magpies can recognize themselves in a mirror, as can human kids around age 2. In fact, chimpanzees seem to develop self-awareness half a year after humans, that is, at about 2 years of age, and both species are preceded by dolphins (Morrison & Reiss, 2018; Lei, 2023). It is interesting that rhesus monkeys can be trained to acquire self-awareness (Chang et al., 2017), which probably means that the mental capacity was present but needed activation through training. Dogs have body awareness (Horowitz, 2009), which is another component of self-awareness (Damasio, 2010). The self-awareness discussed with respect to elephants, dolphins, and monkeys is the third level of self-awareness in the course of human development, that of "identification." Humans acquire the capacity of meta-awareness or self-consciousness around their fourth year, and this goes hand in hand with the development of a theory of mind. Thus, only creatures that combine self-awareness and a theory of mind can acquire self-consciousness. It should be noted that even self-recognition comes in many levels in the sense that self-recognition in fish differs from self-recognition in apes, for example. Remember from the human developmental studies, at the fourth level there is a phase that the ability oscillates and comes and go.

Among the components of self-awareness that have been proposed in the literature, three stand out because of the importance they have for the main object of the book, which is to examine the cognitive abilities required for representations to appear, since, as the reader recalls, self-consciousness is a pre-requisite for having representational capacities. These components are self-recognition, a theory of mind, and meta-cognition. The importance of self-recognition, as a component of self-awareness, has been studied by Anderson and Gallup (2011), and De Vignemont (2011) among others. The importance of a theory of mind for self-consciousness is suggested by the fact that that agents can have it only if they can understand that others have minds, that that they can think about these agents, and they also recognise that what they think of themselves may not be the same as what others think of them. This is a second order theory of mind. This last capacity also requires meta-cognition, that is, having thoughts and cognize about one's own thoughts and the thoughts of others.

At a minimum, since apes have self-recognition, H.E that appeared 5–8 Ma since Hominis split from the great apes certainly had it. The problem is not whether H.E had this capacity, but the extent to which H.E had it. In other words, how much more advanced was this capacity in H.E compared to the self-awareness in the sense of self-recognition of the great apes, given the millions of years that separate them? More specifically, could we attribute to H.E self-consciousness, that is the combination of self-awareness with a theory of mind and metacognition? Chimpanzees and other monkeys have a theory of mind in that they can attribute to the other members of the group mental states that allow them to perceive, feel, and know. In addition, chimpanzees have mastered the notion of deception and can use it as a means for formulating strategies to deceive others (Call & Tomasello, 2008; Le Roux et al., 2013; Overduin-de Vries et al., 2015). Moreover, a host of studies show that great apes have metacognition as well (Call, 2010; Call & Carpenter, 2001; Perdue et al., 2018); Tomasello, 2023). It follows that H.E had all these capacities.

As I have alluded to, and as we shall in some detail in a while, the important factor to consider in attempting to detect the cognitive capacities of a species is the environmental (including social) pressures whose successful negotiation on the part of the species required that some adaptations be selected and transmitted genetically of epigenetically (for the adaptive advantages of self-awareness, for example, see the discussion in Lage et al., (2022)). We shall address this problem at the end of this section, after we have concluded the examination of the cognitive capacities required for the emergence of representations, of the neural underpinnings that could subserve these capacities, and of the way they have been developed in hominins.

The capacity for self-awareness leads also to the awareness of others as distinct beings with whom one can cooperate or compete not at the level of instinct but at the level of thought. Cooperation works better if a member of the group could infer the intentions of the other members of the group so as to ensure that they all aim for the same task. Having the ability to make such inferences, therefore, constitutes an evolutionary advantage that could have led to the selection of this ability. Cooperation requires what Tomasello (2019) calls "joint intentionality", which means that different members of the group have similar intentions on a joint enterprise, and they know that they do so. This allows planning and, thus, best hunting techniques that confer a significant evolutionary advantage. Joint intentionality also allows planning for excursions in places sometime far away to obtain the appropriate stones to construct good quality hand-axes, and, importantly for planning long-distance migrations. Planning is a crucial factor in our discussion because planning certainly requires the usage of a rich representational arsenal; it is the kind of behavior philosophers call "representation hungry." If it could be shown that a species does engage in planning, this is a very good reason to think that this species has representational capabilities which, even though they may not be as advanced, abstract, or sophisticated as the representational capabilities of Homo Sapiens or modern humans, they are relatively speaking sophisticated enough to allow this species to engage in successful planning.

Let me point out that it is useful to distinguish between levels of TOM and, thus, levels, of intentionality. In his classical work Dennett (1971) distinguishes between orders of intentionality. A first-order intentional system is one whose behavior is predictable by attributing (simple) beliefs and desires to it; this being also has a first-order TOM. A first-order intentional system has beliefs and desires (etc.) but no beliefs and desires about beliefs and desires. A second-order intentional system is predictable only if it can attribute beliefs about beliefs, or beliefs about desires, or desires about beliefs, and so forth. Thus, a second-order intentional system is more sophisticated; it has beliefs and desires about beliefs and desires (and other intentional states), both those of others and its own; this being has also a 2-order TOM.

Accordingly, a being, Dennett continues, that can be seen to act on the expectation that you will discover that it wants you to think that it does not want the contested food would be a fifth-order intentional system.

Let us return to the great apes. They have a theory of mind, metacognition, and self-awareness. In fact, they have all the ingredients for self-consciousness that humans do and, yet humans have a joint intentionality that allows them to do things far more advanced than chimpanzees could do. I hasten to note that great apes have social intelligence too and, thus, display some sort of joint intentionality (Seyfarth & Cheney, 2018a, 2018b; Tomasello, 2023; Worden, 1996). Thus, the great apes can understand the goals, intentions, perceptions, and knowledge of others, and they use this knowledge to their advantage in competitive interactions, and they can also engage in deceptive behaviors to achieve their goals (Hirata, 2009). They can make elaborate plans, construct iconic representations, and use and understand symbols culminating in the acquisition of language. The reason that the great apes, unlike Homo sapiens, did not advance these capacities is obviously that their strategies in coping with their environment were successful enough, and, as a result, they did not face any environmental pressures that could have led to adaptations be selected over the course of the evolution of the species. The problem we face is the extend to which H.E had advanced these capabilities so that they be able to entertain representations advanced enough to support Tomasellllo's joint intentionality. Since none of the cognitive components of self-consciousness can be observed directly, one must take recourse to behaviors and try to infer from them the minimum cognitive requirements that could make them possible. Thus, let us turn to examining the H.E behaviors in detail since we have already brought to the fore many aspects of these behaviors.

We know that between 2.1 and 1 Ma H.E travelled long distances, sometimes across oceans to colonize Europe, India, Indonesia, etc. Undertaking such trips and be consistently successful at doing so requires serious planning ahead. This, in turn, requires the ability to integrate past experiences with probable future events in an attempt not only to determine how to proceed but also how to react in unforeseen circumstances. This is crucial for determining how much water and food to take in the trip, how many members will travel, etc. The same level of planning is required by other activities H.E were engaged in, such as stone knapping and the quarrying of the appropriate stone (Hiscock, 2014; Planer, 2017). H.E sometimes had to travel considerable distances to quarry the stones most appropriate for making their tools.

Even, big-game ambush hunting in which H.E were engaged (Ben-Dor et al., 2011; Binford, 2007; Kübler et al., 2015; Pickering & Bunn, 2012; Pickering, 2013) requires careful planning as to where to ambush the animals, what to do with the carcasses, etc. It also requires, as the previous activities do too, teamwork, division of labor, and, in the case of hunting, food-sharing and food preservation. In addition, successful hunting, especially with the short range, small tools used by H.E requires a rather sophisticated knowledge of natural history concerning the behavioral patterns of the prey (Binford, 2007). All these require significant planning ahead and cooperation at many levels of complexity, and all these presuppose, among many other things (such as the capacities for attention and working memory) that we shall discuss later on, the capacity of self-consciousness. Thus, the first condition for being able to have representations was satisfied the latest by *Homo erectus*.

Planning of any sort requires the ability to attend to what the others have to say, whenever means they use to do that—we shall discuss the problem of the linguistic abilities, if any, of H.E later—and to the details of the plan. It also requires working memory (WM)[1] both to synthesize information to produce a plan, and to be able to reactivate the plan at the proper time and place. In fact, the development of awareness to the stage of self-consciousness requires attention and WM. To have self-consciousness means that agents become aware that what they see in front of them is also seen by others, and they can also understand that what they think of themselves may not be the same as what others think of them. This, in turn, requires the capacity to shift attention from oneself to another agent and sustain attention enough to understand the other's intentions and desires. In other words, the neural substrates subserving attention shifts and control must have matured. Since, as the evidence suggests, H.E were able to plan in all these disparate facets of their lives, it follows that they had developed attention and WM sufficiently well; in other words, their systematic behaviors suggest that they had developed these cognitive capacities. This evidence can be disputed, however, and, thus, to be able to examine whether H.E actually had these capabilities to such an extent that would allow them to plan successfully, one must examine the development (ontogenesis) and evolution (phylogenesis) of attention and WM. To be sure, once we have concluded the examination, we shall find ourselves relying on behavioral evidence to draw conclusions, but this, time we shall have considerably augmented our arsenal by adding psychological and neuropsychological considerations.

Let us start with the development of attention. Executive functions describe the range of cognitive processes underpinning complex goal-directed behaviour and cognitive functions. A part of executive functions is attentional set shifting, that is, the ability to switch quickly attention between different rules or instructions, allowing individuals to adjust to changing environments, an ability that is found in other species too (Brown & Tait, 2015; Garon et al., 2008; Rosati, 2017). An attentional set results from a bias to direct attention preferentially to one stimulus or group of stimuli over others. This results in heightened processing of the relevant stimuli, while inhibiting processing of other stimuli. In humans, attentional set shifting capacities grow substantially during childhood and well into adolescence and early adulthood (Benitez et al., 2017; Buttelmann & Karbach, 2017; Cooper et al., 1978).

According to Cooper et al. (1978), the development of attention in general in children follows the following pattern.

Between 0–1 years is the stage of *fleeting attention*: Attention shifts from one thing to another, is involuntary, is caught by basic but salient stimuli, and is exclusively under the control of a reflexive system (Schiller, 1985). Thus, the infant is easily distracted by a new event occurring e.g. a loud noise. Since development is a dynamic phase, one expects to find small developmental shifts within each stage. Such shifts are found at the first stage of attentional development. Though looking and visual fixation stay reflexive for the first 2 months, at the end of the newborn period the posterior orienting system that is involved in the voluntary control of eye movements reaches functional onset. This system shows considerable development from 3 to 6 months of age. During this period, infants begin to show the ability to disengage and shift attention voluntarily (Courage et al., 2006). Areas of the brain involved in the posterior orienting system include posterior parietal areas, pulvinar, and frontal eye-fields (Posner & Peterson, 1990). The posterior parietal areas are believed to be involved in disengaging fixation, and the frontal eye-fields are key for initiating voluntary saccades.

At around 6 months of age, the anterior attention system reaches functional onset as well, and infants begin to develop inhibitory control and executive attention. Not only do infants have better voluntary control over their visual fixations, but they can also now inhibit attention to distractors and maintain attention for more prolonged periods when it is called for (Courage et al., 2006). This suggests the emergence of some rudimentary level of attentional control at around 6 months of age. The cortical areas that underpin these attentional capacities are predominantly the PFC including the inferior and superior PFC and the anterior cingulate (Reynolds et al., 2010).

Between 1–2 years is the stage of *rigid attention*: Children cannot attend to a concrete task of their own choice. Attention is still involuntary. External stimuli are ignored in order to allow focusing on the task at hand. The maturation of PFC that started earlier is progressing and the children can now concentrate on a task, even if it is captured by exogenous factors. Children may concentrate on blocks and will ignore adult intervention both verbal and visual.

Between 2–3 years the stage is called *single channelled attention*: Children of this age cannot attend to other stimuli while focussing on an activity. However, they can shift full attention with adult's help. This means that the control of the child's attention lies with the adult.

Between 3–4 years is the stage of *focussing attention*: The children start to control their own focus, but can still only focus on one activity at a time. They will give full attention to specific activity including both visual and auditory, and can focus and refocus their attention by themselves. Children, for example, playing a game will stop when adults speak, give their full attention to the adult, and then will continue with their game.

4–5 years-old are in the stage of *two channelled attention*: Children can focus on an activity but can also listen to what is being said without stopping their activity and looking at the speaker. However, they can focus on a single aspect of a complex situation. These mean that children can shut out unwanted and unrelated information and concentrate only on the necessary, for the task they are currently performing, features.

Between 5–6 years, finally, is the stage of *integrated attention*: The two channelled attention of the previous stage is well entrenched and the child can implement it across different situations and people. Attention is well established and sustained.

More recent research (Reindl et al., 2023) corroborates that the crucial step between 4 and 6 years is the capacity to shift and refocus attention, and, in addition, compares children's performance in attention shifting tasks with that of chimpanzees. The researchers found that there is a significant difference between 5- and 6-years old children, on the one hand, and between 3- and 4-years old children and chimpanzees, on the other hand. The older children perform better at tasks that require attention shifting from a set of instructions to another set. Reindl et al. (2023) presented 3–6-years old children and chimpanzees with two shelves holding the same set of four boxes. To choose the correct box on each shelf, the participants had to switch attention depending on which shelf they were currently presented with. This is a non-verbal attentional set shifting task. The chimpanzees' performance was comparable to the performance of 3- and 4-years old, but worse than the performance of 5–6 years old.

Pope et al. (2020) found that chimpanzees and baboons were less susceptible to forming a cognitive set compared to humans in the initial learning phase of a sequence learning touchscreen task. Pope et al., think that underlying this development is probably an interplay of "biological" and "cultural" factors that boost children's attentional set shifting abilities. These factors likely involve the biological maturation of fronto-parietal brain regions, which allow for the development of other core cognitive abilities, such as inhibitory control and working memory, and the acquisition and use of cultural tools such as language, rules, values, knowledge, and beliefs that facilitate sustaining and switching attention to stimuli.

Herrmann and Tomasello (2015) also found that attention shifting in 5-years old was significantly better than in chimpanzees in a task in which participants had to monitor and switch attention between two apparatuses, whereas 4-year olds and chimpanzees performed on a comparable level.

These results suggest that chimpanzees and young children share attentional set shifting capacities, and that there are unique changes in the human lineage from 5 years of age. Moreover, whatever mistakes were made by the older children, a larger proportion of them were switching mistakes rather than 'random'errors, a finding showing that the key factor that determined the performance in the task was the capability to switch attention and that this capability undergoes an important development around the ages of 5. Older preschoolers have a greater capacity than younger children and chimpanzees to switch, focus and maintain their attention on the relevant stimulus dimension. In addition, this capacity distinguishes older children from chimpanzees as well, who, apparently, never acquire fully this capacity. Thus, the strengths of the attentional set formed by the chimpanzees in this study seem to be similar to the one of younger children, but weaker than the attentional set formed by the 5-year-olds.

At the same time, these studies show that chimpanzees already have a rudimentary capacity for attention shifting. They can control their own focus, but can still only focus on one activity at a time, which means that they cannot attend to other stimuli while focussing on an activity. If chimpanzees have this capacity, early hominids also had it and, thus, one can explore the development of attention starting from this level. Pope's et al., (2020) suggestion that this development is tied to the development of fronto-parietal brain regions is a starter because one can search for developmental differences among different hominids species in general and in Homo-Sapiens in particular. The guide again is the belief that ontogenesis, that is, the development of attention is infancy could shed some light on phylogenetic studies of the development of attention and working memory systems in our lineage.

As we shall see, social factors played a crucial role in the development of symbolic cognition and language in hominins. Studies on the development of attention in children (Fraser & Scerif, 2024) show that from very early in infancy, attentional selection skills gradually improve and that there is bidirectional interplay between attentional control, learning, and memory, rather than a unidirectional bottleneck or filter on incoming information. However, not all stimuli influence equaly attention, as highly salient social attentional biases (for eye gaze, faces, and people) affect learning, faster recall from memory, and further influence memory-guided attention orienting to a greater extent that other kinds of stimuli. This is because they modify the attentional state of learners across development. In other words, social stimuli play a more influential role in ontogenesis than other kinds of stimuli. From this, one can extrapolate that the same holds for phylogenesis and, thus, argue that social factors were decisive in the development of symbolic thought and language through their predominant effects on the development of attention.

Closely related to attention is WM and many researchers think that attentional control is a significant component of working memory (Amso & Scerif, 2015; Astle & Scerif, 2011; Bauer, 2009; Courage & Howe, 2004; Cowan & Morey, 2006; Klingberg et al., 2002; Reynolds & Romano, 2016). D'Esposito and Postle (2015, 115) argue that information is encoded in WM by allocating attention to internal representations stored in long-term memory (LTM) and perceptual circuits. Thus, it is plausible to suggest that the development of WM should parallel that of attention. The development, and evolution, of working memory is important because to be able to discuss the cognitive capacities required for working memory to be able to play such an important role in the emergence and use of representations, both iconic and symbolic, one need have a picture of what is involved in working memory. The reason is that it is likely that the differences in the size or holding capacity of WM underlie the qualitatively differences in cognitive abilities between humans, earlier hominis, and the great apes because many of these cognitive abilities (such as recursive reasoning, which is a precursor of our modern languages) depend on the holding capacity of WM (see Read et al. (2022) for a discussion and further references).

Recent theories posit that WM consists of two parts, a small, capacity limited part that is the focus of attention (FoA), and a larger part, an activated portion of VLTM that has no capacity limits but is susceptible to temporal decay and interference effects (Cowan, 1995; Oberauer, 2001, 2009). When attention is withdrawn from an item that was put in WM because it was selected by attention, this item remains in WM outside the FoA and is relegated into the active LTM and may be used when attention shifts to it again. Thus, LTM participates in WM since a part of it is activated to serve the depository demands of WM. Oberauer (2001) provides empirical support for the thesis that items stored initially in WM that are subsequently de-selected and removed from FoA remain in the active LTM and can be used again in ongoing information processing. Boettcher et al. (2013) and Schurgin et al. (2018) argue that items that are expected to be used in a task in hand are more accessible or active in LTM than other items in LTM, a phenomenon that Cowan (2008) coins "active long-term memory". Additional evidence for the active involvement of activated LTM in tasks engaging WM is adduced by the finding that WM and LTM tasks draw on the same neural representations (Lewis-Peacocke & Postle 2008). According to these theories, the view that the capacity of WM is limited by a number of slots that contain chunks of information is undermined, since these theories suggest that it is likely that Wolfe's (2014) view that accounts of WM should incorporate aspects of both slots and flexible resource models is correct.

Beaman (2010) elaborates further this view of WM, by proposing that WM consists of three sub-systems. The first is a holding sub-system of buffers for temporarily holding or storing rapidly decaying information as part of the manipulation of that information. The holding sub-system, whether it involves the buffer portion of the phonological loop, the visuospatial, or some other subsystem of working memory, holds activated information in LTM, some subset of which will be subjected to attentional control and processing by the executive function of WM. The second sub-system is the strategic part of WM, because it establishes the goals for guiding behavior. The last sub-system of WM is an attentional system that focuses attention to salient information and also filters out distractors. The last two sub-systems are in effect the central executive of WM posited by Baddeley (2012). Ontogenetic or phylogenetic development in any of these sub-systems entrains changes in the other sub-systems and, also, affects a wide range of cognitive capacities that depend on WM.

Let us, thus, examine the development of working memory in general and of visual working memory (VWM) in particular. Let us expand, first, on the nature of the representations stored in VWM. When a representation stored in implicit, i.e., inactive, VLTM, is activated (either because it receives bottom-up some sensory signals, or because of the demands issued to the viewer, or because of for some other cognitive factor) it becomes part of VWM. Recent research examines the role of various brain areas involved in VWM and the relations among these areas in memory tasks using various neuroimaging techniques and EEG recordings. This line of research sheds light on the nature of the representations employed in memory and, as we shall see, suggests that VWM, the conceptual involvement notwithstanding, probably contains relatively rich iconic representations. Fuster and Alexander (1971) first showed that the Prefrontal Cortex (PFC) neurons exhibited persistent activity during the stage of a delayed-response task in which a monkey had actively to maintain information that was no longer present but was relevant for successfully completing the task. Kubota and Niki (1971) reported a comparable finding during the delay period of a delayed alternation task. The ability of neurons in some high-level brain areas to generate persistent activity in the absence of external stimuli is probably of fundamental importance to the neural basis of WM. VWM-related neurons within the PFC have been reported by other studies too (Funahashi et al., 1989). Courtney et al. (1997) showed that human PFC exhibited persistent neural activity that appeared to be coding task-relevant information during WM tasks. Many characteristics of this activity support the view that PFC maintains representations that are critical for guiding behavior; PFC, thus, is critically connected to VWM.

fMRI studies with humans investigate the nature of representations held in PFC compared to those held in posterior cortical regions that sustain perceptual processing. Delay-period MVPA (multivariate pattern analysis) decoding of stimulus direction was successful only at lateral and medial regions of the occipital cortex that are associated with visual perception. PFC was seen to represent a more abstract level of task performance, whether, for example, a trial was a speed trial or a direction trial (Riggall & Postle, 2012). Another study (Lee et al., 2005) using different stimuli, but a similar procedure found analogous results. On perceptual trials, MVPA decoded stimulus identity in the ventral occipitotemporal cortex but not in the PFC, which suggests that PFC does not represent the visual characteristics of a visual scene. On category trials, MVPA decoded stimulus identity from the PFC, but not from the occipitotemporal cortex (Lee et al., 2005), which shows that PFC represents, in addition to strategies and instructions relevant to the task, abstract representations of object-categories, information which is not represented in perceptual processing that takes place in the visual areas of the brain in the occipitotemporal cortex.

These findings are consistent with prior studies demonstrating that the lateral PFC preferentially encodes and maintains arbitrary and abstract representations of object category over representations of visual similarity (Freedman et al., 2003; Meyers et al., 2008). Further support for the distinction between stimulus-selective lateral PFC representations and sensory representations comes from another fMRI study that required subjects to remember over a short interval either faces or scenes or both categories of information (Reynolds & Romano, 2016; Sreenivasan et al. 2014b). The findings of these fMRI studies suggest that what PFC represents higher-order information, such as task rules, goals, or abstract representations of the categories, as compared to what is represented in the extrastriate cortex, which is more stimulus specific (e.g., the identity of specific faces). In this vein, Constantinidis et al. (2001) showed that PFC exhibits coarse selectivity for objects and features maintained in VWM. It seems, thus, that PFC plays less or no role in information storage; instead, it plays an active role in providing top-down control over all other brain areas where information is actually stored (D'Esposito et al. 2000; Postle 2015; Sreevinasan et al. 2014a).

To recapitulate the discussion about the relation between PFC and VWM, the persistent activity recorded in specific groups of neurons in several, mainly non-visual cortical areas during VWM tasks is thought to be the main neuronal correlate of VWM. In particular, a substantial fraction of neurons in the prefrontal cortex (PFC) show elevated activity selectively and persistently during the delay period of visuo-spatial WM tasks in which participants have to remember a visual cue for several seconds. This persistent activity has to be created internally, because during the delay period the stimulus is absent. The standard view of the role of PFC in VWM is that PFC activity reflects the existence of many intrinsic states of the PFC neuronal circuits, each of which is an attractor of the network dynamics (Wang et al., 2006, 2022). Each state is characterized by a specific group of active neurons that sustain their activity through their recurrent interactions. The transiently presented cue triggers the selection of a specific attractor, that in whose basin of attraction the cue falls. During the delay period, the network remains in this attractor after the cue has disappeared until the behavioral task is performed. Therefore, during the delay, the neuronal activity in PFC encodes a memory trace of the cue feature.

These considerations make clear the close relation between PFC and memory and, specifically, the crucial role PFC top-down activity plays in sustaining representations in visual areas, and also suggest that PFC provides top-down control of visual areas where the perceptual information relevant to the memory task is stored. IT is one of those areas, as is well known that complex stimuli are encoded there.

PFC is being implicated as a source of top-down signals that can influence processing in other cortical and subcortical brain regions (Braver et al., 2008; Duncan, 2001; Fuster, 2008). One type of PFC top-down signal likely provides direct feedback to posterior cortical regions that process incoming visual input. For example, when a person is looking into a number of people in a room, the visual scene presented to the retina includes a vast array of visual information. Suppose someone is searching the crowd for a person wearing a red hat; in this case some top-down mechanism must exist that allows one to suppress irrelevant visual information, while enhancing task-relevant information, allowing for an effective search. In this way, the maintenance and representation of the goal (find the person who wears a hat) by PFC serves as bias signals. Indeed, given that PFC represents rules and goals at multiple levels of abstraction, it is in an ideal position to influence in a top-down manner the processes in those sensory brain regions that receive its anatomical projections.

Since fMRI cannot directly determine the source of top-down signals, to obtain evidence that PFC is indeed the source of top-down signals that modulate the visual association cortex, one must examine the physiological responses of the visual association cortex after disruption of PFC function because if PFC controls top-down the visual association cortex, its disruption should affect the processing in the visual cortex. Fuster (2008) examines the effect of PFC inactivation in monkeys by cooling on the spiking activity in IT cortex neurons during a delayed-match-to-sample color task and found that the activity in IT is strongly affected by top-down signals from PFC, signals that likely are necessary for maintaining robust sensory representations in the absence of bottom-up sensory activity.

Miller et al. (2011) investigated further the effect of disrupted PFC function on the selectivity of category representations (faces or scenes) in the temporal cortex. Since different object categories, such as faces and scenes, are represented by spatially distributed and overlapping assemblies in the extrastriate visual cortex (Haxby et al., 2001), the disruption of PFC function should lead to a decrease in category selectivity by inducing higher spatial correlations between scene- and face evoked activity in the extrastriate cortex. Indeed, following disruption of PFC function, stimulus-evoked activity in the extrastriate cortex exhibited less distinct category selectivity to faces and scenes (more spatial overlap). In a follow-up study (Lee & D'Esposito, 2012), showed that the decreased tuning of the extrastriate cortex response coincided with decrements in WM performance. This work, which extended the findings of Fuster et al. (1985) from monkeys to humans, suggests that PFC may sharpen the representations of different object categories in the extrastriate cortex by increasing the distinctiveness of their distributed neural representations. This evidence supports the view that PFC is the source of top-down signals that act via gain and selectivity mechanisms and enhance processing in visual areas in the brain. Since the neural correlate of VWM consists in the patterns of persistent activity recorded in specific groups of neurons in several cortical areas during the delay period of a VWM task, it follows that VWM includes both higher-level cortical areas such as PFC, and visual areas at posterior regions of the brain.

Finally, Gazzaley et al. (2004) using fMRI and ERPs found significant functional connectivity between FFA (a face selective region within the fusiform gyrus) and PPA (a scene selective region in the parahippocampal place area), on the one hand, and the parietal cortex during the delay period of the task, which supports the idea that higher-order association cortices interact with posterior sensory regions to facilitate the active maintenance of perceptual representations held in memory in visual posterior areas.

The foregoing discussion shows the close collaboration between cognitive centers and visual areas in the brain in WM, since the higher-level cognitive centers guide attention through which they sustain the perceptual representations in visual areas during memory tasks. The perceptual information used in memory tasks is most likely represented in visual areas and, in this sense, memory recruits the representations in these areas to achieve its goals, a process in which PFC that stores highly abstract information, and the goals plays a prominent role. This is the basic premise of sensorimotor recruitment models of WM, a class of models that hold that the systems and representations engaged to perceive information can also contribute to the short-term retention of that information in WM (D'Esposito & Postle, 2015). It seems, thus, that WM, and VWM in particular, provide an essential link between perception and higher cognitive functions, allowing for the active maintenance of information about stimuli no longer in view. Sustained activity in higher-order prefrontal, parietal, inferotemporal, and lateral occipital areas supports visual maintenance, and may account for the limited capacity of working memory, linking these areas to VWM.

The higher order areas lack the visual selectivity of early sensory areas, and, thus, the activity of these areas during VWM tasks cannot account for the fact that viewers can remember specific visual features with minimal decay in performance over delays of many seconds. It is likely that this information is stored in the visual areas of the brain and is recruited via top-down activation to participate in memory tasks. The posterior sensory regions form representations that are informationally rich enough to support successful performance, and these representations are maintained during the delay period in the absence of a sensory stimulus by means of a top-down control from a wide network of higher-order brain areas.

Both behavioral evidence (Magnussen, 2000; Magnussen & Greenlee, 1999; Magnussen et al., 2003; Zaksas et al., 2001), and evidence concerning the neural signatures of activities in visual areas in various mnemonic tasks (Dijkstra et al. 2019; Emrich et al., 2013; Harrison & Tong, 2009; Lee et al., 2013; Nelissen et al., 2013; Riggall & Postle, 2012; Serences et al., 2009: Sreenivasan et al., 2014) support the view that neuronal assemblies in early visual areas, where specific visual details are encoded, provide the necessary detailed information observed in VWM and VLTM tasks despite the long delays. In seems, thus, that the activity in the early visual areas persists over time even after the stimulus is no longer present.

These studies reveal that VWM recruits specific perceptual information that can be stored only in early visual areas. Thus, perceptual representations in early visual areas remain available for use several seconds after the disappearance of the sensory stimulus. What remains unclear is the way these perceptual representations are recruited by VWM, because fMRI studies cannot specify what mechanism underlies persistent neural activity. There are two possibilities. The first is that these representations remain active, which means that the neuronal assemblies implementing them show a persistent activity during the delay period, an activity revealed by MVPA analyses. Or that these representations are stored in the connection weights of these assemblies and remain there inactive until the network is reactivated.

The evidence presented thus far suggests the crucial role of PFC in WM and this entails that the development of PFC and WM parallel each other. Let us examine now the development of WM in infancy before we turn to examine the development of PFC. Most of the research on working memory in infancy has focused on tasks similar to the Piagetian A-not-B task that involve some delayed response (DR) with the correct response requiring some level of attentional control. The A-not-B and other DR tasks typically involve the presentation of two or more wells. While the participants watch, an attractive object is placed in one of the wells and the participantss view of the object is then occluded. Following a brief delay, the participants are allowed to retrieve the object from one of the wells. In the A-not-B task, after multiple successful retrieval trials, the location of the hidden object is reversed, and the reversal is observed by the participants. The classic A-not-B error occurs when the participant continues to reach for the object in the original hiding location even after observing the reversal of the hiding location.

Studies utilizing looking versions of the DR task, show that significant development occurs from 5 to 12 months of age. With increasing age, infants show higher rates of correct responses, and infants can tolerate longer delays and still demonstrate successful responses (Cuevas & Bell, 2010; Pelphrey et al., 2004). Cuevas and Bell (2011) measured EEG responses in looking versions of the A-not-B task in a systematic. They found power changes from baseline to task in the 6–9 Hz EEG frequency band that correlate with successful performance for 8-month-old infants. Furthermore, higher levels of frontal-parietal and frontal-occipital EEG coherence as well as decreased heart rate from baseline to task are all associated with better performance on the looking version of the A-not-B task (Bell, 2012).

Several studies concerning the development of WM (Read et al., 2022) converge to the view that WM becomes active at about 7 months after birth (Diamond, 1990) and then increases in size to 7 ± 2 that characterizes the size of adult humans in puberty or shortly thereafter. There is linear growth pattern of WM from 7 months when WM becomes active up to 15-year-olds, although at about 9 years of age the growth pattern of WM slightly decreases. There are variations in the development of the three components of WM posited by Baddeley (2012), namely the central executive, the visuospatial sketchpad, the phonological loop, and the episodic buffer. Gathercole et al. (2004) data show that the three measures for each WM component have strikingly similar linear patterns, except that the phonological loop shifts to a lower growth rate after 6 years of age, and the rate of growth in WM holding capacity slows down and stabilizes at 11 years of age for the central executive and visuospatial sketchpad.

These findings also provide support for the role of a PFC network in WM tasks in infancy which is consistent with findings from neuroimaging studies with older children and adults showing recruitment of dorsolateral prefrontal cortex (DLPFC), ventrolateral prefrontal cortex (VLPFC), intraparietal cortex, and posterior parietal cortex (Crone et al., 2006; Klingberg et al., 2002). Crone et al. (2006) utilized fMRI during an object working memory task with children and adults and found that VLPFC was involved in maintenance processes for children and adults, and DLPFC was involved in manipulation of items in working memory for adults and children older than 12. The youngest group of children (8–12 years of age) did not recruit DLPFC during item manipulation, and did not perform as well as the other group on the task. If, as it was mentioned above, working memory and focused attention are interrelated, in view of the PFC's crucial role in attention, it comes as no surprise that PFC is also a crucial component underpinning the function of working memory.

Káldy and Leslie (2005) argue that the significant improvements on WM tasks between 6 and 9 months of age are related to further development of medial temporal lobe structures (i.e., enthorhinal cortex, parahippocampal cortex) which allows older infants to continue to hold objects in working memory in the presence of distractors. Káldy and Leslie (2005) proposed an alternative model of working memory development which emphasizes the importance of medial temporal lobe structures more so than PFC. However, Buss et al. (2014) utilized functional near infrared spectroscopy fNIRS to image cortical activity associated with visual working memory capacity in 3- and 4-year-old children. They applied receptors over frontal and parietal locations. Frontal and parietal channels in the left hemisphere showed increased activation when working memory load was increased from 1 to 3 items, a result supporting the view that young children do indeed utilize a frontal-parietal working memory circuit similar to adults. Both of these findings from fNIRS studies provide preliminary support for the role of PFC in working memory during early development.

The exact contribution of PFC to working memory functions in early development remains unclear. Even though PFC plays a critical role in integrating sensorimotor traces in working memory to guide future behavior, the A-not-B task may overestimate the functional maturity of the PFC in infant participants because it does not require the accurate integration of sensorimotor traces in working memory. Instead, the integration of sensorimotor traces should be considered a core process in working memory. What is clear is that infants beyond 5–6 months of age are capable of demonstrating basic yet immature aspects of working memory, and significant improvement in these basic functions occurs from 5–6 months. What is also clear is the role of the PFC in the executive function of WM is undeniable.

The improvements in WM performance which occur after 5–6 months of age are likely influenced by further development of the attention systems, especially Beaman's (2010) working memory attentional system. The majority of the WM studies examine visuospatial working memory and performance on these working memory tasks involves voluntary eye movements and controlled scanning of the stimuli involved. Thus, functional maturity of the posterior orienting system would be instrumental for successful performance. This system shows significant development from 3 to 6 months of age (Courage et al., 2006; Reynolds et al., 2013).

This timing coincides with the time frame at which infants begin to demonstrate above chance performance on working memory tasks. Successful performance on working memory tasks, however, also involve attentional control and inhibition. These cognitive functions are both associated with the anterior attention system (Posner & Peterson, 1990), which shows significant and protracted development from 6 months on.

In view of the evidence, one can assume that the development of WM performance with age parallels the maturation of the prefrontal cortex (Spear, 2000), and the frontal lobes (Conklin et al., 2007). Casey et al. (2005, 104) argue that "the sequence in which the cortex matures parallels cognitive milestones in human development." Furthermore, the maturation of the connectivity between the prefrontal region and the parietal region also correlates positively with increases in WM performance (Olesen et al., 2003). Conklin et al. (2007) show that that the temporal order in which performance was mastered for WM tasks parallels the temporal order for the developmental maturation of the frontal substrates associated with the performance of these WM tasks.

Thus, the PFC and its recurrent connections to the occipital cortex and to other areas in the frontal lobe are crucial for both attention and VWM. I also said that studies in developmental psychology show that working memory and attention develops gradually in children and suggested that this might reflect the course of evolution of working memory and attention in a phylogenetic scale. In addition, it is plausible to assume that the developmental trends in attention and working memory are underwritten by the development of PFC. The development of attention and working memory at the same time underlies the development of higher-order cognitive capacities such as problem solving, planning, knowledge-transfer, and others. Let us now examine the evidence concerning the ontogenetic development of PFC. Then, we could explore the possibilities that ontogeny offers for exploring the phylogenetic development of the PFC, since it is obvious that the evolution of the PFC is interlinked with the development of representational capacities in phylogeny.

A few words about the PFC first. Brodmann (1909) has identified a large frontal region in multiple primate species that includes areas occupying the anterior-most lateral, dorsal, medial, and orbital surfaces of the hemisphere. This region became known as the "PFC," "granular frontal cortex, or" "frontal association cortex." As we shall see, the PFC is heavily interconnected with hierarchical lower and higher brain regions and in position to synthesize inputs from various brain areas, including the somatosensory and motor cortex. For example, the parietal visual areas of macaques and humans have extensive interconnections with dorsolateral and dorsomedial PFC (DLPFC and DMPFC) (Remple et al., 2007).

PFC consists of distinct areas that seem to have different and overlapping functionality. The PFC in humans and nonhuman primates is divided into a collection of structurally and functionally different subregions that are positioned anterior to the motor cortex. These are, the medial (mPFC), lateral prefrontal cortex (lPFC) and orbitofrontal cortex (oFC). The development of the PFC is important for the acquisition of many cognitive abilities including socio-emotional and executive functioning in humans and other primates and apes. The PFC is involved in higher-level cognitive processes that are usually referred to as "executive functions" and are supported by the dorsolateral areas BA9/10/46 (in Brodmann's classification). These executive functions crucially involve attention and working memory, the involvement of PFC in which was attested in the studies discussed in previous paragraphs. The PFC also contains Brodmann's areas 44/45 that underlie language, the orbitofrontal cortex areas 47, 11, and 13, which, together with area 10, support social-emotional processing. In general, the lPFC is mostly involved in language and executive processing, while the oFC and mPFC are known to contribute to cognitive functioning and emotional control (Fuster, 2000a, 2000b; Goldman-Rakic, 1987). The mPFC can be further subdivided into the infralimbic (IL), the prelimbic (PL) and anterior cingulate cortex (ACC). The most ventral subdomain of the mPFC is the infralimbic cortex (IL) and is involved in coping with chronic stress eventually leading to structural changes and prefrontal dysfunction (Dias-Ferreira et al., (2009; Moench et al., 2016; Shansky et al., 2009). All these cognitive, social, and emotional capacities are important, both for humans and primates, in successfully negotiating a complex and constantly changing natural and social world. Successfully interaction with the world requires detecting the patterns and their variation in the world and this activity is largely subserved by the PFC (Teffer & Semendeferi, 2012; 192).

Turning now to the development of the PFC, the PFC is the last part of the brain to mature (Fuster, 2000a, 2000b, 2002). Its development starts in utero during the third week and continues after birth until adolescence. The frontal lobe and much of the PFC matures late relative to the remainder of the cortex, although regions of the temporal cortex that are also higher-order association areas that integrate inputs from sensorimotor and other lower-order regions develop late too. The last parts of the PFC to mature are those on the lateral part that subserve executive functions and are connected to the occipital temporal, and parietal cortices. By receiving input from these disparate areas, the later PFC is able to synthesize somatosensory signals and use them to guide executive functions, and language (Miller & Cohen, 2001). The orbital and medial regions of the PFC are thought to be involved in the processing of emotional signals and in the regulation of emotional behavior.

It is believed that in humans the PFC has evolved disproportionally large, and it is thought to be the last region of the brain to gain full maturity (Donahue et al., 2018; Fuster, 2000a, 2000b, 2002, 2008; Petanjek et al., 2011). During early childhood, the brain quadruples in size and grows to roughly 90% of the adult volume by age 6 (Knickmeyer et al., 2008). During this period of brain development, dorsolateral and medial PFC expands nearly twice as much as some other regions, including medial occipital and insular cortex (Hill et al., 2010). Throughout childhood and adolescence, brain development is characterized by both growth and then decline in gray matter volume (the latter is called pruning), and increases in white matter volume, that is, in the connectivity of the various brain regions, as well as in ganglia cells. The frontal lobe myelinates last, and in its most rostral regions myelination can continue well into the third decade of life (Sowell et al., 1999). From ages 7–16, the frontal lobe experiences an increase in white matter volume that goes beyond what is expected from overall brain growth during these ages (Sowell et al., 2002). The anterior-medial aspect of the frontal lobe, which is involved in attention and self-referential tasks (Zysset et al., 2003), is one of the last regions, along with the temporal lobe, to myelinate postnatally (Barkovich, 2005).

Developmental increase in the prefrontal cortex is being measured both by changes in grey and white matter volumes and myelination of axon fibers (Casey et al., 2005; Zhou et al., 2016), and by increase in the OFC activity using fMRI (Casey et al., 2005). Regarding the association between WM maturation and maturation of the prefrontal cortex, there is a positive association between increase in WM performance and maturation of grey and white matter (more about that in a while). Since WM development parallel mastering WM tasks, it seems that "cortical activity and structure are closely related to current WM, while the subcortical white matter tracts and activity in the caudate are preceding these changes and predict future WM capacity" (Darki & Klingberg, 2015: 1594).

Before we turn to examine the phylogenetic development of WM, let us say a few things about the ontogenetic development of WM (and the PFC) in monkeys. The reason is that, as we saw above, the cognitive capacities that chimpanzees and the other apes may have form the baseline from which the phylogenetic evolution of these capacities in the hominins started.

There are significant differences in the pattern and pace of cognitive development between humans and chimpanzees (Wobber et al., 2014), even though human and chimpanzee toddlers are comparable with respect to their cognitive capacities. This discrepancy is probably due to a more limited WM capacity for chimpanzees compared to humans. Though WM of humans and non-human primates show parallels, there are differences, such as the ability of humans to maintain actively information in WM through rehearsal using the "phonological loop" (Baddeley, 2003). Brady and Hampton (2018) argue that it has been assumed that while rhesus monkeys can keep familiar stimuli in WM through rehearsal, the same is not the case for novel stimuli since they lack the language facility to enable the verbal recoding and rehearsal through a phonological loop the way humans do to keep novel stimuli active in WM. However, monkeys can keep novel stimuli active in WM through encoding that does not depend on language, which means that it is possible that mechanisms supporting the maintenance of novel items in WM evolved before language and at least 32 Ma when a common ancestor of humans and rhesus monkeys lived.

The pattern of increase in WM performance with age in humans also occurs in monkeys, with WM performance already in place before 1 year of age (Diamond, 1990) and increasing with age to 36 months and then through puberty until early adulthood (Zhou et al., 2016). Alexander and Goldman (1978) show that, in the rhesus monkeys, there is a parallel between increase in WM performance with age and increase in prefrontal cortex development and activity from 12 to 36 months, which suggests that PFC development parallels WM development as in humans. Zhou et al. (2016) measured, at different ages, their performance on a 12 s delayed response task, both under normal conditions and when the activity of the dorsolateral prefrontal cortex (DLPFC) had been depressed by induced hypothermia. They found no effect in the 9 –16-month-old monkeys, a reduction of 7–8% in the 19–31-month-old monkeys, and a reduction of 21–25% in the 36-month-old monkeys. This development parallels the pattern found in humans from adolescence to adulthood.

When compared to human infants and using the same task for both human infants and monkey infants, macaque performance at 4 months matches human toddler performance at 12 months (Diamond, 1990), indicative of the development extension that is characteristic of humans. The protracted maturation of the PFC in humans relates to neuronal plasticity that allows for increased effect of post-natal experiences on brain functioning (Johson, 2001). The same extension in time to maturation occurs with the chimpanzees, which suggests that it would also be present in an ancestor common to chimpanzees and humans (Sakai et al., 2012), though there are differences in prefrontal cortex maturation rates between humans and chimpanzees.

Prefrontal cortex develops at a faster rate in humans than in chimpanzees (Somel et al., 2011). Sakai et al. (2012) show by means of a longitudinal study of prefrontal development in chimpanzees from 6 months to 6 years using magnetic resonance image scans that the prefrontal white matter volume is still immature and has not reached adult levels during prepuberty. In humans, in contradistinction, prefrontal development extends from 1 month to 10.5 years and is part of the development extension that is characteristic of our species.

Read et al. (2022) show, at most, that chimpanzees (and possibly the other great apes) can, to a certain extent solve new problems by inventing new solutions, plan for future needs, and have at least a limited theory of mind. They have cognitive skills that rely on the power of WM to store and manipulate mental objects and compute plans, but it is not evident that WM $> 2 \pm 1$ is necessary. Instead, these studies show the limited capacity of apes to draw up and execute plans and to switch attention from the perspective of oneself to that of others. It follows that there seem to be fundamental differences between chimpanzees' and ours WM. Read et al. (2022) suggest that the WM capacity of chimpanzees is limited compared to that of modern humans in that the size of WM in chimpanzees is 2 ± 1 versus Miller's 7 ± 2 in humans (Read et al., 2022). The chimpanzee WM is, thus, comparable to that of 3.5-year-old children. The problem is to see how the WM of H.E may have evolved from this baseline.

Let us turn to examine the evolution of PFC, that is, the phylogenetic development of PFC. The first piece of hard scientific evidence comes from comparative studies of humans, primates, and great apes that establish the cognitive baseline for our species and suggest ways the evolution could have proceeded in view of the fact that ontogenesis may be helpful in examining phylogenesis. The second piece of evidence that concerns directly extinct hominis comes from anatomical and cranial studies and from what conclusions we can draw from the comparative studies of humans and apes concerning the development of the PFC.

At the start, hominin WM and chimpanzee WM were essentially the same at the level of 2 ± 1 for the adult members of the two species. Modern adult Homo sapiens had a holding WM capacity of 7 ± 2. This is so because the last common ancestor for hominins and the great apes dates to around 6–8 Ma, hence WM capacity would be, at most, the same as for chimpanzees today, that is 2 ± 1 for the last common ancestor. Since Homo sapiens' WM holding capacity is 7 ± 2, this means that H.E had a WM capacity between 2 ± 1 and 7 ± 2. H.E appeared about 1.8 Ma and, thus, evolution had at least 5 million years to work, as it were, on the development of

WM capacities of H.E. Given that WW develops in human infants until it reaches the 7 ± 2 threshold, a development that mostly depends on, and, in turn, influences, post-natal experiences and social learning, this is another reason to assume that H.E WM capacities have evolved from the 2 ± 1 base-line and, as I shall argue in a while, for the same reasons, even though the role of social learning in humans is more prominent because the post-natal maturation of H.E infants was much faster than in humans. Assuming a linear increase in WM capacity through time (recall that I assume a gradualist model of evolution), Read et al. (2022, 22) estimate a WM capacity of around 5 for H.E, which places them at the fifth level of Carruthrer's (2013) scale and endows them with the capacity for rehearsal and a limited ability to transform WM representations into problem solving. Read et al. (2022) also note that there is striking correlation between the stepwise increase in WM capacity and major events in the hominin phylogeny leading to Homo sapiens. The storing WM capacity of H.E also explains, with respect to the H.E tool construction technique, their ability to execute sequences of actions that were planned ahead in their minds and were guided by the overall goal, which was the final form of the tool that they had to construct. It also explains their ability to create mental models that allowed them to plan ahead their ambush hunting because as the WM capacity increases, the agents can construct more elaborate plans by analyzing the possible situations that they may encounter both in width and depth and, thus, determine in advance possible courses of action should these situations arise. In other words, an increased WM holding capacity, and the attentional development that accompany it, gradually made possible both the formation of hierarchical systems of representations, and the executive control exercised over them so that they can be brought to instigate a sequence of actions towards achieving a prespecified goal held simultaneously in mind.

Consider how the artifacts made by H.E dating from around 1.8–0.5 Ma, hence before the origin of syntactically structured language that probably occured about 200 Ka (Knight, 2009), relate to WM. Coolidge et al., (2012, 38) claim that about 1.8 Ma, H.E strike a rock against another while holding a mental image of the shape they want to impart on the stone in their mind. Through a series of skillful choices of angles and strikes, they impose that shape on the stone to produce a hand axe. No other species impossible could conceive and execute the complex planning involved in this series of actions. This means that HE had the ability to construct and carry out increasingly elaborate plans of action without possessing a language. To the extent that WM underlies these capacities, it flows that H.E had a much more advanced WM that chimpanzees and the other great apes that diverged earlier from hominis, and H.E had these abilities despite the lack of speech and language.

Curving a hand-axe out of a piece of stone involves a planning scenario that is striking in its complexity in comparison to a simple scenario for chimpanzees making termite sticks (Read & Andersson, 2019). Although researchers (Mithen, 1996, being a prominent example) usually distinguish knowledge relating to the physical world and the forces that govern this world (physical cognition or natural history intelligence) from knowledge of other minds (social intelligence), there is an aspect of cognition feeding both that is at play in the scenario for making a hand-axe, or, else the two domains interact by feedback connections and inform each other, a sort of cognitive fluidity, which allows the making of the hand-axe. If, as it is likely, WM underlies both kinds of intelligence, when a developmental threshold is reached through expansion of WM, a qualitatively new way of processing information is enabled across all cognitive abilities, a way that allows the combination of information from different domains of intelligence (natural world and social intelligence in our case).

Since WM can be activated in a top-down way by mentally entertaining the mental states we attribute to other persons as well as one's own, and the agent can distinguish between these representations, it is possible to shift from one representation to the other to extract information relevant to making appropriate decisions in the social domain. The simultaneous activation of alternative representations occurs when knapping a hand-axe. The artisans have in mind not only a representation of the intended hand-axe and its desired features, but may also have in mind an alternative artifact that they could possibly make, while simultaneously attending to the piece of stone they are working with. At this moment, the artisans have in mind a mental representation of the intended hand-axe, of the piece of stone they are working with and its possible transformations, and of the cutting tool that is also being used by the artisans.

All these imagined representations and scenarios can be held in the artisans' minds as they proceed to construct the tool only when WM is powerful enough to entertain multiple mental objects for which information is updated as the artisan proceeds (because as the construction proceeds the properties of the objects, both real and mental, that they work with do change as a result of the artisans' actions), and each of which can be mentally accessed by the artisan at will. In this sense, social intelligence with its multiple representations and use of episodic and prospective memory in the preparation of tools, parallels the properties of WM and this suggests that prospective (prospective memory is a form of long-term memory and refers to the memory for performing or carrying out future-oriented intentions) and episodic memory probably appeared roughly at the same time and developed hand in hand.

Thus, there is an increase in WM capacity that can be related to archaeological evidence for qualitative changes marking different stages in the design and technological complexity of tool manufacture (Read & van der Leeuw, 2008), beginning with tools used by chimpanzees and extending through the stone tools that are critical to hominin evolution leading to Homo sapiens. With the hominins, stone tool manufacturing evolves from the earliest technology for removal of flakes used as tools 3.3 Ma through the Oldowan and the Acheulean technologies (the latter being used by H.E 1,76 Ma) to the complex blade technologies eventually appearing in most parts of the world (50,000 Ka). There is an interrelated pattern incorporating cranial, technological, and behavioral changes associated with increase in effective WM capacity that is a central part of hominin evolution leading to Homo sapiens.

Read et al. (2022) argue that the striking difference for making a hand-axe and its dependence on an already elaborated WM in H.E compared to that of a chimpanzee-like ancestor involves elaboration of WM that does not derive from speech rehearsal, because H.E did not have a language. Read and van der Leeuw's (2008, 2015) data about hominin activity before the origin of syntactic forms of language also show that language is not the basis for more elaborated WM in hominin ancestors to Homo sapiens. Read's et al. (2022) target is Carruther's (2013) claim that inner speech, or mental rehearsal of speech, as he calls it, is necessary to explain the cognitive differences between humans and non-human primates, despite the fact that, as Carruthers thinks, have the same WM capacities. Read et al. (2022, 22) argue that if Carruthers is right, prior to the development of language, hominin WM and chimpanzee WM were essentially the same. This claim, Read et al. (2022) rightly argue is false, because, as we have seen, H.E's tool industry would be possible only in H.E has advanced WM capacities. Notice that Read et al. (2022) jump from the premise that WM development does not rely on having syntactic forms of language, because H.E had well-developed WM capacities but did not have syntactically structured languages, to the conclusion that language is not the basis for elaborate WM. This leaves out the possibility that H.E had a protolanguage that possible allowed for mental rehearsal or inner speech and this contributed to the development of WM. This last remark brings us back to the claim made by Coolridge et al. (2012) and others that H.E had all these capacities without possessing speech and language. I postpone discussing this problem until Sect. 7.4, where I address the problem of whether H.E had a protolanguage, even though I have little to add to Planer and Stereleny's (2021) seminal book on this topic.

To summarize, the storing WM capacity of H.E also explains, with respect to the H.E tool construction technique, their ability to execute sequences of actions that were planned ahead in their minds and were guided by the overall goal, which was the final form of the tool that they had to construct.

Let us move to the anatomical evidence. There are three areas of the PFC that are of particular interest. The first is the limbic frontal cortex BA 13, which is found in the posterior part of the orbitofrontal cortex and is part of the neural substrate underlying emotional reactions to social stimuli. It is found in the posterior orbitofrontal cortex and has been identified across humans, chimpanzees, bonobos, gorillas, orangutang, and gibbons. BA 13's structure, and features such as size of cortical layers, density of neurons, and space available for connections are similar across hominoids, with only subtle differences present. Area 13 occupies a smaller percentage of the human and bonobo brains than it does in other hominoid species; the absolute size of BA 13 is quite similar among humans and all of the great apes, except the bonobos (Semendeferi et al., 1998).

BA10, which lies at the most anterior aspect of the PFC, belongs to the association cortex and is involved in higher cognitive functions, such as planning future actions and decision-making (Fuster, 2008). BA 10 has been identified across humans and most of the apes (Semendeferi et al., 2001) and has similar cytoarchitectonic features among hominoids. BA10 is larger in humans than in apes both in absolute terms and relative to the rest of the brain. The expected volume for BA 10 in an ape brain of human size is little more than half of the observed volume, which means that BA10 has expanded disproportionally in humans. Semendeferi et al. (2001) show that the absolute size of the human area 10 exceeds the size of area 10 in any other anthropoid and does so dramatically. When the volume of the human polar PF cortex is compared with the brain as a whole, it is twice as large as in great apes. Moreover, compared with the PFC of chimpanzees, the human PFC is vastly expanded in relation to the indirect inputs from the primary visual cortex, which means that in humans the PFC has expanded much more that in the great apes (Donahue et al., 2018). This suggests that the neural substrates supporting cognitive functions associated with this part of the cortex enlarged and became specialized during hominid evolution. In the great apes, BA10 expanded from its restricted orbital location and came to occupy the entire frontal pole. As suggested before, this expansion was the result of responding to various socio/environmental pressures.

Finally, another part of the PFC, Broca's area, or BAs 44 and 45, comprises part of the inferior frontal gyrus in the human brain. These regions are involved in language production, sequencing, planning, syntax, and phonological processing (Damasio et al., 2004). However, Broca's area is also involved in the processing of hierarchically structures strings of information irrespective of their modal origin and in domains other than language, for example, music or arithmetic (Fadiga et al., 2009). Based on cytoarchitectonic criteria, both regions have been identified in the inferior frontal gyrus of chimpanzees, bonobos, gorillas, and orangutans (Schenker et al., 2008) and display similar cytoarchitectonic characteristics. There are no relative volumetric differences in Broca's area between humans and the apes. In humans, there is a distinct trend for both BA 44 and 45 to be larger in the left hemisphere than the right. This asymmetry is not present in chimpanzees (Schenker et al., 2010), suggesting that Broca's area in the left hemisphere expanded in relative size during human evolution, possibly as an adaptation for our species' language abilities.

Bruner (2004) and (2017) argue that, compared to Neanderthal and chimpanzee brains, human brains have an enlargement of the PFC and of the medial parietal cortex. The increase of the frontal and parietal lobe in Homo Sapiens was associated with an increase in encephalization that saw a rapid increase from 400 Ka. The *relative* rates of the growth of the brain and body are not matched between modern humans and chimpanzees (Rice, 2002). This means that even if a chimpanzee or macaque monkey had as long a period of ontogeny as modern humans, they would not have the same ratio of brain to body weight as humans. They represent different grades, which suggests that there was a change in the relative rates of the development of the brain and body in the ancestral hominins.

The same can be said about the proportion of the neocortex that is taken up by association cortex, including the PFC. The development of the primary sensory and motor areas of the neocortex parallels that of the growth of the body. Chugani et al. (1987) using functional activation showed that the striate and sensorimotor cortex matured early. Other areas such as the PFC and premotor cortex reached maturity much later. This means that during the prolonged period of human postnatal development, the areas that are neither primary, sensory, nor motor take up an increasing proportion of the brain. The areas that are most expanded are the association areas including the PF cortex. Significantly, the areas that show the greater expansion during human ontogeny are those that can be shown to be expanded when the chimpanzee brains are compared to the adult human brain. Thus, it is highly likely that the changes during ontogeny are similar to the changes that occurred during evolution (Passingham, 2021). Hill et al. (2010) concur with this assessment arguing that the cortical expansion displays the same patterns during human development and evolution.

In another study, Passingham and Smaers (2014) compared the size of the PFC with the inferior temporal cortex. Their results indicate that the human PFC cortex is expanded relative to the temporal association cortex and relative to the middle and inferior temporal cortex, and, thus, there is more tissue in the human PFC for re-representing the input from the ventral visual stream. The PFC also lies at the top of the motor hierarchy. The consequence of the expansion of the PFC compared with the premotor and motor cortex is that there is relatively more tissue for generating goals, as opposed to directing the movements that are needed to achieve them (Passingham, 2021).

Preuss and Wise (2022) claim that one would never expect a great ape to have a brain as large as ours. In body size, humans overlap with the chimpanzees and other great apes but our brains, at about 1450 cc on average, are 3–4 times larger than theirs. One would expect humans to have a brain size similar to that of a chimpanzee (~ 400 cc). Although most human primary sensorimotor areas are only marginally larger than those of chimpanzees, certain parts of the association cortex—including the dorsolateral, frontopolar, and anterior orbital parts of the granular PFC—are enormously larger (see also Wei et al., 2019).

The evidence we have examined (see Passingham, 2021 for an overview and discussion of this evidence) suggests that the PFC is significantly expanded in the human brain compared with the great apes. Indeed, it has commonly been accepted that the granular PFC, and the association cortex in general, underwent enormous expansion in human evolution, both absolutely and relative to the amount of primary sensorimotor cortex (Donahue et al., 2018; Glasser, 2014; Smaers et al., 2017). This view has been challenged, however, as some studies dispute this finding. Based on scaling studies, Barton and Venditti (2013) have argued that even though PFC is absolutely much larger in humans than in other primates, humans have the expected amount of PFC for a primate of our brain size. So, those researchers argue that human PFC is not exceptionally large. Similarly, Semendeferi et al. (2002) measured the size of the frontal lobe as a proportion of the neocortex as a whole. The mean value for humans was 38% compared with 35% for chimpanzees. Given variation in the values between individuals within each species, this difference was not statistically significant. Furthermore, Semendeferi and Damasio (2000) found that the proportions of the temporal and parieto- occipital cortex were also the same in the human and chimpanzee brain. Passingham (2021, 346–351) does a nice job in debugging these and other objections to the thesis that the human PFC is particularly expanded compared to other apes and argues that the human PFC is indeed larger relatively and absolutely compared to that of the other apes.

Beyond differences in relative size, it has been argued by Sallet et al. (2013) and Neubert et al. (2014), based on studies of the structural-connectivity MRI parcellation of frontal areas, that the cortex on the lateral aspect of the frontal pole in humans has no clear counterpart in macaques, a result which, given the involvement of that region in higher cognitive function, such as generalized relational reasoning, is telling about the cognitive differences from our closest relatives.

Be that as it may, even those who question the claim that the human PFC and the PFC in apes differ in relative size, it is unanimously held that there are significant differences in the organization and connectivity of the PFC in humans in comparison to the PFC of apes. Recent research shows that the neurons in layer III of the PFC are significantly more widely spaced throughout the PFC in humans than they are in great apes, while spacing in sensorimotor and visual cortex is similar in humans and apes. One interpretation of the functional significance of absolutely wider mini columns is that they are associated with being more generalized processors (Gustafsson, 2004). Mini columns largely comprise pyramidal neurons in layer III, along with their myelinated axons and apical dendrites (DeFelipe, 2005). The human PFC exhibits more complex dendritic branching than the visual cortex (Elston et al., 2006). Pyramidal cells in the human PFC are more branched and spinous than those in the temporal and occipital lobes and are also more branched and spinous than those in the PFC of macaques and marmosets (Elston et al., 2001). In the human PFC, layer III projections possess more branched and spinous dendritic arbors than in temporal, occipital, or parietal cortex (Elston et al., 2001; Petanjek et al., 2008). The long-range cortico-cortical projections of layer IIIc neurons are thought to be critical to working memory and other higher-order cognitive processes in primates (Elston et al., 2006; Fuster, 2000b; Lewis et al., 2002), suggesting that the reported differences in dendritic tree structure are related to cognitive differences (Zeba et al., 2008). These findings suggest some degree of reorganization of the human PFC in general, and possibly the frontal lobe as a whole, that is not found in other apes. Kolk and Rakic (2021) discuss extensively the development of inter- and intra-connectivity of PFC and its subregions, as well as the role of neurotransmitter systems in the PFC, and conclude that 'in this way, a unique and higher-order functional network capable of emotional processing and complex cognitive abilities is established' (Kolk & Rakic, 2021, 44–45).

Overall, studies comparing the PFC in humans and other apes suggest that the human PFC differs from that of closely related primate species both in relative size but, also and most importantly, in the organization of its subparts and the inter- and intra-connectivity patterns. It is clear, thus, that during both ontogenetic development and phylogenetic evolution the PFC undergoes several reorganizational events in its neural circuitry. These events in phylogeny may have taken place either as a consequence of adjusting to increases in brain size or as adaptive responses to specific selection pressures. I return later to address this critical issue in relation to the cognitive abilities and the underlying neural mechanisms, particularly the PFC, of *Homo erectus*.

Even though the size of the PFC is important, so is the connectivity of the PFC with other regions lower or higher, and the interconnectivity of the various PFC regions. The former is measured by the size of the gray matter, while that latter by the size of the white matter. Schoenemann et al. (2005) measured the extent of the grey matter and white matter in the frontal lobe anterior to the corpus callosum. Compared with the grey matter, there was significantly more PFC white matter in the human brain than in the brains of great apes or monkeys. These data are consistent with the evidence from tracer studies suggesting that the different areas of the PFC are highly interconnected (Petrides & Pandya, 2002). These internal interconnections are important because they allow the PF cortex to generate the appropriate goal within a given the current context and the desired outcome. It is important to note that in view of the fact that the dendrites in the polar PFC are larger and have more spines in the human than chimpanzee brain (Bianchi et al., 2013), it is likely that the human PFC has an increased capacity for integrating a wide range of inputs from sensorimotor and other association areas in the brain.

Our excursion to the phylogenetic evolution of PFC and, by extension to that of WM, thus far concerns comparisons of modern human brain, which has remained unchanged the last 70 Ka or so, with the brain of our closest cousins, that is, the great apes. What matters for the theme of this book, however, is the brain of H.E and whether its neuronal circuits could subserve the cognitive capacities necessary for H.E having mental representations, either iconic or symbolic, and, furthermore, whether H.E could construct and understand external iconic and symbolic representations. Before we turn to H.E, let us concern ourselves with the hominin cognitive baseline as it relates to WM, that is, with the WM related capacities of the great apes. There is overwhelming evidence that not only chimpanzees but also other apes, such as baboons, have good WM (Claidiere et al., 2014). As we have seen, research suggests that the WM capacity of chimpanzees is limited compared to that of modern humans in that the size of WM in chimpanzees is 2 ± 1 versus Miller's 7 ± 2 in humans (Read et al., 2022). The chimpanzee WM is, thus, comparable to that of 3.5-year-old children. The problem is to see how the WM of H.E may have evolved from this baseline.

We can say something about the H.E brain based on the examination of their sculls, and we can also induce certain hypotheses from the phylogenetic evidence concerning the evolution of modern human brains from the apes bearing in mind that H.E intersects the trajectory from apes to humans since their appearance about 2.1 Ma. To the extent that ontogeny sheds light onto phylogeny, we may also be able to say a few things concerning the H.E brain based on the typical development of the brain in children from birth to maturity. However, most of the hypotheses concerning H.E brain comes necessarily from extrapolations from their cognitive capacities.

The problem, of course, is the status of our knowledge concerning these capacities; how could one tell what H.E were cognitively speaking capable of? The answer to that is that one could start from the behavioral evidence we have about their activities, because if, for instance, one is able to plan ahead and thus, extrapolate from past experiences and make estimations of what the future situation may be, they must have the requisite cognitive abilities. And since having these abilities requires certain neural circuits, then, it is likely that they had these neural circuits as well. Such a discussion is fraught with dangers, of course, because inferring cognitive capacities from behavioral trends is not easy, a fact that is attested by the many disagreements concerning, for example, whether H.E had some sort of proto-language. It follows that one cannot deduce cognitive capacities from behavior; however, one can form educated hypotheses trying to come with one that best fits the data and hope that this hypothesis could be the best game in town, for the time being, of course, as new evidence accumulates. This is important, because who could imagine 20 years ago that we would be talking about H.E and Neanderthals nursing the sick or injured members of their groups, or about H.E and Neanderthals constructing iconic representations irrespective of how rudimentary they were.

Let us examine, therefore, what we can say about the H.E brain from the evidence we have from all the aforementioned sources. Since the crucial cognitive capacities for engaging in planning and, thus, for having any sort of representations are, as we have seen, attention and WM and since the basic brain structure subserving these capacities is the PFC and the fronto-lateral parietal regions, let us concentrates on these brain areas.

Let us start with the H.E sculls. The average volume of the H.E skull is about 1000 cc, and the typical variation is 546–12651 cc. The average size of Homo Sapiens varies between 1250 and 1730 cc with a mean value of 1330 cc. Recently Clark and Henneberg (2022) argued that the brain size of Asian H.E over the last 600,000 years overlaps significantly with modern human populations. A difference between the two hominin species is that the brain of Homo-Sapiens has a taller frontal area and the parietal areas bulge; indeed, the human skull shows a unique globalization (Lombard, 2024).

Studies of the hyoid bone that modulates the vocal tract and controls pitch, and volume suggest that H.E lacked a human like vocal apparatus and, thus, could not have the sound capabilities of modern humans. Whether this means that H.E did not have some sort of language is another problem with which I shall deal later in Chap. 7. It should be noted that studies show that H.E infants did not have an extended childhood, as a 1-year old infant had an endocranial capacity at 72–84% of an average adult H.E. (the modern human brain represents only 25% of its adult size at birth), which means that they did not have much time to master some sort of language and its rules (Coquegniot et al., 2004). This, as a matter of course, has implications for the cognitive abilities of the species because most of them are developed or tuned during human growth in a relatively rich environment. Coquegniot et al. (2004) conclude that these data suggest that in H.E only a brief period of brain maturation took place in the environment outside the womb. Thus, it is unlikely that H.E had cognitive skills comparable to those of modern Homo Sapiens, and it also entails that complex spoken language emerged relatively late in the course of human evolution, since language learning requires a prolonged infancy of the sort that characterized only Homo-Sapiens. As we shall see, H.E had well developed Broca and Wernicke's brain areas that subserve modern language capabilities, but whether this entails that they had some sort of language will be discussed in the last chapter of the book. Showing that they did have some sort of language requires further argumentation.

Let us see what glimpses we can get into the H.E brain from the ontogeny and phylogeny of the human brain. H.E are situated between apes and Homo Sapiens; since we know from developmental studies that the PFC and the fronto-lateral areas are among the last to mature, and since researchers, as we have seen, think that, in this respect at least, we have good reasons to believe that this mirrors phylogeny as well, one could induce that the PFC of the H.E had developed considerably since the split of hominis from the apes, and continue to develop during the about 2 million years that H.E existed, on a genetic or epigenetic basis. (The last traces of H.E, specifically, H.E soloensis have been found from 117 Ka in Java, but it is believed that H.E was replaced by Homo Heidelbergensis by 300 Ka (Rizal et al., 2020)).

The main cause driving development, of course, is environmental pressure that leads to the selection of adaptations that successfully meet the challenges imposed by the environment. Did the H.E felt any environmental pressures either due to climate changes, or changes in their physical habitat due to migrations, or social challenges due to changes in the social fabric of their groups, pressures that were such that required adaptations that could be established through natural selection, and which favoured the development of new or rearranged neural circuits that resulted, in turn, in the acquisition of new cognitive capacities? The answer to that is emphatically in the affirmative.

First, during the 2 million years of their existence, H.E faced severe climate instability, including global cooling. They faced 6 different ice ages from 900 to 290 Ka the most severe of which occurred between 860 and 635 Ma, the so-called Cryogeniann. Second, H.E colonized almost the entire world and at each new place they arrived they encountered physical environments sometimes radically different from their origin place of departure in Africa. This means different living-conditions, different food sources, different game to hunt, different foes against whom the group had to protect itself, etc. All these posed challenges that had to be dealt with successfully for the migrants to survive.

Passingham (2021, 360) provides a list of behavioral responses to these challenges and the cognitive functions subserved by the PFC that would enable these responses. Climate instability leads to clothing that requires analogical reasoning, since one needs to think that as fur protects animals from the cold, artificially made skins could do the same. A behavioral solution to global cooling is fire for cooking, which requires semantic associations, because the hominis that first used fire for this purpose got the idea upon seeing forest fires that the dry periods of the ice age make abundant. (There is some evidence for the use of fire about 1.5 Ma, but the evidence is more compelling at 800Ka. It may be that in the earliest sites fire was used opportunistically, say by getting some pieces of wood on fire from some wildfire, and by about 800 ka fire was domesticated and controlled.) The semantic association is needed because one has to think that the power of the fire to melt things also means that it could soften the raw meet. The new environments the emigrants encountered led to the use of fire for protection, and this requires imagination or simulation, because one has to imagine that the fire will keep possible foes away. The need to gather new foods leads to group defence, which requires metacognition. Facing new predators needs social cooperation, which requires social and moral rules because it is conditioned on the understanding that each part of the group has social responsibilities. The hunt of new kinds of prey is facilitated by spoken language because it allows for more efficient planning through a more effective communication. All the aforementioned challenges, in addition to new social systems, probably led to the invention of new tool industries, which require a theory of mind because the know-how of making the tools must be transmitted to the new generation and this means that the teacher and the pupil think that others have mental states, that each can form opinions on the others based on these mental states, that each understand that the other has some purposes, etc. All these require a theory of mind and, in fact, an advanced theory of mind since one must be able to read the mind of others. Some of the cognitive capacities described above are needed for facing most of the challenges encountered by H.E.

One may not agree with all parts of Passingham's analysis, but the fact is that all these cognitive abilities are subserved by the PFC and, thus, their existence entails that the PFC of H.E was developed in such a way that could allow H.E to display these cognitive capacities. This does not mean that the PFC of H.E was as developed as that of modern Homo-Sapiens; the taller frontal lobe and the bulged parietal cortex of the modern Homo-Sapiens suggest that most likely it was not. As we have seen, Lombard (2024) argued that the development of the precuneus, along with the unique globularisation of the human skull characterizes Homo sapiens and can be traced in the Homo sapiens fossil record since about 160 thousand years ago, reaching the modern range by around 100 thousand years ago within a continuum of brain modification. This, apparently, was the result of Homo-Sapien's neuro-genetic adaptations since their split from the Denisovans and Neanderthals. These brain modifications suggest differences in attention as a cognitive trait between recent big-brained humans. It follows that the cognitive abilities of H.E that relied on attentional circuits were underdeveloped compared to those of modern Homo-Sapiens. This, however, does not entail that H.E lacked the cognitive capabilities, and the underlying subserving neural structures, that could support the adaptive behaviors discussed above. Moreover, the linguistic abilities of modern humans, the G3 languages with recursive operations that modern humans speak and write require a much more sophisticated cognitive apparatus, which, in turn must be subserved by a more sophisticated PFC and other association areas in the brain.

Few researchers in the field question the view that one way that hominins adapted to the problems they faced involved cooperation within their social group (Tomasello, 2016). This brings into focus the theory of mind. Most of the behaviors discussed above require cooperation. To cooperate, Hominins would have gained an advantage if they were able to infer the intentions of other members of their group, which is, as we have seen, Tomasello's (2019) "joint intentionality", meaning that different individuals have similar intentions on a joint enterprise and each one could understand that the others did so. The requirement to cooperate would also have promoted the need to understand social rules and conventions (Tomasello, 2016), but it is not possible to have a system of social rules without imposing sanctions on those who disobey them. It follows that those who obeyed would have had a better chance of being integrated in the group, and thus to survive, which shows how such social systems appeared in the course of evolution.

Another reason H.E had probably a sophisticated system of social rules is found in the fact that they were likely monogamous. Sexual dimorphism, that is, the physical differences between males and females, is an important source of variation within species, and in primates can be an indicator of reproductive strategy and group dynamics. The large amount of size variation observed within H.E makes it difficult to estimate average levels of dimorphism. The H. E fossil record provides clear evidence of a large range of skeletal size variation, at least equivalent to that observed in living human populations. Thus, it is plausible to assume that males were not systematically larger than females to a greater extent than they are today. If H.E did not have more sexual dimorphism than Homo sapiens, one could infer that male competition for mates was not more dependent on body size than it is today and that, thus, probably H.E practiced monogamy.

Because *H. erectus* males and females are thought to have been about the same size compared to other great apes (exhibit less size-specific sexual dimorphism), it is generally hypothesised that they lived in a monogamous society, as reduced sexual dimorphism in primates is typically correlated with this mating system (Plavcan, 2012). However, it is unclear if *H. erectus* did in fact exhibit humanlike rates of sexual dimorphism (Spoor et al., 2007). Vilmoare et al. (2019) studying footprints from Kenya argue that East African H.E 1.5 Ma was more dimorphic than modern *Homo sapiens*, although less so than highly dimorphic apes. Finally, Garvin et al. (2017) argue that *H.E* and more recent *Homo* species were characterized by reduced mass dimorphism. Whether monogamy was practiced is especially important because as we shall see when, in the last chapter, we discuss the appearance of symbolic representations and symbolic systems, one of the reasons adduced to explain the need for the emergence of symbolic thought is the social practices associated with monogamy.

Finally, let us see what inductions we can make based on the behavioral trends of H.E. This behavior is closely related to the environments in which H.E lived throughout their presence on earth, because it is the environment, both physical and social, that shapes behavior as organisms are trying to respond to the challenges the environment imposes on them. In turn, and in order to support the appropriate behavior, cognitive abilities area shaped, and with them the brain structures that can subserve these abilities are formed. This is how organisms adapt to environmental pressures both on a genetic or epigenetic scale. Kolk and Rakic (2021, 47) nicely summarize all these:

> Cognitive abilities are shaped by experience over time and seem to be in synchrony with PFC structural changes such as synaptogenesis and pruning. Following the 'use it or lose it' principle, the developing PFC dynamically rearranges incoming and outgoing wiring depending on usage and need. Specific for the PFC, the non-coding microRNAs mir-128b and mir-30a-5p have shown to be involved in prefrontal-dependent cognitive maturation by affecting epigenetic mechanisms. The constantly developing cognitive and executive capabilities occur parallel to the neurophysiological changes within the PFC and its connected areas and seem to reach a plateau in teenagers (around 12 years in human, around P50 in rodents). Adolescence is typically characterized by changes in social interactions and cognitive abilities in order to gain independence and adult skills and competences. In nonhuman primates this is characterized by risk-taking, novelty seeking, and increased vigilance; whereas in rodents by play behavior, increased exploratory activity and impulsivity are peaking. Higher order cognitive functions, in which PFC plays a prominent role, such as language and intelligence, continue to develop into adulthood.

I have argued that the most likely explanation of the behavioral traits of H.E is that they had the cognitive capacities that could support these behaviors; in other words, without these cognitive capacities, such behaviors would be impossible. These cognitive capacities, in turn, can exist only if the appropriate neural circuits could subserve them. Since in all apes, and rodents too (but this is another story), the main structure that subserves such cognitive capacities is the PFC, one can induce that H.E has a well-developed, PFC, which, however, fell well short of the PFC of Homo Sapiens, especially after 100 Ka.

To be able to represent iconically is conditioned on the development of the capacity to recognize that an object encountered in the past looks like or resembles a perceptual representation, a drawing, a picture, etc. This means that the brain stores in the long run the objects and the properties encountered in the past. This is a form of knowledge and the memory that stores it is called semantic long-term memory, which is memory for experiences with a specific spatial and temporal context. Chimpanzees learn a vast amount of information about their environment and use this information to forage successfully and they certainly develop mental maps of their environment. In addition, tamarins show accurate familiarity semantic memory across context changes. Thus, apes seem to have the cognitive arsenal underlying semantic memory (Neiworth et al., 2023). Since H. E evolved from these apes, they certainly had this capacity too. Moreover, Homo Habilis and *Homo erectus* were meat eaters, starting from scavenging carcasses to hunting animals eventually (Lewis-Williams, 2004; Mithen, 1996). Both activities require information about the possible locations at which to find meat, information that would allow the forming of hypotheses concerning possible meat sources that could be acted upon accordingly. This activity requires a rich semantic memory. Thus, H.E had this capacity too.

5.2 The Emergence of Iconic Representations

Granted that in some way representations emerged, the next question to ask is what is the form of the representations that are formed first. The answer is easy, and we have already brought it to the fore; perception is our first contact with the world and, thus, the first representations to be formed must be produced in perception and its contents. If the first internal representational vehicles are parts of perceptual states, and, eventually, perceptual contents themselves become representational contents, what is their form? The discussion on iconic representations in the third chapter entails that since perceptual representations are iconic (as the reader recalls that I assume that early vision has purely iconic representations, while late vision has hybrid iconic and symbolic representations), the first representational form to appear is that of iconicity; indices are succeeded by icons. Therefore, the first agents to have representations certainly had iconic mental perceptual representations whose content presented the world to them as being in some way. This means that they were conscious of the contents of their perceptual states, that is, they had self-consciousness, but this is not surprising because, as we saw, to be able to have representations of the sort discussed in the book, that is, representations that are in principle eventually available to cognitive access, requires self-consciousness.

Before I proceed, a few comments about iconic representations that pertain to our discussion. According to Peirce (1977), icons are non-arbitrary representational signs that resemble what they stand for. They physically evoke through their resemblance what they refer to, and they are not arbitrary constructions whose meaning is agreed upon through some convention. If the first iconic representations are the representations formed in perception, Peirce is right that icons resemble what they refer to because perception is designed to ensure that we extract from the environment and represent the information required for our successful interaction with it (evolution has seen to it), and this can be achieved only if perception depicts accurately with an acceptable margin of error the physical attributes in the environment with which we have interacted in our evolutionary past. Even though, as we discussed in the third chapter, a class of iconic representations resemble their referents only in the sense that there is a 2nd order structure mapping between the sign and the referent, and there is no requirement that there be any physical resemblance (that is, any resemblances concerning their qualitative properties), iconic perceptual representations do resemble in qualitative ways many of the properties (including spatial properties) of the objects in the visual scenes they represent, as these properties look from the perspective of the viewer. This view can be extended to other pictorial, external representations. Recall that according to Ayers (2019, 77–78), pictorial in general representations convey information about what is represented 'by exploiting the content of the experience of seeing it, that is, how the object or scene looks from a particular point of view'. This is a good example of what Peirce meant by writing that icons 'physically evoke through their resemblance what they refer to'.

Let me interject here that I assume that there are two main stages of visual processing (Raftopoulos, 2009). The first is early vision that lasts to about 120–150 ms. after stimulus onset and the other is late vision. The former is a purely visual stage that involves iconic representations. The latter is a hybrid stage in which cognitive contents interface with purely visual contents and in which the representations are purely visual iconic or visual iconic + symbolic hybrid representations.

The early visual iconic representations are intertwined with the context of their specific visual encounter. First, their content is always tied to the perspective of the viewer; the same visual scene looks different from different perspectives. Second, the denoting terms in the visual representational content are indexical in that they have direct referential links (that is, links that are not conceptually mediated) to the objects in the visual scene. It would take us far afield to give the details of this view (see Raftopoulos, 2009), so I restrict myself to noting that in perceiving a visual scene, the perceptual demonstratives, that is, the terms in the perceptual content that purport to refer to the objects in the visual scene, have a mode of presentation. This mode of presentation of the referent of a perceptual demonstrative allows early vision to individuate objects in conceptually unmediated, non-descriptive ways. Spatial and motion information regarding an object mainly constitute the mode of presentation of the object in a perceptual demonstrative, but these are not attached to the object, and they are not encoded as its properties, and this mode fixes the reference of the demonstrative by individuating the referent. Thus, one should not associate the sense of the demonstrative, that is, the referent of the demonstrative, with a description of the object's features and location. Since spatial and motion information is highly contextual, the perceptual demonstratives are essentially indexical and, hence, their referents vary with the context. It follows that early vision iconic contents are highly contextual. This entails the restriction mentioned above, that the perceptual iconic representations of early vision can be used only in the context of their formation. We shall explore the repercussions of this sort of context sensitivity for external icons in a while.

There is a further reason iconic signs were the first representational forms and relates to communication and, in fact, communication associated with action. The first forms of communication, researchers agree, were vocalization, and imitation and rehearsal that rely mostly on gesturing. In fact, many argue that communication expanded through the expansion in the number and precision of gestural signs. As we have already discussed indices, the first vocalizations were not representational. However, imitation through gestures or through sounds, and rehearsal involve iconic signs because they are based on the resemblance between the gesture/sign and its intended referent. When someone flips their hands in the air, in the context of a communication exchange concerning possible hunting targets, they intend the communicate the information that birds should be hunted. The hand movement resembles the bird's fly and indicates 'bird' because only birds can fly. The same holds when someone imitates a series of movements in an attempt to, say, exercise at some skill; the training exercise resembles the action sequences required by the actual act. Note that the cognitive baggage required for resemblance to work is minimal, if any at all, and this as it should be because we are discussing the emergence of icons that supposedly happened before the agents involved have the capacity for symbolic thought, and, thus, for the use of symbols. For a receiver to understand the meaning of flipping hands all that is required is that the iconic signal resembles some act in the receiver's behavioral repertoire, that is, that the receiver has seen in the past a flying bird and has stored this in a mental template. The appreciation on the part of the receiver of this sort of resemblance requires no cognitive apparatus as it takes place at the perceptual level by means of a simple template matching process (Raftopoulos, 2019).

Communication, of course, requires a cognitive baggage. For example, the sender of information must know that the receiver will perceive the sender's motions, that the receiver understands that the sender wishes to convey a message such as the sender and other pragmatical facts. However, simple forms of TOM and perception can warrant this understanding of the pragmatics of communication, and this sort of TOM and the requisite perceptual abilities are possessed by animals that do not have any conceptual capacities, at least in the wild. For example, there is evidence that great apes can produce various kinds of iconic gestures (Perlman et al., 2014) even though iconicity seems not to play a prominent role in the communication among great apes through gesturing (Planer & Sterelny, 2021, 105). So, iconic signs were first used before they were effective means for communication, long before the agents involved possessed any conceptual/symbolic capacities. (I hasten to note that I do not discuss the way these iconic signs are embedded in a culture or learned because this is orthogonal to my project in this book. The reader should consult Planer and Sterelny's (2021) discussion in Chap. 5 of their book.)

This sort of communication is coined "resemblance-based communication' by Planer and Sterelny (2021, 100–101), who also discuss further ways in which iconic signs can become established. The reader has noticed that both imitation and rehearsal involve in a significant degree actions, the flipping of hands in the air in imitation, or the mental training sequence in rehearsal. This should not come as a surprise. We saw above when we discussed the frog and the fly that initially there were purely casual sensorimotor action schemes that were holistic and non-representational. Representations emerged out of the gradual detachments of these schemes from the here-and-now when agents were able to detach the perceptual content associated with a fly from the affordances associated with perceiving a fly, and use it as an internal copy of a fly that may be evoked even in the fly's absence; that is, when agents were able to achieve displaced reference. Since action is in a last analysis what matters for survival, the action component remained at central stage in the gradual appearance of the ability for displaced reference and, thus, the first detached signs, i.e., representations, were having a strong action-related component, which is exactly what imitation and rehearsal do have.

Granting that iconic signs/representations emerged owing to, first, the fact that they were already part of the perceptual arsenal of some hominins that once they acquired self-consciousness they could access the contents of their perceptual states, and, second, owing to the fact that they enhanced communication within a group, the next problem to address is how these icons could be combined to produce representations of more complex situations. In other words, how could structure emerge using iconic representations? In the case of perceptual icons, the structure is already there since a visual scene is already a highly structured arrangement of particulars and their properties and the problem of how this structure gets perceptually represented is solved by the perceptual system. What about icons in communication? How could one compose icons to be able to communicate more complex situations?

We discussed above the example of hand flipping to represent iconically birds through gesturing. Suppose that someone (the sender) wishes to convey to another person (the receiver) that a bird is behind a bush. All the sender has to do is to point to the bush and flips his/her hands. Pointing by itself is an index and, thus, its use was already well entrenched in the group. Combining the index with the iconic sign conveys the intended information but notice that this complex representation is context dependent on account of the index. No special cognitive capacities are required for this composition of signs once iconic representations are possible. All it takes to allow one to use pointing to indicate a location one wishes to draw attention to is a rudimentary knowledge (if it is knowledge at all) of how pointing works. In addition, the sender must understand certain things about the visual experience of others. Experiences being mental states, this ability rests on the level of TOM possessed by the sender. However, this need not be a sophisticated TOM since having the ability requires that the sender know that the receiver can see what is his/her visual field, and that sometimes looking at something does not necessarily mean being consciously aware that it is there. Great apes have the requisite TOM to support these capacities and, thus, hominis did certainly have them as well.

Suppose, now, that a sender wishes to convey to a receiver a wish to go hunting birds in a lake, when no birds or a lake are in the immediate context. This immediately precludes the possibility of using pointing. To convey the information, the sender flips his/her hands and also uses an iconic sign for the lake. Or suppose that the sender wishes to ask the receiver to pick up a piece of fruit and give it to a third person in the scene. This is a triadic predicate relation and, thus, requires third degree predicate of the form Gives (o1, x, o2), where "G" stands for the action of giving (the predicate), x stands for the object to be transferred, and o1 and o2 are the two persons involved. To indicate his/her intention, the sender can use a sequence of three signs, a mixture of icons and indices. These actions do not require much of cognitive sophistication either and, thus, a minor increase in the ability to reason causally and in TOM compared to the baseline suffices to support this kind of composing iconic signs. Notice that in the first case, one of the icons plays the role of the predicate (being in a lake), and the other plays a nominal role indicating the target.

Putting two icons, or a pointer and an icon in a row, or even combing gestures create a composite sign. At a first level, the order of signs does not seem to play any role in determining the meaning of the composite sign, especially when both sender and receiver share the same spatio-temporal context; thus, no syntax is required. In more complicated situations, however, when cooperation for some group activity requires sophisticated forms of communication, the ability to use and interpret more complex sign formations in which the order of the signs plays a semantic role is advantageous to the effectiveness of communications and, thus was selected for. The semantic ordering of the signs allows the creation of more complex structures and, thus, expands the communicative capabilities. The first form of ordering relation that probably mattered was the temporal succession of events, which, to be expressed by signs requires that a temporal order be imposed on the composition of the signs so that the first sign represents an event or item that precedes a second event or item. This means that the ordering of the signs must map onto the temporal sequence of the events/items being represented which, very often, reveals the causal relation between the two events/items as the preceding events may be the cause of the second, whereas the reverse is not possible. For this reason, to understand the composite sign, the receiver must be sensitive to the ordering of the signs. As we shall see in Chap. 7, old world monkeys and the great apes already have this capacity or, at least, have the requisite cognitive mechanisms that, under the proper circumstances, can support this cognitive capacity, and, thus, early hominis had it too.

We discussed the case in which two icons are combined to form a linear syntactic structure in which one icon is used in a nominal role and the other as a predicate, say the icon of a bird drawn next to an icon of a lake. There is an additional complication in the sense that this case involves the activity of drawing an external iconic representation and there are different cognitive demands supporting using mental, internal iconic representations to communicate and using external iconic representations for the same purpose. I shall address this problem next, but for the time being I wish to concentrate on the nature of iconic composition. As we saw when we discussed iconic representations, icons cannot be combined to express logical relations. Therefore, one cannot, for example, simply add a picture to another picture to get the conjunction of the two meanings expressed by each picture. So, what two icons set side by side can express? One way would be to superimpose the icon of a bird onto the icon of a lake and, thus, effectively, place the bird in the lake. This way makes it clear that the bird-sign is used as nominal and the lake-sign as a predicate. The reader should recall that in iconic representations actions and transformations on the representational vehicles mirror in a natural way transformations in the represented domain. That is, the transformations that occur in the represented medium must be isomorphic to the transformations in the representing medium if the latter is iconic. Thus, to represent iconically a bird siting on a lake, one must put an iconic sign of a bird on an iconic sign of the lake by superposition.

The problem with this view of composing icons is that it has very limited range in the sort of state of affairs it can be used to indicate because of the expressive limitations of icons when it comes to representing complex situations that result from logical combinations of their constituent parts, a limitation that results from the way icons can be composed and which we extensively discussed in Chap. 4. Icons cannot be used to express and, thus, cannot be used in discussions about, hypothetical situations, but reasoning on the basis of hypothetical situations is critical in planning. Combining pointers and icons to extend expressive capabilities has its own limitations as far as the use of pointers is highly contextual and, thus, it cannot be used when displaced reference is required. This means that hominins and certainly H.E, whose activities, as we have seen, show that they were engaged in such planning, had to rely on other forms of representation, namely, symbols, in order to be able to communicate effectively to draw plans for their future activities. We shall discuss symbols in Chap. 7.

If, as I have argued, H.E were capable of having self-consciousness, and, thus, had the capacity to represent the world, the first representations that they had were the iconic representations of their perceptions, and the various gestures and vocalizations that are iconic representations as well. The next question to ask relates to H.E and external iconic representations. Could they interpret external iconic representations? That is, if they saw, say a figure of a person carved on a piece of stone could they recognize it as a person? As we shall see, it takes four years for children to be able to tell the difference. Moreover, could they understand that the person whose figure is curved on the stone is not the actual person depicted? Finally, could H.E construct external iconic representations?

What about external icons, that is, iconic representations carried by external physical vehicles, and not by mental states as the iconic representations in perception? The Neolithic Sumerian cuneiform and Egyptian Hieroglyphics relied on pictorial, iconic representations as well. Sumerian writing was first developed as a means of accounting material goods, such as sheep or grain. The quantity or number of sheep were incised in clay tablets or seals. This presupposes two sorts of signs; one for the number and the other for the sort of goods that the number counts. Since these seals were also used for foreign commercial exchanges along vast regions with peoples speaking different languages, the signs used should be such that people who did not speak Sumerian could understand what goods were referred to in the seals or tablets. Sumerian may be different from Elamite, but the cows were the same and this is why the signs for cows and other goods resembled the physical form of the cow so that everyone could understand the significance of the seal independent of the linguistic conventions of each civilization or local community. This is why iconic representations and not symbols were the first to appear in written languages, even though any language requires symbols because it must ensure communication across different contexts. Different civilizations need not share any conventions in order to be able to understand the meaning of an icon; it was enough that they had the same visual systems. The pictograms, thus, resembled the physical form of their referents. This explains why iconic representations were first associated with the notion of physical resemblance that is grounded in qualities that representations and their representata share. Interestingly, the pictograms were not initially related to the names of the goods they signified, the way 'Cat'' is associated with the animal it refers to through the name of the animal and, thus, though the concept *CAT*, pictograms referred directly without any conceptual intermediary to some physical entity. Moreover, and gradually, iconic linguistic symbols paved the way to abstract symbolic signs that substituted them and bear no resemblance to their referents as is evidenced by the evolution of both Sumerian cuneiform from archaic to Babylonian to Assyrian, and Egyptian hieroglyphics (Schmandt-Besserat, 2014). Thus, iconic signs tended to become simpler and more abstract by gradually losing their resemblance to what they represent and, eventually, ceased to be icons and became symbols that had to be learned and not just be perceived to establish their referents.

Icons are not arbitrary as symbols are; there is no set of conventions that dictate how the perceptual system represents visual scenes. It does so through a purely physical (and neurophysiological) set of laws. In this sense, iconic perceptual representations are not invented either and are naturally produced. It is not quite clear, though, that all these theses hold true for all sorts of iconic representations. One could, for example, adopt two different sets of perspectives to draw a picture and even though both perspectives obey the laws of optics and geometry, still, it is a matter of decision on the part of the agent which one to use.

Even though Ayer is generally speaking right that the cognitive processes governing the construction and interpretation of pictures, paintings, and drawings parallel the perceptual processes at work in normal visual perception, this applies to the pictorial representations that use the rules of linear perspective and linear geometrical projection that are at work in visual perception. However, linear perspective in picture, painting, and drawing construction is not the only game in town. One could equally use, and for certain purposes it is advisable to use, parallel perspective and parallel geometrical projection (Giardino & Greenberg, 2015, 20). In linear perspective, the projection lines (that is, the lines that start from the depicted object and terminate to the viewpoint) converge on a single viewpoint. In parallel perspective, the projection lines are all perpendicular to a single plane, and are, thus, parallel to each other, rather than converging to a single point. (It is evident now why visual perception uses linear perspective; the fact that the eye is at a distance from the source from which the light emanates, and is, usually, smaller in size that the items in the visual scene entails that the projection lines terminate to a single point rather than on a plane.)

Adopting a different perspective leads to different drawings and the difference originates with the painter's decision. This, in turn, is influenced by social, economic, artistic, or other cultural factors that are to a certain extent matters of convention. In addition, and since the perceptual system uses a liner perspective, adopting to use parallel perspective leads to a drawing that does not represent a visual scene in the way normal vision naturally does.

The role of context in selecting a particular perspective leads us back to the problem of the context sensitivity of iconic representations. The discussion in the third and fourth chapters suggested that there are basic differences between icons and symbols. On account of these differences, icons are more restricted in their application compared to symbols. The first characteristic of iconic perceptual representations is that they are essentially tied to the context in which they are formed. Symbols are not because when the conceptual framework is fixed, they refer to the same entity irrespective of the context of occurrence. Thus, there is a sense in which symbols too depend on context, because even though they are by arbitrary conventions this presupposes a context in which the participants agreed on the conventions, but this is quite different kind of contextuality that does not impose the restrictions accompanying the contextuality of perceptual iconic representations.

External iconic representations, unlike, iconic perceptual representations are not tied in the same way to the context of a perceptual encounter even though they represent through evoking some perceptual content and this is why, say, drawings can convey their meaning across different contexts. After all, I just said that Sumerians drew icons of cows to refer to cows exactly to be able to express this information across different linguistic and perceptual contexts. Since iconic representations are interpreted by using the viewer's visual modality, pictorial representations convey information about what is represented by exploiting the content of the experience of seeing it from a particular point of view, it follows that pictorial representations in general should inherit the context dependency of the early visual modality.

Before I turn to this problem, let me say that there is another sense icons are context dependent, which I shall use in the discussion. One could have decided to use the world/symbol "cat" to refer to dogs just by changing the linguistic convention. However, one could not decide to use the image of a cat to refer to dogs because that would require changing the perceptual system, which brings to the fore the importance of the fact that perceptual iconic representations do not refer by convention but through physical resemblance in a given context. How does this help address the problem at hand? It does because it entails that viewers with similar perceptual systems when presented with the icon of a cow will see a specific shape, color, texture, size, etc. It is true that the percept depends on the viewing conditions, illumination, and the viewer's perspective, which means that different viewers may form different iconic representations of the same visual scene. However, these different percepts are statistical variants of the same template, since they resemble to a certain extent their referent despite variations, and they eventually fall within this template and their referent is recognized as a cow. Admittedly, when the viewing conditions are radically different viewers may fail to recognize the present entity as a cow. However, this does not detract from the fact that the role of the perceptual system is to form a representation that resembles the referent of the representation. Besides, a bizarre drawing of a cow will probably result in recognition failure.

Allow me to elaborate on this. Even though viewers who see a cow may have used different context-dependent information to individuate the cow from the background and the other objects in the visual scene, which means that the contents of their early visual states are highly contextual and, thus, may differ, once the percept (that is, the mental image of a cow) has been formed, these viewers will see and recognize a cow. The reason is that the percept is formed in late vision where the images of early vision interface with the iconic templates stored in memory. In this case, the iconic template of "cow" is activated and is compared to the representation of that object in early visual content following the process of hypothesis testing that takes place in late vision, and which I described a bit earlier. For a match to be achieved, so that the viewers see a cow, similarities and differences between the image in vision and the activated template are assessed. The information stored in the template is statistically extracted from previous perceptual encounters with cows. Obviously, these encounters have happened under different viewing conditions and, thus, each time different information may have been retrieved. However, the template contains the statistical average of the perceptual information concerning cows, that is, the typical properties of cows. If the object in early visions falls within the statistical average, it is seen as a cow. Statistical averages have the property that they abstract from the context of each specific visual encounter with cows, and this is why the resulting percept is context independent, while it is, still, an iconic representation. This explains why all the different iconic representations of the animal in early vision, which are due to the context dependency of the representations in early vision, when the percept is formed converge to the image of a cow.

To put the previous discussion in context and underline the role of mental iconic perceptual representations in interpreting external iconic representations, such as drawings etc., the reader should bear in mind that we are concerned with the iconic representations that appeared first in the prehistory of the human kind, that is, that we are concerned with the origins of pictorial representations. These were based on a strong notion of resemblance, in the sense that in order to be understood these icons should invoke what they represent by means of a sufficient physical resemblance to it. The perceptual visual system provides this resemblance. Occasionally, resemblance played a pivotal role in the construction of the representational vehicle of an icon, that is, in choosing the material with which to construct a representation. The rock painters in the caves 45 Ka used some of the protrusions and cavities of the rocks to paint animals because these protrusions and cavities resembled some of the animals' body parts (Clottes, 2016; Curtis, 2006; Rigal, 2016). As we shall shortly see, H.E probably did the same thing. Recall that according to Ayers (2019, 77–78), the pictorial representations convey information about what is represented 'by exploiting the content of the experience of seeing it, that is, how the object or scene looks from a particular point of view'. It follows that H.E were quite likely able to interpret external iconic representations, that is, to understand what is depicted. This is reinforced by the findings of the Venuses that H.E constructed (which we shall discuss in the next paragraph). If, as is suggested, H.E recognized in some pieces of stones bodily characteristics and highlighted them by curving the stone to make the resemblance more prominent, this entails that H.E could interpret external iconic representations by relying, obviously, on their experiences of seeing these bodily characteristics on actual persons. This answers the first question, that is, if H.E could interpret external iconic representations.

When did our ancestors start producing external iconic representations? What is the evidence we have for the use of iconic representations in our evolutionary past? The most unequivocal piece of evidence consists of the numerous figurines and rock-paintings on the walls of caves (and not only caves) that appeared throughout the world about 45 Ka. However, the evidence extends much more back in time. The oldest known iconic representation is the series of petroglyphs discovered in the 1990s in two ancient quartzite caves in India: the Auditorium Cave at Bhimbetka and a rock shelter at Daraki-Chattan. This cave art consists of numerous cupules, that is, hemispherical cup-shaped depressions. Research indicates that they must date from at least 290 Ka and there is some evidence that they may be much older, originating as early as 700 Ka. The early Stone Age figurine of the Venus of Berekhat Ram (Birkat Ram) discovered in hills in Israel may be the oldest piece of figurines, dating back from 230 to 700 Ka. Initially some paleontologists preferred to see it more as a product of natural erosion rather than a deliberate human act of creativity. Although these figurines were mostly formed by natural processes, they were modified to resemble human form by emphasizing certain grooves to resemble hairline limbs, and eyes. There is almost a consensus nowadays that it is manmade, as a second similar and contemporaneous figurine, the Venus of Tan-Tan, was discovered in Morocco. This Venus shows traces of pigments suggesting that it was colored (d'Errico & Nowell, 2000; Morriss, 2010). The earlier dates are within the reign of *Homo erectus*, who, remember spread out almost throughout the world. The later dates are within the range of Archaic Homo sapiens in Africa and China and Homo Heidelblbergensis in Europe. There is, of course, the problem concerning the functionality, practical or ceremonial, of these icons, but this need not concern us here. However, it should be noted that many hand-axes—in fact, they were mass-produced—constructed by H.E were found intact without any trace of use. They may have been left unused for several reasons, but the copious quantities found suggest that they had some social or ceremonial significance, and they were not designed to be used for cutting meat or other menial tasks (Kohn & Mithen, 1999). These dates are within the lifespan of H.E. This discussion provides an answer to the second question, namely whether H.E could produce external iconic representations.

This leaves us with the third question. Could H.E tell the difference between the content of the external iconic representation, that is the object as depicted in the image, from the actual object? At a first glance, the answer is in the affirmative. I said above that he rock-painters in the caves 45 Ka years ago or perhaps even earlier used some of the protrusions and cavities of the rocks to paint animals because these protrusions and cavities resembled some of the animals' body parts. This entails that icons are not always intentionally produced, but when perceived they can be recognized as such, that is, as representing something with which the viewer is familiar. This entails, in turn, that the creators or collectors or users of icons, be they Homo Sapiens, or Neanderthals, or *Homo erectus*, grasped the relation between form (the representational vehicle) and meaning (the thing represented), and this means that they knew the difference between the thing depicted in the icon and the icon itself (an ability that infants acquire after their fourth year of age).

Let me elaborate on this. Having the ability to represent iconically, that is, to understand and use images as representations, and also having the ability to construct external iconic representations, requires that one can tell the difference between an iconic representation *qua* vehicle, the thing as depicted in the icon, and the thing represented. DeLoache (2004) coins this characteristic of iconic representations the "dual nature" of symbolic objects. A picture is both a concrete object (the vehicle) and a representation of something other than itself (the representatum). To use effectively an image, one has to perceive the image and interpret its relation to its referent.

Turning to children, psychological research suggests that children younger than four-years old fail to understand the nature of pictures or drawings. This is evidenced by the fact that they often try to retrieve objects portrayed in the picture or the drawing, and by the fact that turning around a picture of a cup, leads them to expect that the liquid in the cup will spill out, etc., (Deloache & Burns, 1994). Moreover, when shown pictures of the same object from different perspectives, which are different pictures, 3-years-old children saw them as the same picture because the pictures showed the same stuff (DeLoache, 2004; Liben, 2003). Evidently, these children do not have the capacity to understand that the icon does not contain the thing itself, but it is only a representation of it. In other words, they cannot draw the distinction between the intentional object and the object of intention, which is not surprising since several philosophers also fail to do. On other tasks using models, which are representations of physical systems, 3 years-old children succeed in understanding them as representations, while 21/2 years-old fail. It is interesting that performance on such tasks depends also on the physical properties of the symbols/representations. The more salient these features, the more children tend to focus on the symbol itself and ignore its representational function. Increasing the salience of the symbol, even 3 years-old children, which, otherwise, in general succeed in tasks requiring grasping the dual nature of representational items, fail the tasks. By the age of 4, children in general understand the dual nature of icons, which means that the capacity has been consolidated. This back and forth in performances and the significant role that the experimental context plays in success or failure suggests that the developing ability relies on a developing neural system that displays the properties of dynamic systems (Kelso, 2007; Rolls, 2008).

To be able to understand the dual nature of icons and symbols in general, one must be able to understand that the content of a mental state differs from the object it purports to represent. This, in turn, requires that someone have cognitive access to the contents of their mental states, that is, that they have self-consciousness. As we saw, children solidly acquire self-consciousness around the age of 4, which is exactly the age at which the capacity to understand images in the sense that they know that the depicted object is not the actual object that the image purports to represent consolidates as well.

Since hominis diverged from chimpanzees 7–10 Ma and since it is natural to think that the cognitive abilities of Hominins related to understanding icons gradually evolved starting form a base-line determined by the abilities of the common ancestor, examining the cognitive capacities of Chimpanzees related to icon understanding is useful because it allows us to establish this base-line. Of course, chimpanzees have evolved too, but since the habitats in which they live have not changed that much (recall that they live in almost the same places in Africa they used to live millions of years ago) one can assume that they faced much less evolutionary pressures and, thus, that their cognitive capacities have not changed that much since they did not have to face serious adaptation challenges, as Hominis did. In addition, if other monkey species that diverged from Hominis earlier than did the chimpanzees have related capabilities, these are, too, parts of the cognitive baseline from which hominis started.

Pokorny and de Waal (2009) and Perret and Mistlin (1990) showed that capuchins understand the representational nature of pictures. Chimpanzees, in general, can recognize line drawings of familiar humans and apes (Itakura, 1990). Since, obviously, these drawings differ from the referents, the chimpanzees could not have taken the drawings to be the items in the environment and this means that the recognition of the line drawings relies on understanding the representational nature of the drawings and, thus, their referential role. Savage-Rumbaugh et al. (1980) showed that chimpanzees can learn symbols to refer to the categories of food and tools, and then use these symbols to label photographs of foods and tools, a capacity that clearly requires the ability to distinguish between the representation and its content. In all these cases, the monkeys or apes succeeded in the relevant tasks only after training or practice (Parron et al., 2008). Notice, however, that even in humans, the acquisition of the capacity to understand the dual nature of representations relies on maturation, but also on training and practice with pictures and videos (DeLoache, 2004; Fagot et al., 2010). Going back in time when pictures and videos were not available, no one would claim that the children at those times never acquired the capacity to understand icons and, thus distinguish, say between objects as they see them and objects as they really are. Thus, it seems that the maturation of the appropriate neural circuits suffices for the capacity to appear. In the case of monkeys, earlier evidence suggested that learning or practice are required for the capacity to develop. This means that the relevant cognitive processes and the neural mechanisms that realize these processes are available and need fine tuning to express themselves. However, newer studies (Zhao et al., 2020) suggest that monkeys in the wild can recognize facial expressions from photographs and react accordingly. If these findings are confirmed by further research, they will show that apes or some among them can understand iconic representations as representations. In any case, the literature we examined reveals that the cognitive capacities underlying the understanding of the representational function of iconic representations are parts of the cognitive baseline from which hominis evolved.

We saw that *Homo erectus* did have self-consciousness. However, self-consciousness is necessary for understanding imagistic representations, but it is not sufficient to allow image-making. This returns us to the problem of whether H.E could draw external iconic representations and if they did (or did not), what cognitive ability did they have (or did not have) to be able to draw icons (or be unable to draw them)? They certainly had perceptual icons and were aware of their contents. The cognitive capacities required for their ability to plan ahead suggests that they knew that the contents of their perceptual states differed from the objects of their perceptual states. However, could they construct icons? Image-making demands cognitive abilities of a more advanced kind; in fact, this ability presupposes a more advanced stage of consciousness, a topic to which I return in the next chapters. H.E were constructing sophisticated stone tools and transmitted their knowledge of how to make these tools to the next generation. This means that H.E had mental images of their tools and they knew how to transform these images to real tools, which means that they recognized both that there are two dimensional iconic representations of three-dimensional objects, and that there is a connection between mental images and three-dimensional physical objects, since one could use the former to knap stone to construct the latter. Besides, if the view that the early Venuses we discussed above are indeed iconic representations of human forms is correct, this settles the issue; H.E were able to construct iconic representations. As we have seen, H.E curved some stones so as to take advantage of some of their natural properties that made them resemble H.E figures in order to emphasize these characteristics to increase the similarity of the piece of stone with an actual H.E figure. This means, first, that H.E could recognize the contours on a stone as portraying a H.E figure which, however, was not a good match to the actual figure and needed reworking, This presupposes that H.E could tell the difference between the figure as presented on the stone and the actual figure, which means that they understood the representational function of the stone/icon. Second, by curving the stone to increase the similarity to an actual H.E figure, they surely knew that their actions were performed on the stone and did not affect the actual H.E figure portrayed, which also entails that they knew that the figure on the stone was not a real H.E figure.

Since H.E had mental images of their tools and used them both to construct their tools and teach the techniques for constructing them, could they draw an icon of such a tool and understand that the there is a difference between the tool in the icon and the actual tool in the real world? Lewis-Williams (2004, 193–194) claims that even in the Upper Paleolithic (about 40 Ka) the painters of the rocks in caves did not think that icons stood for real animals (one cannot go around them, it is too small to be a real animal, etc.), but the paintings depicted the contents of their visual states, the intentional objects. Thus, the Upper-paleolithic Homo Sapiens could tell the difference between the object in their imagination and the real one. Could H.E do the same thing? The only way to attempt to answer this question is to speculate about the cognitive abilities of H.E based on the available empirical evidence (stone tools, travels, etc.)

I have already discussed some of these cognitive abilities and I shall expand more on them in the next chapters. As we shall see, Planer (2017, 212–13) proposes that H.E possessed the following cognitive abilities. First, an informational theory of mind, which means that an agent can represent other agents as possessing information-carrying states, as well as states that specify an environmental condition that the other agent is motivated to effect. Second, H.E possessed the ability to construct multifaceted mental representations or models of a situation, in which they integrate different kinds of information, be it spatial, temporal, causal, etc. Third the cognitive ability to imitate acts that others have performed either towards oneself or towards other agents. This ability depends on the ability to construct abstract representations of social interactions), that is, representations that represent interactions in terms of abstract roles, each of which can be filled by some agent or other. The existence of abstract representations is important when collaboration among the members of a groups is required, because collaboration is based on an understanding of distinct roles in an activity, as well as understanding of the possibility of role reversals. Adding to that that H.E have developed working memory and autobiographical memory and they were able to interpret and foresee natural regularities (as they knapping techniques, the way they quarried for stones often undertaking very long-distance trips to the source of the stones, their ambush hunting, and their successful migrations suggest), I think, to reiterate a point made earlier, that it is very plausible to assume that they could understand the function of icons as representations.

However, if as I have also suggested, they could make icons, and there is evidence that they did so, why there is no evidence of, say, stone icon-like carvings or other forms of icons of their tools that could have survived? The answer is that they saw no need to do so, which means that there were no environmental pressures that would lead them to do so. The fact that they probably constructed icons of Venuses means that these icons played some social role, which is unknown to us, as is the social role of the mass of deliberately hand-axes that were left unused. Perhaps, icons of tools had no social role to play, and this is why they were not constructed. One might argue that such icons could have been proved useful in instructing tool-making techniques. This is true, but remember that this instruction, partly because of lack of a sophisticated grammatical language, relied mostly on trial and error. Trial and error means that the apprentices learned by trying their hands on the stone and their teacher directed this activity by showing the appropriate hand movements, etc. Thus, no icons of tools were required.

However, even if H.E could make rudimentary iconic representations by taking advantage of the naturally occurred contours of pieces of stones that elicited in H.E to see a person and complete the image, one must be careful what iconic drawing capacities one attributes to H.E. Back to the problem of why they did not make icons of their tools or other artifacts they used in their everyday life, especially since they probably and mental images of these tools, one has to remark that passing from a having a mental image of something to construct an external iconic representation of this something requires the ability to coordinate the hand movements with the image in the head, that is, to coordinate both consciously and unconsciously body and mind. As Lombard (2024) argues, a fully inextricable relation between mind and body may have originated 60–80 Ka with Homo Sapiens, and Lombard relates it to the development of attentional capacities of Homo Sapiens subserved by evolutionary changes in their brains. These capacities allowed Homo Sapiens to practise for the first time bow-hunting. If true, this means that H.E were incapable of using their mental images of tools to construct an image of a tool. When they found materials whose physical make-out remind them of, say, persons, they could finish the job by adding some coarse details that would enhance the similarity, but their capacities for constructing images stops there. Another factor that may have contributed to the inability of H.E to transfer their mental images to the world of image making, with the exception noted above, is the so-called functional modularity of the mind of hominids that we shall explore in Chapter six.

Let us revisit a problem that we encountered before. H.E were able to construct the means of transportation across oceans to reach other continents. They certainly constructed some kind of raft that could transport a fair number of passengers along with their provisions. Obviously, to construct the raft, they must have formed first a mental image of the raft even if this image was constructed as a result with an encounter of an object that resembled a raft that gave them the idea in the first place. Moreover, H.E could transform their mental images of hand-axes to the actual tools. These entail that H.E has certainly achieved affair amount of mental-body coordination. So, why did they fail to construct images of their tools, assuming they did not, of course, because there is the possibility of future findings. To answer that, one need consider the difference between constructing a tool based on a mental image and constructing an iconic representation of a tool on stone, or wood, or bone. Instead of taking a piece of stone and curve it to make a tool, one could curve it to draw a tool on it, the way they curved stones to make the Venuses. Of course, in the latter case they used a piece of stone that already bore some resemblance to a woman and, thus, to draw the analogy one must assume that to do the same with tools they should have found a piece of stone that resembled a tool. But, if they found a such a piece of stone, why not finish it to make an actual tool? If it were not of the appropriate quality to make an actual tool, then one has to wonder what reason did they have to make an icon of a tool, as opposed to making an icon of a woman. The answer must be sociological, given the significance of Venus figures across almost all eras and cultures. Thus, it may be that the reason H.E did not make images, if they did not, is that they saw no need to do so with the notable exception of Venuses, and not because they lacked the requisite cognitive capacities.

Note

1. In discussing MW I do not distinguish between storage-specific WM, or, pure working memory size, and process relating short-term working memory, or effective working memory size (Cowan, 2001, 2010), as this distinction does not impact on my project in the book.

References

Alexander, G. E., & Goldman, P. S. (1978). Functional development of the dorsolateral prefrontal cortex: An analysis utilizing reversible cryogenic depression. *Brain Research, 143*, 233–249.

Amso, D., & Scerif, G. (2015). The attentive brain: Insights from developmental cognitive neuroscience. *Nature Review Neuroscience, 16*, 606–619. https://doi.org/10.1038/nrn4025

Amsterdam, B. (1972). Mirror self-image reactions before age two. *Developmental Psychobiology, 5*(4), 297–305.

Anderson, J. R., & Gallup, G. G. (2011). Which primates recognize themselves in mirrors? *PLoS Biology, 9*(3), Article e1001024. https://doi.org/10.1371/journal.pbio.1001024

Astle, D. E., & Scerif, G. (2011). Interactions between attention and visual short-term memory (VSTM): What can be learnt from individual and developmental differences? *Neuropsychologia, 49*, 1435–1445. https://doi.org/10.1016/j.neuropsychologia.2010.12.001

Ayers, M. (2019). *Knowing and seeing: Groundwork for a new empiricism*. Oxford University Press.

Baddeley, A. (2003). Working memory: looking back and looking forward. *Nature Review Neuroscience, 4*(10), 829–839.

Baddeley, A. D. (2012). Working memory: Theories, models, and controversies. *Annual Review of Psychology, 63*, 1–29.

Barkovich, A. J. (2005). Magnetic resonance techniques in the assessment of myelin and myelination. *Journal of Inherited Metabolic Disease, 28*, 311–343.

Barton, R. A., & Venditi, C. (2013). Human frontal lobes are not relatively large. *PNAS, 110*(22), 9001–9006.

Bauer, P. J. (2009). The cognitive neuroscience of the development of memory. In M. Courage & N. Cowan (Eds.), *The development of memory in infancy and childhood* (pp. 115–144). Psychology Press.

Beaman, C. P. (2010). Working memory and working attention. What could possibly evolve? *Current Anthropology, 51*(Suppl. 1), S21–S38.

Beets, M. W., Cardinal, B. J., & Alderman, B. L. (2010). Parental social support and the physical activity–related behaviors of youth: a review. *Health Education Behavior, 37*(5), 621–644.

Bell, M. A. (2012). A psychobiological perspective on working memory performance at 8 months of age. *Child Development, 83*, 251–265. https://doi.org/10.1111/j.1467-8624.2011.01684.x

Ben-Dor, M., Gopher, A., Hershkovitz, I., & Barkai, R. (2011). Man the fat hunter: the demise of *Homo erectus* and the emergence of a new hominin lineage in the Middle Pleistocene (ca. 400 kyr) Levant. *PLOS ONE 6*(12), e28689.

Benitez, V. L., Vales, C., Hanania, R., & Smith, L. B. (2017). Sustained selective attention predicts flexible switching in preschoolers. *Journal of Experimental Child Psychology, 156*, 29–42. https://doi.org/10.1016/j.jecp.2016.11.004

Bianchi, S., Stimpson, C. D., Bauernfeind, A. L., Schapiro, S. J., Baze, W. B., McArthur, M. J., Bronson, E., Hopkins, W. D., Semendeferi, K., Jacobs, B., Hof, P. R., & Sherwood, C. C. (2013). Dendritic morphology of pyramidal neurons in the chimpanzee neocortex: Regional specializations and comparison to humans. *Cerebral Cortex, 23*, 2429–2436.

Binford, L. (2007). The diet of early Hominins: Some things we need to know before 'reading' the menu from the archaeological record. In W. Roebroeks (Ed.), *Guts and brains* (pp. 185–222). Leiden University Press.

Block, N. (2007). Consciousness, accessibility, and the mesh between psychology and neuroscience. *Brain and Behavioral Sciences, 30*, 481–548.

Boesch, C. (1994). Cooperative hunting in wild chimpanzees. *Animal Behavior, 48*, 653–667.

Boesch, C. (2002). Cooperative hunting roles among Taï chimpanzees. *Human Naure, 13*, 27–46.

Boettcher, S. E., Drew, T., & Wolfe, J. M. (2013). Hybrid search in context: How to search for vegetables in the produce section and cereal in the cereal aisle. *Visual Cognition, 21*(6), 678–682.

Brady, R. J., & Hampton, R. R. (2018). Nonverbal working memory for novel images in Rhesus monkeys. *Current Biology, 28*(24), 3903–3910.

Brauer, J., & Call, J. (2015). Apes produce tools for future use. *American Journal of Primatology, 77*, 254–263.

Braver, T. S., Gray, J. R., & Burgess, G., C. (2008). Explaining the many varieties of working memory variation: dual mechanisms of cognitive control. In A. R. A., Conway, C. Jarrold, M., J. Kane, A. Miyake, & J. N. Towse, (Eds.), *Variation in working memory* (pp. 76–106). Oxford University Press.

Brodmann, K. (1909). Vergleichende Lokalisationslehre der Grosshirnrhinde. Leipzig: Barth (Reprinted as Brodmann's *Localisation in the cerebral cortex*, translated and edited by L. J. Garey, London: Smith-Gordon, 1994).

Brown, V. J., & Tait, D. S. (2015). Attentional set-shifting across species. In T. W, Robbins & B. J., Sahakian, (Eds.), *Translational neuropsychopharmacology: Current topics in behavioral neurosciences* (Vol. 28, pp. 363–395). Springer.

Bruner, E. (2004). Geometric morphometrics and paleoneurology: Brain shape evolution in the genus Homo. *Journal of Human Evolution, 47*(5), 279–303.

Bruner, E., Peuss, T. M., Chen, X., & Rilling, J. K. (2017). Evidence for expansion of the precuneus in human evolution. *Brain Structure and Function, 222*, 1053–1060.

Burge, T. (2022). Perception. [illegible] University Press.

Buss, A. T., Fox, N., Boas, D. A., & Spencer, J. P. (2014). Probing the early development of visual working memory capacity with functional near-infrared spectroscopy. *NeuroImage, 85*, 314–325. https://doi.org/10.1016/j.neuroimage.2013.05.034

Buttelmann, F., & Karbach, J. (2017). Development and plasticity of cognitive flexibility in early and middle childhood. *Frontiers Psychology, 8*, 1–6. https://doi.org/10.3389/fpsyg.2017.01040

Call, J. (2010). Do apes know that they could be wrong? *Animal Cognition, 13*(5), 689–700.

Call, J., & Carpenter, M. (2001). Do apes and children know what they have seen? *Animal Cognition, 3*(4), 207–220.

Call, J., & Tomasello, M. (2008). Does the chimpanzee have a theory of mind? 30 years later. *Trends in Cognitive Sciences, 12*(5), 187–192.

Carey, S. (1995). Continuity and discontinuity in cognitive development. In E. Smith & D. Osherson (Eds.), *An invitation to cognitive science, volume 3: Thinking* (pp. 101–129). MIT Press.

Carey, S., & Xu, F. (2001). Infant's knowledge of objects: Beyond object files and object tracking. *Cognition, 80*, 179–213.

Carruthers, P. (2013). Evolution of working memory. *Proceedings of the National Academy of Science, U. S. A., 110*, 10371–10378.

Casey, B. J., Tottenham, N., Liston, C., & Durston, S. (2005). Imaging the developing brain: What have we learned about cognitive development? *Trends in Cognitive Sciences, 9*(3), 104–110.

Casile, A., & Giese, M. (2006). Nonvisual motor training influences biological motion perception. *Current Biology, 16*(1), 69–74.

Chang, L., Zhang, S., & Mu-ming Poo, M. (2017). Spontaneous expression of mirror self-recognition in monkeys after learning precise visual-proprioceptive association for mirror images. *PNAS, 114*(12), 3258–3263.

Chugani, H. T., Phelps, M. E., & Mazziotta, J. C. (1987). Positron emission tomography study of human brain functional development. *Annual Neurology, 22*, 487–497.

Claidiere, N., Smith, K., Kirby, S., & Fagot, J. (2014). Cultural evolution of systematically structured behaviour in a non-human primate. *Proceedings of the Royal Society B, 281*, 20141541.

Clark, G., & Henneberg, M. (2022). Interpopulational variation in human brain size: Implications for hominin cognitive phylogeny. *Anthropological Review, 84*(4), 405–429.

Clotte, J. (2016). *What is Paleolithic art: Cave paintings and the dawn of human creativity?* Chicago University Press.

Coltheart, M. (1980). The persistences of vision. *Philosophical Transactions of the Royal Society of London: Series B, 290*, 269–294.

Conklin, H. M., Luciana, M., Hooper, C. J., & Yarger, R. S. (2007). Working memory performance in typically developing children and adolescents: Behavioral evidence of protracted frontal lobe development. *Developmental Neuropsychology, 31*(1), 103–128.

Constantinidis, C., Franowicz, M. N., & Goldman-Rakic, P. S. (2001). The sensory nature of mnemonic representation in the primate prefrontal cortex. *Nature Neuroscience, 4*, 311–316.

Coolidge, F. L., Wynn, T., & Overmann, R. S. (2012). The evolution of working memory. In T. J. P Alloway, & F. G. Alloway (Eds.), *Working memory: The connected intelligence* (pp. 37–40). Psychology Press.

Cooper, J., Moodley, M., & Reynell, J. (1978). *Helping language development: A developmental programme for children with early learning handicaps*. Edward Arnold.

Coqueugniot, H., Hublin, J. J., Veillon, F., Houët, F., & Jacob, T. (2004). Arly brain growth in *Homo erectus* and implications for cognitive ability. *Nature, 431*(7006), 299–302.

Courage, M. L., & Howe, M. L. (2004). Advances in early memory development research: Insights about the dark side of the moon. *Developmental Review, 24*, 6–32. https://doi.org/10.1016/j.dr.2003.09.005

Courage, M. L., Reynolds, G. D., & Richards, J. E. (2006). Infants' attention to patterned stimuli: Developmental change from 3 to 12 months of age. *Child Development, 77*, 680–695. https://doi.org/10.1111/j.1467-8624.2006.00897.x

Courtney, S, M., Ungerleider, L. G., Keil, K., & Haxby, J. V. (1997). Transient and sustained activity in a distributed neural system for human working memory. *Nature, 386*, 608–11.

Cowan, N. (1995). *Attention and memory: An integrated framework*. Oxford University Press.

Cowan, N. (2001). The magical number 4 in short-term memory: A reconsideration of mental storage capacity. *Behavioral and Brain Sciences, 24*, 87–185.

Cowan, N. (2008). What are the differences between long-term, short-term, and working memory? *Progress in Brain Research, 169*, 323–338.

Cowan, N. (2010). The magical mystery four: How is working memory capacity limited, and why? *Current Directions in Psychological Science, 19*(1), 51–57.

Cowan, N., & Morey, C. C. (2006). Visual working memory depends on attentional filtering. *Trends Cognitive Science, 10*, 139–141. https://doi.org/10.1016/j.tics.2006.02.001

Crone E. A., Wendelken C., Donohue S., van Leijenhorst L., & Bunge S. A. (2006). Neurocognitive development of the ability to manipulate information in working memory. *Proceedings of the National. Academy of Science, U S A, 103*, 9315–9320. https://doi.org/10.1073/pnas.0510088103.

Cuevas, K., & Bell, M. A. (2010). Developmental progression of looking and reaching performance on the A-not-B task. *Developmemtal Psychology, 46*, 1363–1371. https://doi.org/10.1037/a0020185

Cuevas, K., & Bell, M. A. (2011). EEG and ECG from 5 to 10 months of age: Developmental changes in baseline activation and cognitive processing during a working memory task. *International Journal of Psychophysiology, 80*, 119–128. https://doi.org/10.1016/j.ijpsycho.2011.02.009

Curtis, G. (2006). *The cave painters: Probing the mysteries of the world's first artists*. Anchor Books.

D'Errico, F., & Nowell, A. (2000). A new look at Berekhat Ram figurine: Implications for the origin of symbolism. *Cambridge Archaeological Journal, 10*(1), 123–167.

D' Esposito, M., Postle, B. R., & Rypma, B. (2000). Prefrontal cortical contributions to working memory: Evidence from event-related fMRI studies. *Experimental Brain Research, 133*, 3–11.

D'Esposito, M. D., & Postle, B. R. (2015). The cognitive neuroscience of working memory. *Annual Review of Psychology, 6*, 115–142.

Damasio, A. (2010). *Self comes to mind: Constructing the conscious brain*. Pantheon Books.

Damasio, H., Tranel, D., Grabowski, T., Adolphs, R., & Damasio, A. (2004). Neural systems behind word and concept retrieval. *Cognition, 92*, 179–229.

Darki, F., & Klingberg, T. (2015). The role of fronto-parietal and fronto-striatal networks in the development of working memory: A longitudinal study. *Cerebral Cortex, 25*(6), 1587–1595.

De Vignemont, F. (2011). Embodiment, ownership and disownership. *Consciousness and Cognition, 20*(1), 82–93.

DeFelipe, J. (2005). Reflections of the structure of the cortical minicolumn. In M. Casanova (Ed.), *Neocortical modularity and the cell minicolumn* (pp. 57–92). Nova Biomedical.

DeLoache, J. S. (2004). Becoming symbol-minded. *Trends in Cognitive Science, 8*(2), 66–70.
DeLoache, J. S., & Burns, N. M. (1994). Early understanding of the representational function of pictures. *Cognition, 52*, 83–110.
Dennett, D. C. (1971). Intentional systems. *The Journal of Philosophy, 68*(4), 87–106.
Di Lollo, V. (1980). Temporal integration in visual memory. *Journal of Experimental Psychology. General, 109*, 75–97.
Diamond, A. (1990). Developmental time course in human infants and infant monkeys, and the neural bases of, inhibitory control in reaching. *Annual New York Academy of Science, 608*, 637–669.
Dias-Ferreira, E., Sousa, J., Melo, C., Morgado, P., Mesquita, A. R., Cerqueira, J. J., et al. (2009). Chronic stress causes frontostriatal reorganization and affects decision-making. *Science, 325*, 621–625.
Dijkstra, N., Bosch, S., & van Gerven, M. (2019). Shared neural mechanisms of visual perception and imagery. *Trends in Cognitive Sciences, 23*, 423–434.
Donahue, C. J., Glasser, M. F., Preuss, T. M., Rilling, J. K., & Van Essen, D. C. (2018). Quantitative assessment of prefrontal cortex in humans relative to nonhuman primates. *Proceedings of the National Academy of Sciences, USA, 115*, E5183–5192.
Dretske, F. (2000). *Perception, knowledge, and belief: Selected essays*. Cambridge University Press.
Duncan, J. (2001). An adaptive coding model of neural function in prefrontal cortex. *Nature Review Neuroscience, 2*, 820–829.
Edwards, S. W. (1978). Nonutilitarian activities in the lower Palaeolithic: A look at the two kinds of evidence. *Current Anthropology, 19*(1), 135–137.
Elston, G. N., Benavi-Despiccione, R., & DeFelipe, J. (2001). The pyramidal cell in cognition: A comparative study in human and monkey. *Journal of Neuroscience, 21*(RC163), 1–5.
Elston, G. N., Benavides-Piccione, R., Elston, A., Zietsch, B., Defelipe, J., Manger, P., et al. (2006). Specializations of the granular prefrontal cortex of primates: Implications for cognitive processing. *Anatomical Record Part a: Discoveries in Molecular Cellular and Evolutionary Biology, 288A*, 26–35.
Emrich, S. M., Riggall, A. C., Larocque, J. J., & Postle, B. R. (2013). Distributed patterns of activity in sensory cortex reflect the precision of multiple items maintained in visual short-term memory. *Journal of Neuroscience, 33*, 6516–6523.
Fadiga, L., Craighero, L., & D'Ausillo, A. (2009). Broca's area in language, action, and music. *Annals of the New York Academy of Sciences, 1169*(1), 448–458.
Fagot, J., Thompson, K. R., & Parron, C. (2010). How to read a picture: Lessons from nonhuman primates. *PNAS, 197*(2), 519–520.
Fraser, A., & Scerif, G. (2024). Attention in infancy and childhood. In K. C. Kadosh (Ed.), *The Oxford handbook of developmental cognitive neuroscience* (pp. 431–454). Oxford University Press.
Freedman, D. J., Riesenhuber, M., Poggio, T., & Miller, E. K. (2003). A comparison of primate prefrontal and inferior temporal cortices during visual categorization. *Journal of Neuroscience, 23*, 5235–5246.
Funahashi, S., Bruce, C. J., & Goldman-Rakic, P. S. (1989). Mnemonic coding of visual space in the monkey's dorsolateral prefrontal cortex. *Journal of Neurophysiology, 61*, 331–349.
Fuster, J. M., & Alexander, G. E. (1971). Neuron activity related to short-term memory. *Science, 13, 173*(3997), 652–4. https://doi.org/10.1126/science.173.3997.652. PMID: 4998337.
Fuster, J. M. (2000). Executive frontal functions. *Experimental Brain Research, 133*, 66–70.
Fuster, J. M. (2000). Prefrontal neurons in networks of executive memory. *Brain Research Bulletin, 52*, 331–336.
Fuster, J. M. (2002). Frontal lobe and cognitive development. *Journal of Neurocytology, 31*, 373–385.
Fuster, J. M. (2008). *The prefrontal cortex*. Elsevier.
Fuster, J. M., Bauer, R. H., & Jervey, J. P. (1985). Functional interactions between inferotemporal and prefrontal cortex in a cognitive task. *Brain Research, 330*, 299–307.

Garon, N., Bryson, S. E., & Smith, I. M. (2008). Executive function in pre-schoolers: A review using an integrative framework. *Psychological Bulletin, 134*, 31–60. https://doi.org/10.1037/0033-2909.134.1.31

Garvin, H. M, Elliott, M. C., Lucas K., Delezene, L. K., Hawks, J., Steven E. Churchill, S. E., Berger, L. R., & Holliday, T. W. (2017). Body size, brain size, and sexual dimorphism in Homo naledi from the Dinaledi chamber. *Journal of Human Evolution, 111*, 119–138.

Gathercole, S. E., Pickering, S. J., Ambridge, B., & Wearing, H. (2004). The structure of working memory from 4-15 years of age. *Developmental Psychology, 40*(2), 177–190.

Gazzaley, A., Rissman, J., & D'Esposito, M. (2004). Functional connectivity during working memory maintenance. *Cognitive and Affective Behavioral Neuroscience, 4*, 580–599.

Giardino, V., & Greenberg, G. (2015). Introduction: Varieties of iconicity. *Review of Philosophy and Psychology, 6*, 1–25.

Glasser, M. F. (2014), Trends and properties of human cerebral cortex: Correlations with cortical myelin content. *Neuroimage, 93*(2), 165–175.

Goldman-Rakic, P. S. (1987). Circuitry of the primate prefrontal cortex and the regulation of behavior by representational memory. In F. Plum (Ed.), *Handbook of physiology, the nervous system, higher functions of the brain* (pp. 373–417). American Physiological Society.

Gómez-Robles, A. (2019). Dental evolutionary rates and its implications for the Neanderthal–modern human divergence. *Science Advances, 5*(5), eaaw1268.

Gustafsson, L. (2004). Comment on Disruption in the inhibitory architecture of the cell minicolumn: Implications for autism. *The Neuroscientist, 10*, 189–191.

Hare, B., Call, J., Agnetta, B., & Tomasello, M. (2000). Chimpanzees know what conspecifics do and do not see. *Animal Behavior, 59*, 771–785.

Harmand, S., Lewis, J. E., Feibel, C. S., Lepre, C. J., Prat, S., Lenoble, A., Boës, X., Quinn, R. L., Brenet, M., Arroyo, A., & Taylor, N. (2012). 3.3-million-year-old stone stools from Lomekwi 3, West Turkana, Kenya. *Nature, 521*(7552), 310–315.

Harrison, S. A., & Tong, F. (2009). Decoding reveals the contents of visual working memory in early visual areas. *Nature, 45*, 632–635.

Haxby, J. V., Gobbini, M. I., Furey, M. L., Ishai, A., Schouten, J. L., & Pietrini, P. (2001). Distributed and overlapping representations of faces and objects in ventral temporal cortex. *Science, 293*, 2425–2430.

Herrmann, E., & Tomasello, M. (2015). Focusing and shifting attention in human children (Homo sapiens) and chimpanzees (Pan troglodytes). *Journal of Comparative Psychology, 129*, 268–274. https://doi.org/10.1037/a0039384

Hill, J., Inder, T., Neil, J., Dierker, D., Harwell, J., & Van Essen, D. (2010). Similar patterns of cortical expansion during human development and evolution. *Proceedings of the National Academy of Sciences of the United States of America, 107*, 13135–13140.

Hirata, S. (2009). Chimpanzee social intelligence: Selfishness, altruism, and the mother–infant bond. *Primates, 50*, 3–11.

Hiscock, P. (2014). Learning in lithic landscapes: a reconsideration of the hominid "toolmaking" niche. *Biological Theory, 9*, 27–41.

Hommel, B. (2004). Event files: feature binding in and across perception and action. *Trends in Cognitive Science, 8*(11), 494–500.

Hommel, B., Müsseler, J., Aschersleben, G., & Prinz, W. (2001). The Theory of Event Coding (TEC): A framework for perception and action planning. *Behavioral and Brain Sciences, 24*(5), 849–878.

Horowitz, A. (2009). Disambiguating the "guilty look": Salient prompts to a familiar dog behaviour. *Behavioural Processes, 81*(3), 447–452.

Humphreys, G. W., & Riddoch, M. J. (2007). How to define an object: Evidence from the effects of action on perception and attention. *Mind and Language, 2*(5), 534–547.

Irwin, D. E., & Yeomans, J. M. (1986). Sensory registration and informational persistence. *Journal of Experimental Psychology. Human Perception and Performance, 12*, 343–360.

Itakura, S. (1990). Recognition of line-drawing representations by a chimpanzee. *Journal of General Psychology, 121*, 190–197.

Johson, M. H. (2001). Functional brain developemnt in humans. *Nature Reviews Neuroscience, 2*, 475–483.

Káldy, Z., & Leslie, A. M. (2005). A memory span of one? Object identification in 6.5-month-old infants. *Cognition, 97*, 153–177. https://doi.org/10.1016/j.cognition.2004.09.009

Kaminski, J., Call, J., & Tomasello, M. (2008). Chimpanzees know what others know, but not what they believe. *Cognition, 109*, 224–234.

Karg, K., Schmelz, M., Call, J., & Tomasello, M. (2015). The goggles experiment: Can chimpanzees use self-experience to infer what a competitor can see? *Animal Behavior, 105*, 211–221.

Kelso, J. A. S. (2007). *Dynamic patterns: The self-organization of brain and behavior*. Cambridge, M.A: MIT Press.

Klatzky, R. L., Wu, B., & Stetten, G. (2010). The disembodied eye: Consequences of displacing perception from action. *Vision Research, 50*(24), 2618–2626.

Klingberg, T., Forssberg, H., & Westerberg, H. (2002). Increased brain activity in frontal and parietal cortex underlies the development of visuospatial working memory capacity in childhood. *Journal of Cognitive Neuroscience, 14*, 1–10. https://doi.org/10.1162/089892902317205276

Knickmeyer, R. C., Gouttard, S., Kang, C. Y., Evans, D., Wilber, K., Smith, J. K., et al. (2008). A structural MRI study of human brain development from birth to 2 years. *Journal of Neuroscience, 28*, 12176–12182.

Knight, C. (2009). Introduction: Perspectives on the evolution of language in Africa. In R. Botha & C. Knight (Eds.), *The cradle of language* (pp. 1–15). Oxford University Press.

Kohn, M., & Mithen, S. (1999). Handaxes: Products of sexual selection? *Antiquity, 73*(281), 518–526.

Kolk, S. M., & Rakic, P. (2021). Development of prefrontal cortex. *Neuropsychopharmacology, 47*, 41–57.

Kripke, S. A. (1980). *Naming and necessity*. Harvard University Press.

Krupenye, C., Kano, F., Hirata, S., Call, J., & Tomasello, M. (2016). Great apes anticipate that other individuals will act according to false beliefs. *Science, 354*, 110–114.

Kubota, K., & Niki, H. (1971). Prefrontal cortical unit activity and delayed alternation performance in monkeys. *Journal of Neurophysiology, 34*, 337–347.

Kübler, S., Owenga, P., Reynolds, S. C., Rucina, S. M., & King, G. C. P. (2015). Animal movements in the Kenya Rift and evidence for the earliest ambush hunting by hominins. *Scientific Reports, 5*, Article Number 14011.

Lage, C. A., Wolmarans, D. W., & Mograbi, D. C. (2022). An evolutionary view of self-consciousness. *Behavioural Processes, 194*, Article 104543.

Laland, K. N., Uller, T., Peldman, L. W., Sterelny, K., Muller, G. B., Moczek, A., Jablonka, R., & Olding-Smee, J. (2015). The extended evolutionary synthesis: Its structure, assumptions and predictions. *Proceedings of the Royal Society B: Biological Sciences, 282*(1913), 20151019.

Le Roux, A., Snyder-Mackler, N., Roberts, E. K., Beehner, J. C., & Bergman, T. J. (2013). Evidence for tactical concealment in a wild primate. *Nature Communications, 4*, 1462–1466.

Lee, S. H., Kravitz, D. J., & Baker, C. I. (2013). Goal-dependent dissociation of visual and prefrontal cortices during working memory. *Nature Neuroscience, 16*, 997–999.

Lee, S. H., Simpson, G. V., Logothetis, N. K., & Rainer, G. (2005). Phase locking of single neuron activity to theta oscillations during working memory in monkey extrastriate visual cortex. *Neuron, 45*, 147–156.

Lee, T. G., & D'Esposito, M. (2012). The dynamic nature of top-down signals originating from prefrontal cortex: A combined fMRI-TMS study. *Journal Neuroscience, 32*, 15458–15466.

Lei, Y. (2023). Sociality and self-awareness in animals. *Frontiers Psychology, 1*, 106538.

Lewis, A., Berntsen, D., & Call, J. (2019). Long term memory of past events in great apes. *Current Directions Psychological Science, 28*(2), 117–123.

Lewis, D. A., Melchitzky, D. S., & Burgos, G. G. (2002). Specificity in the functional architecture of primate prefrontal cortex. *Journal of Neurocytology, 31*, 265–276.

Lewis-Peacock, J. A., & Postle, B. R. (2008). Temporary activation of long-term memory supports working memory. *Journal of Neuroscience, 28*, 8765–8771.

Lewis-Williams, J. D. (2004). *The mind in the cave*. Thames & Hudson.

Liben, L. B. (2003). Beyond point and shoot: children's developing understanding of photographs as spatial and expressive representations. In R. Keil (Ed.), *Advances in child development and behavior* (Vol. 31, pp. 1–42). Academic Press

Lombard, M. (2024). *Paying attention: The neurocognition of archery, middle Stone Age bow hunting, and the shaping of the sapient mind*. Advance Online Publication. https://doi.org/10.1007/s11097-024-09980-z

Magnussen, S. (2000). Low-level memory processes in vision. *Trends in Neuroscience, 23*, 247–251.

Magnussen, S., & Greenlee, M. W. (1999). The psychophysics of perceptual memory. *Psychological Research Psychologische Forschung, 62*, 81–92.

Magnussen, S., Greenlee, M. W., Aslaksen, P. M., & Kilbedo, O. (2003). High fidelity long-term memory revisited—and confirmed. *Psychological Science, 14*(1), 74–76.

Marquet, J.-C., Freiesleben, T. H., Thomsen K. J., Calligaro, M., Macaire, J.-J. (2023). The earliest unambiguous Neanderthal engravings on cave walls: La Roche-Cotard, Loire Valley, France. *PLoS One, 18*(6), e 0286568.

Martin-Ordas, G., Haun, D., Colmenares, F., & Call, J. (2010). Keeping track of time: Evidence for episodic-like memory in great apes. *Animal Cognition, 13*, 331–340.

Melis, A. P., Warneken, F., Jensen, K., Schneider, A., Call, J., & Tomasello, M. (2011). Chimpanzees help conspecifics to obtain food and non-food items. *Proceedings Royal Society, B: Biological Science, 278*, 1405–1413.

Melis, A. P., Hare, B., & Tomasello, M. (2006). Engineering cooperation in chimpanzees: Tolerance constraints on cooperation. *Animal Behavior, 72*, 275–286.

Mellars, P. (1996). *The Neanderthal legacy: An archaeological perspective from Western Europe*. Princeton University Press.

Mendes, N., Steinbeis, N., Bueno-Guerra, N., Call, J., & Singer, T. (2018). Preschool children and chimpanzees incur costs to watch punishment of antisocial others. *Nature Human Behavior, 2*, 45–51.

Meyers, E. M., Freedman, D. J., Kreiman, G., Miller, E. K., & Poggio, T. (2008). Dynamic population coding of category information in inferior temporal and prefrontal cortex. *Journal of Neurophysiology, 100*, 1407–1419.

Miller, B. T., Vytlacil, J., Fegen, D., Pradhan, S., & D'Esposito, M. (2011). The prefrontal cortex modulates category selectivity in human extrastriate cortex. *Journal of Cognitive Neuroscience, 23*, 1–10.

Miller, E. K., & Cohen, J. D. (2001). An integrative theory of prefrontal cortex function. *Annual Review of Neuroscience, 24*, 167–202.

Mithen, S. (1996). *The prehistory of the mind: A search for the origins of art, religion and science*. Thames and Hudson Ltd.

Moench, K. M., Maroun, M., Kavushansky, A., & Wellman, C. (2016). Alterations in neuronal morphology in infralimbic cortex predict resistance to fear extinction following acute stress. *Neurobiological Stress, 3*, 23–33.

Moncel. M. H., Chiotti, L., Gaillard, C., Onoratini, G., & D. Pleurdeau, D. (2012). Non-utilitarian lithic objects from the European Paleolithic. *Archaeology, Ethnology and Anthropology of Eurasia 40*(1), 24–40.

Morriss, G. M. (2010). The evolution of human artistic creativity. *Journal of Anatomy, 216*(2), 158–176.

Morrison, R., & Reiss, D. (2018). Precocious development of self-awareness in dolphins. *PLoS One, 13*(1). e0189813. https://doi.org/10.1371/journal.pone.0189813

Mulcahy, N. J., & Call, J. (2006). Apes save tools for future use. *Science, 312*, 1038–1040.

Müsseler, J., & Hommel, B. (1997). Detecting and identifying response-compatible stimuli. *Psychonomic Bulletin & Review, 4*, 125–129.

Neiworth, J. J., Thall, M. E., Liu, S., Leon-Mofy, E., Rankin, M., LoRussos, M. A., Thandi, S., & Garay-Hernandez, J. (2023). A recognition test in monkeys to differentiate recollection from familiarity memory. *Scientific Reports, 13*, 17579. https://doi.org/10.1038/s41598-023-44804-1

Nelissen, N., Stokes., M, Nobre, A. C., & Rushworth, M. F. (2013). Frontal and parietal cortical interactions with distributed visual representations during selective attention and action selection. *Journal of Neuroscience, 33*, 16443–16458.

Neubert, F. X., Mars, R. B., Thomas, A. G., Sallet, J., & Rushworth, M. F. (2014). Comparison of human ventral frontal cortex areas for cognitive control and language with areas in monkey frontal cortex. *Neuron, 81*, 700–713.

Oberauer, K. (2001). Removing irrelevant information from working memory: A cognitive aging study with the modified Stenberg task. *Journal of Experimental Psychology: Learning, Memory, and Cognition, 27*, 948–957.

Oberauer, K. (2009). Design for a working memory. *Psychology, Learning, and Motivation, 51*, 45–100

Olesen, P. J., Nagy, Z., Westerberg, H., & Klingberg, T. (2003). Combined analysis of DTI and fMRI data reveals a joint maturation of white and grey matter in a fronto-parietal network. *Cognitive Brain Research, 18*, 48–57.

Överduin-de Vries, A. M., Spruijt, B. M., de Vries, H., & Sterck, E. H. M. (2015). Tactical deception to hide sexual behaviour: Macaques use distance, not visibility. *Behavioral Ecology and Sociobiology, 69*(8), 1333–1342.

Parron, C., Call, J., & Fagot, J. (2008). Behavioural responses to photographs by pictorially naïve baboons (Papio anumbis), gorillas, (Gorilla gorilla) and chimpanzees (Pan troglodytes). *Behavioral Processes, 28*, 351–357.

Passingham, R. E. (2021). *Understanding the prefrontal cortex: Selective advantage, connectivity, and neural operations*. Oxford Psychology Series (Oxford, 2021; online edn, Oxford Academic, 19 Aug. 2021). https://doi.org/10.1093/oso/9780198844570.001.0001

Passingham, R. E., & Smaers, J. B. (2014). Is the prefrontal cortex especially enlarged in the human brain allometric relations and remapping factors. *Brain Behavioral Evolution, 84*, 156–166.

Peirce, C. S. (1977). *Semiotics and significs*. In C. Hardwick (Ed.), Indiana University Press.

Pelphrey, K. A., Reznick, J. S., Davis, G. B., Sasson, N., Morrow, J., Donahoe, A., et al. (2004). Development of visuospatial short-term memory in the second half of the 1st year. *Developmental Psychology, 40*, 836–851. https://doi.org/10.1037/0012-1649.40.5.836

Perdue, B. M., Evans, T. A., & Beran, M. J. (2018). Chimpanzees show some evidence of selectively acquiring information by using tools, making inferences, and evaluating possible outcomes. *PLoS ONE, 13*(4), 1–20. https://doi.org/10.1371/journal.pone.0193229

Perlman, M., Clark, N. B., & Tanner, J. E. (2014). *Iconicity and Ape Gesture*. https://doi.org/10.1142/9789814603638_0029

Perret, D. I., & Mistlin, A. J. (1990). Perception of facial characteristics by monkeys. In W. C. Stebbins & M. A. Berkeley (Eds.), *Comparative perception* (Vol. 2, pp. 187–215). Wiley.

Petanjek, Z., Judas, M., Kostovic, I., & Uylings, H. B. M. (2008). Lifespan alterations of basal dendritic trees of pyramidal neurons in the human prefrontal cortex: A layer-specific pattern. *Cerebral Cortex, 18*, 915–929.

Petanjek, Z., Judas, M., Simic, G., Rasin, M. R., Uylings, H. B., Rakic, P., et al. (2011). Extraordinary neoteny of synaptic spines in the human prefrontal cortex. *Proceedings of the National Academy of Science USA, 108*, 13281–13286.

Petrides, M., & Pandya, D. N. (2002). Comparative cytoarchitectonic analysis of the human and macaque ventrolateral prefrontal cortex and corticocortical connection pattern in the monkey. *European Journal of Neuroscience, 16*, 291–310.

Pickering, T. R. (2013). *Rough and tumble: Aggression, hunting, society, and politics*. Oxford University Press.

Pickering, T. R., & Bunn, H. (2012). Meat foraging by Pleistocene African hominins: Tracking behavioral evolution beyond baseline inferences of early access to carcasses. In M. Dominguez-Rodrigo (Ed.), *Stone tools and fossil bones* (pp. 152–173). Cambridge University Press.

Planer, R. J. (2017). Talking about tools: Did early Pleistocene Hominins have a protolanguage? *Biological Theory, 12*, 211–221.

Planer, R. J., & Sterelny, K. (2021). *From signal to symbol: The evolution of language*. MIT Press.

Plavcan, J. M. (2012). Body size, size variation, and sexual size Dimorphism in Early Homo. *Current Anthropology, 53*(S6), 309–423.

Pokorny, J. J., & de Waal, F. B. M. (2009). Monkeys recognize the faces of group mates in photographs. *Proceedings of the National Academy of Sciences, USA, 106*, 21539–21543.

Pope, S. M., Fagot, J., Meguerditchian, A., Watzek, J., Lew-Levy, S., Autrey, M. M., & Hopkins, W. D. (2020). Optional-switch cognitive flexibility in primates: Chimpanzees' (Pan troglodytes) intermediate susceptibility to cognitive set. *Journal of Comparative Psychology, 134*, 98–109. https://doi.org/10.1037/com0000194

Posner, M. I., & Peterson, S. (1990). The attention system of the human brain. *Annual Review of Neuroscience, 13*, 25–42. https://doi.org/10.1146/annurev.neuro.13.1.25

Postle, B. R. (2015). The cognitive neuroscience of visual short-term memory. *Current Opinions in Behavioral Science, 1*, 40–46.

Preuss, T. M. M, & Wise, S. P. (2022). Evolution of prefrontal cortex. *Neuropsychopharmacology, 47*, 3–19.

Putnam, H. (1983). Reference and truth. In *Realism and reason: Philosophical papers, volume* 3. Cambridge University Press.

Raftopoulos, A. (2019). *Cognitive penetrability and the epistemic role of perception*. Palgrave, McMillan.

Raftopoulos, A. (2008). Perceptual systems and realism. *Synthese, 164*(1), 61–91.

Raftopoulos, A. (2009). *Perception and cognition: How do psychology and neural sciences inform philosophy*. MIT University Press, A Bradford Book.

Raftopoulos, A. (2015). Cognitive penetrability and consciousness. In J. S. Zeimbekis & A. Raftopoulos (Eds.), *Cognitive effects on perception: new philosophical perspectives* (pp. 268–298). Oxford University Press.

Read, D. W., & Andersson, C. (2019). Cultural complexity and complexity evolution. *Adaptive Behavior, 28*(5), 1–30.

Read, D. W., Monrique, H. M., & Walker, M. J. (2022). On the working memory of humans and great apes; strikingly similar or remarkably different? *Neuroscience & Biobehavioral Reviews, 134*(1–29), Article 104496.

Read, D. W., & van der Leeuw, S. (2008). Biology is only part of the story. *Philosophical Transactions of the Royal Society Series B, 363*, 1959–1968.

Read, D., & van der Leeuw, S. (2015). The extension of social relations in time and space during the Palaeolithic and beyond. In F. Wenban-Smith, F. Coward, R. Hosfield, & M. Pope (Eds.), *Settlement, society, and cognition in human evolution* (pp. 31–53). Cambridge University Press.

Reindl, E., Völter, C. J., Civelek, Z., Duncan, L., Lugosi, Z., Felsche, E., Herrmann, E., Call, J., & Seed, A. M. (2023). The shifting shelf task: A new, non-verbal measure for attentional set shifting. *Proceedings of the Royal Society, B, 290*, 20221496. https://doi.org/10.1098/rspb.2022.1496

Remple, M. S., Reed, J. L., Stepniewska, I., Lyon, D. C., & Kaas, J. H. (2007). The organization of frontoparietal cortex in the tree shrew (Tupaia belangeri): II. Connectional evidence for a frontal-posterior parietal network. *Journal of Comparative Neurology, 501*, 121–149.

Reynolds, G. D., & Romano, A. C. (2016). The development of attention systems and working memory in infancy. *Frontiers in Systems Neuroscience, 10*(15). https://doi.org/10.3389/fnsys.2016.00015

Reynolds, G. D., Courage, M. L., & Richards, J. E. (2010). Infant attention and visual preferences: Converging evidence from behavior, event-related potentials and cortical source localization. *Developmental Psychology, 46*, 886–904. https://doi.org/10.1037/a0019670

Reynolds, G. D., Courage, M. L., & Richards, J. E. (2013). The development of attention. In D. Reisberg (Ed.), *Oxford handbook of cognitive psychology* (pp. 1000–1013). Oxford University Press.

Rice, S. H. (2002). The role of heterochrony in primate brain evolution. In N. Minugh-Purvis & K. J. MacNamara (Eds.), *Human evolution through developmental change* (pp. 154–172). The Johns Hopkins University Press.

Rigal, G. (2016). *Le Temps Sacré des caverns*. Editions Corti.

Riggall, A. C., & Postle, B. R. (2012). The relationship between working memory storage and elevated activity as measured with functional magnetic resonance imaging. *Journal of Neuroscience, 32*, 12990–12998.

Rizal, Y., Westaway, K. E., Zaim, Y., et al. (2020). Last appearance of *Homo erectus* at Ngandong, Java, 117,000–108,000 years ago. *Nature, 577*, 381–385. https://doi.org/10.1038/s41586-019-1863-2

Rochat, P. (2003). Five levels of self-awareness as they unfold early in life. *Consciousness and Cognition, 12*, 717–731.

Rolls, E. T. (2008). *Memory, attention, and decision-making*. Oxford University Press.

Rosati, A. G. (2017). The evolution of primate executive function: From response control to strategic decision-making. In J. Kaas, & L. Krubitzer (Eds.) *Evolution of nervous systems* (Vol. 3, pp. 423–437, 2nd edn). Elsevier.

Sakai, T., Mikami, A., Tomonaga, M., Matsui, M., Suzuki, J., Hamada, Y., Tanaka, M., Miyabe-Nishiwaki, T., Makishima, H., Nakatsukasa, M., & Matsuzawa, T. (2012). Differential prefrontal white matter development in chimpanzees and humans. *Current Biology, 21*(16), 1397–1402.

Sallet, J., Mars, R. B., Noonan, M. P., Neubert, F. X., Jbabdi, S., O'reilly, J. X., et al. (2013). The organization of dorsal frontal cortex in humans and macaques. *Journal of Neuroscience, 33*, 12255–12274.

Savage-Rumbaugh, E. S., Rumbaugh, Smith, S. T., & Lawson, J. (1980). Reference: The linguistic essential. *Science, 210*, 922–925.

Schenker, N. M., Buxhoeveden, D. P., Blackmon, W. L., Amunts, K., Zilles, K., & Semendeferi, K. (2008). A comparative quantitative analysis of cytoarchitecture and minicolumnar organization in Broca's area in humans and great apes. *The Journal of Comparative Neurology, 510*, 117–128.

Schenker, N. M., Hopkins, W. D., Spocter, M. A., Garrison, A. R., Stimpson, C. D., Erwin, J. M., et al. (2010). Broca's area homologue in chimpanzees (Pantroglodytes): Probabilistic mapping, asymmetry, and comparison to humans. *Cerebral Cortex, 20*, 730–742.

Schmant-Besserat, D. (2014). The evolution of writing. In J. Wright (Ed.), *International Encyclopedia of social and behavioral sciences*. Elsevier. Downloaded from https://sites.utexas.edu/dsb/files/2014/01/evolution_writing.pdf

Schiller, P. H. (1985). A model for the generation of visually guided saccadic eye movements. In D. Rose & V. G. Dobson, *Models of the Visual Cortex* (pp. 62–70). New York, NY: Wiley.

Schoenemann, P. T. (2022). Evidence of grammatical knowledge in apes: an analysis of Kanzi's performance on reversible sentences. *Frontiers in Psychology, 13*. https://doi.org/10.3389/fpsyg.2022.885605

Schoenemann, P. T., Sheenan, M. J., & Glotzer, L. D. (2005). Prefrontal white matter volume is disproportionately larger in humans than in other primates. *Nature Neuroscience, 8*, 242–252.

Schurgin, M. W., Cunningham, C. A., Egeth, H. E., & Brady, T. F. (2018). Visual long-term memory can replace active maintenance in visual working memory. https://doi.org/10.1101/381848

Semendeferi, K. & Damasio, H. (2000). The brain and its main anatomical subdivisions in living hominoids using magnetic resonance imaging.

Semendeferi, K., Armstrong, E., Schleicher, A., Zilles, K., & Van Hoesen, G. W. (1998). Limbic frontal cortex in hominoids: A comparative study of area 13. *American Journal of Physical Anthropology, 106*, 129–155.

Semendeferi, K., Armstrong, E., Schleicher, A., Zilles, K., & Van Hoesen, G. W. (2001). Prefrontal cortex in humans and apes: A comparative study of area 10. *American Journal of Physiological Anthropology, 114*, 224–241.

Semendeferi, K., Lu, A., Schenker, N., & Damasio, H. (2002). Humans and great apes share a large frontal cortex. *Nature Neuroscience, 5*, 272–276.

Serences, J. T'., Ester, E. F., Vogel, E. K., & Awh, E. (2009). Stimulus-specific delay activity in human primary visual cortex. *Psychological Science, 20*, 207–214.

Seyfarth, R., & Cheney, D. (2018). Pragmatic flexibility in primate vocal production. *Current Opinion in Behavioral Sciences, 21*, 56–61.

Seyfarth, R., & Cheney, D. (2018). *The social origins of language*. Princeton University Press.

Shansky R. M., Hamo, C., Hof, P. R., McEwen, B. S., & Morrison, J., H. (2009). Stress-induced dendritic remodelling in the prefrontal cortex is circuit specific. *Cerebral Cortex, 19,* 2479–84.

Sligte, I., G., Wokke, M. E., Tesselaar, J. P., Scholte, H. S., & Lamme, V. F. A. (2011). Magnetic stimulation of the dorsolateral prefrontal cortex dissociates fragile visual short-term memory from visual working memory. *Neuropsychologia, 49*, 1578–1588.

Smaers, J. B., Gómez-Robles, A., Parks, A. N., & Sherwood, C. C. (2017). Exceptional evolutionary expansion of prefrontal cortex in great apes and humans. *Current Biology, 27*, 714–720.

Somel, M., Liu, X., Tang, L., Yan, Z., Hu, H., et al. (2011). MicroRNA-Driven Developmental Remodeling in the Brain Distinguishes Humans from Other Primates. *PLOS Biology, 9*(12): e1001214.

Sowell, E. R., Thompson, P. M., Holmes, C. J., Batth, R., Jernigan, T. L., & Toga, A. W. (1999). Localizing age-related changes in brain structure between childhood and adolescence using statistical parametric mapping. *NeuroImage, 9*, 587–597.

Sowell, E. R., Trauner, D. A., Gamst, A., & Jernigan, T. L. (2002). Development of cortical and subcortical brain structures in childhood and adolescence: a structural MRI study. *Developmental Medical Child Neurology, 44*(1), 4–16.

Spear, L. P. (2000). The adolescent brain and age-related behavioral manifestations. *Neuroscience Biobehavoral Review, 24*, 417–463.

Spoor, F., Leakey, M. G., Gathogo, P. N., Brown, F. H., Antón, S. C., McDougall, I., et al. (2007). Implications of new early Homo fossils from Ileret, east of Lake Turkana, Kenya. *Nature, 448*(7154), 688–691.

Sreenivasan, K. K., Vytlacil, J., & D'Esposito, M. (2014). Distributed and dynamic storage of working memory stimulus information in extrastriate cortex. *Journal of Cognitive Neuroscience, 26*, 1141–1153.

Teffer, K., & Semendeferi, K. (2012). Human prefrontal cortex: evolution, development, and pathology. In M. A. Hofman & D. Falk (Eds.), *Progress in brain research* (Vol. 95, pp. 191–218).

Tipper, S. P. (2010). Editorial. *Quarterly Journal of Experimental Psychology, 63*, 1–2.

Tomasello, M. (2008). *Origins of human communication*. MIT Press.

Tomasello, M. (2016). The ontogeny of cultural learning. *Current Opinion in Psychology, 8*, 1–4.

Tomasello, M. (2019). *Becoming human*. Harvard University. Press.

Tomasello, M. (2023). Social cognition and metacognition in great apes: A theory. *Animal Cognition, 26*, 25–35.

Tucker, M., & Ellis, R. (1998). On the relations between seen objects and components of potential actions. *Journal of Experimental Psychology: Human Perception and Performance, 24*(3), 830846.

Vandenbroucke, A. R. E., Sligte, I. S., & Lamme, V. A. F. (2011). Manipulations of attention dissociate fragile visual short-term memory from visual working memory. *Neuropsychologia, 49*, 1559–1568.

Villmoare, B., Hatala, K. G. & Jungers, W. (2019). Sexual dimorphism in *Homo erectus* inferred from 1.5 Ma footprints near Ileret, Kenya. *Scientifc Reports, 9*, 7687. https://doi.org/10.1038/s41598-019-44060-2

Wang, M., Yang, P., Zhang, T., Li, W., Zhang, J., Jin, Z., & Li, L. (2022). Working memory biases early object discrimination and parietal activity during attentional selection. *Cortex: A Journal Devoted to the Study of the Nervous System and Behavior, 157*, 53–64.

Wang, Y., Markman, H., Goodman, P. H., Berger, T. K., Ma, J., & Golman-Rakic, P. S. (2006). Heterogeneity in the pyramidal network of medial prefrontal cortex. *Nature Neuroscience, 9*, 534–542.

Wei, Y., De Lange, S. C., Scholtens, L. H., Watanabe, K., Ardesch, D. J., Jansen, P. R., et al. (2019). Genetic mapping and evolutionary analysis of human-expanded cognitive networks. *Nature Communications, 10*, 4839.

Wobber, V., Herrmann, E., Hare, B., Wrangham, R., & Tomasello, M. (2014). Differences in the early cognitive development of children and great apes. *Developmental Psychobiology, 56*(3), 547–573.

Wolfe, J. M. (2014). Introduction to the special issue on visual working memory. *Attention, Perception Psychophysics, 76*(7), 1861–1870.

Worden, R. P. (1996). Primaye social intellegence. *Cognitive Science, 20*(4), 579–616.

Xu, F., & Carey, S. (1996). Infant's metaphysics: The case of numerical identity. *Cognitive Psychology, 30*, 11–153.

Zaksas, D., Bisley, J. W., & Pasternak, T. (2001). Motion information is spatially localized in a visual working-memory task. *Journal of Neurophysiology, 86*, 912–921.

Zeba, M., Jovanov-Milosevic, N., & Petanjek, Z. (2008). Quantitative analysis of basal dendritic tree of layer IIIc pyramidal neurons in different areas of adult human frontal cortex. *Collegium Antropologicum, 32*, 161–169.

Zhao, H., Li, J., Wang, X., Pan, R., Wang, Ch., Ren, Y., Wang, Y., & Li, B. (2020). Facial expression recognition in golden snub-nosed monkeys. *Current Zoology 66*(6), 695–697.

Zhou, X., Zhu, D., Qi, X. L., Li, S., King, S. G., Salinas, E., Stanford, T. R., & Constantinidis, C. (2016). Neural correlates of working memory development in adolescent primates. *Nature Communications, 7*, 13423.

Zysset, S., Huber, O., Samson, A., Ferstl, E. C., & von Cramon, D. Y. (2003). Functional specialization within the anterior medial prefrontal cortex: A functional magnetic resonance imaging study with human subjects. *Neuroscience Letters, 335*, 183–186.

Chapter 6
The Road to Symbols; Why Symbols?

I discuss the factors that made necessary, the emergence of symbols. I note that the limitations of iconic representational systems paved the way for the development of symbolic thought when the need for this kind of thinking arose. I argue that the main reason why symbolic thought became advantageous to hominins is the need to establish and maintain social relationships that benefit a group. I also propose two additional reasons for symbolic thought. The need for social learning, especially as it relates to teaching tool-construction techniques, and the need for the group to come up with an effective management of intragroup social tensions and stress that resulted from the increase of the size of the group, and the demand for more cooperative activities. Finally, I discuss the relation between language and the cognitive capacities underlying symbol use. This relation can be explained by those theories that posit that the organizational features of language are a response to the pre-existing cognitive capacities of the human mind and, thus, languages are adapted to our minds rather than vice versa. Alternatively, it can be explained by theories that hold that language and cognitive capacities coevolved because although language requires certain cognitive capacities, the increasing socio-cultural pressures for more sophisticated language led to the cognitive changes that could support more sophisticated language-forms.

Symbols is the most advanced form of signs. It is of paramount importance because acquiring the capacity to think with symbols paved the way to the emergence of language (Deacon, 1997; Everett, 2017; Mithen, 1996; Torrey, 2017). As Gärdenfors (1995) argues, language use entails the capacity to communicate across contexts and for things that are not spatio-temporally present. This presupposes that language users can think about things that are not present, although the latter capacity by itself is not a sufficient condition for language acquisition. The ability to think symbolically, therefore, predates linguistic abilities. As the reader recalls, spatio-temporal detachment is one of the main characteristics of symbols and explains why they exhibit

A. Raftopoulos, *The Origins of Representations: Cognitive and Brain Development*, Studies in Applied Philosophy, Epistemology and Rational Ethics 73,
https://doi.org/10.1007/978-3-031-91213-9_6

displaced reference. Furthermore, symbol-usage underlies advanced linguistic abilities because it allows syntactic structuring, that is, the construction of hierarchically organized linguistic structures. This requires, in turn, that agents can have hierarchically structured thoughts in the first place. In Sect. 7.3, we shall have the opportunity to discuss the view that hierarchically organized thoughts became possible by the internalization of the tool-making activities of H.E.

This relation between language and cognitive capacities underlying symbol use can be explained by those theories about language that posit that the organizational features of language are a response to the pre-existing cognitive capacities of the human mind and according to which, thus, languages are adapted to our minds rather than vice versa (Christiansen & Chater, 2016). These, in turn, belong to the class of theories that view critical forms of human culture as depending on the prior evolution of their cognitive enablers. Tomasello (1999, 2014, 2016), for example argues that cumulative culture evolution is connected to cultural learning, which, in turn, depends on specific and sophisticated cognitive capacities. Even though, as we shall see next, these views are debatable and language may not be strictly necessary for cumulative cultural learning, it is unquestionably an immensely important aid to cultural learning. Furthermore, critical features of hominin and human cultural life would be impossible without language.

This interrelations between language and cognition can also be explained by another set of theories that hold that both in language and in theorizing about the forms of human culture, as far as language is concerned, language and cognitive capacities coevolved because although language requires certain cognitive capacities, the increasing socio-cultural pressures for more sophisticated language, which enable thinking about the ever-increasing complexity of human societies and effective communication, led to the cognitive changes that could support more sophisticated language-forms. As Sterelny (2020, 776) remarks 'adaptation is almost certainly mutual rather than unidirectional.' As far as the more general theories on the development of social skills and culture are concerned, there is an increasing literature suggesting that that the critical forms of human culture do not depend on the prior evolution of their cognitive enablers, but, rather, on the social organization of a learning community. Thus, the driving force underlying social development is the society itself rather than the individual capacities of the persons in the society (see Sterelny (2020, 773) for a lucid discussion). We shall the opportunity to return to some of these points in the last three chapters of the book.

Why and when did our ancestors start using symbols? It would help in our discussion to revisit some of the limitations of iconic representations. One major difference between icons and symbols is that iconic representations cannot represent logical connectives and quantifiers. 'One important thing that pictures cannot do, then, is represent certain sorts of relations between ideas. They cannot represent, for example, those relations which we express using the words *if… then…* and *not*' (Crane, 2003, 19). This severely restricts the kinds of objects or events that icons can represent, which, as we shall see, is an important factor that led to the "discovery" of symbols. This restriction is most readily seen in the case of negation. '[T] here is no way of simply using a picture alone to deny what it represents. You can only negate or disjoin the content of a picture by using some non-pictorial symbol... Without these non-pictorial symbols, it makes little sense to say that the content of the picture can be something which can be asserted, denied, negated, or disjoined' (Crane, 2009, 459–460).

Sainsbury (2005, 242) also argues that negation applies to propositional content in a unique and truth-value reversing way, and that there is no such thing as the negation of a picture; 'Although ad hoc conventions can relate aspects of a picture to negative facts, there is no standard way of effecting this, no operation that can convert a complete picture into "its negation".' Similar views are held by Heck (2007) and Kulvicki (2006). Heck (2007, 126), for instance, discussing cognitive maps (but his discussion can be easily extended to pictorial representations in general) writes:

> There is no map that is the negation of my cognitive map of Boston; there is no map that is the disjunction of my map and my wife's; and so forth. If the content of a cognitive map is a structured proposition, why shouldn't there be maps with such contents?

If icons cannot represent logical connectives, in view of the fact that in order to build complex symbols one need use logical connectives to bind atomic symbols together in a structured way, binding iconic representations becomes problematic, which is why iconic representations do not display the systematicity and recombinability and, in general, the recursive structure, of symbolic representations; as we saw in Sects. 4.8 and 4.9, icons have a flat structure.

A second limitation of iconic representations stems from the fact that they do no satisfy the *Generality Constraint (GS)*. Remember that according to it, if one understands, say, Fa & Qb, one also understands Fb & Qa, because all these symbols preserve their meanings in the two different contexts and, thus, if one knows what they mean in the first sentence one also knows what they mean in the second and, thus, knows what Fb & Qa means. The GC allows semantic compositionality where the composition occurs by means of some formal rules, i.e., those of grammar. This property is also known as the systematicity of symbolic systems, that is, that the ability to entertain a given thought entails the ability to entertain thoughts that are semantically related to the given thought in the sense that the thoughts have semantically related contents. Semantic compositionality means that if you know the meanings of the constituents of a complex sentence and the meaning of the grammatical connectives used, you know the meaning of the complex sentence. For many, semantic and syntactic compositionality are what make language possible, in the first place, and learnable. GC is based on one of symbols basic properties, namely their context independence; a symbol retains its meaning across different contexts.

Compare this property of symbols with iconic representations, a theme to which I return in the next chapter. Suppose that a signaling flag, such as those used in ships for communication between ships, means "animals on board" and another flag means "customs inspectors welcome". If one wishes to express the proposition "animal inspectors welcome" one cannot combine the two flags, something that would be very easy with the corresponding sentential expressions. One has to design a new flag to express this meaning and if another person does not know the meaning of the new flag there is no combination of the first two flags that could express this meaning.

A third limitation of the initial iconic representations is that, owing to the fact that they physically resemble the objects or events they purport to represent, they are firmly attached to the observable properties of the objects in visual scenes. The reason I talk about the "initial icons" is that in the modern literature iconic representations need not resemble the physical form of their representata and, in addition, they can represent higher-order structures.) They can also present the spatial structure of scene, but they have no capacity to represent non-observable or 2nd or higher order structure. For example, they cannot represent the notion of marriage, and the social conventions associated with it, as well as other abstract concepts. I mention this example because one of the many, largely converging, arguments concerning the evolutionary need to produce symbolic representations is premised, among other things, on the social pressures to manage effectively the social relations in the constantly growing larger populations of the groups of our ancestors. I shall revisit this point in the next section when I discuss the transition from indices and icons to symbols.

A fourth limitation, closely related to the second one, stems from the fact that iconic representations are context-dependent and non-arbitrary. This means that iconic signs can have a limited use as linguistic symbols. If, as we have seen, and we shall return to this in the next chapter, communication was initially based on gestures that are predominantly iconic, the requirement for advanced forms of communication demanded by more complex social systems led to the expansion of the gestural repertory and eventually to the use of symbols that forego the representational restrictions of the iconic gestural representations. Since language, even simple languages with minimal grammatical structure (if at all), require symbols, and since, as I mentioned above, H.E had developed linguistic abilities that allowed them to do some relatively sophisticated planning in order to migrate over long distances and across oceans, to knap and quarry stones, to engage in ambush hunting, etc., one can plausibly assume that *Homo erectus* used symbols.

The foregoing discussion suggests that symbol use and linguistic abilities are closely interrelated because they are both tied to communication. I mentioned above that one of the reasons that led to the need for the ability to use symbolic representations was the need to express abstract social relations. To that one should add that symbols were necessary for effective communication required for social learning, and especially, for teaching the tool-making techniques. Instruction through communication of directives, advices, and error correction is much more effective than instruction through watching the actions of another agents as they construct tools even if it is accompanied by some sort of gestures. This is an additional selective pressure for cognitive symbolic capacities to appear.

To answer, thus, the question concerning the causes that led to the development of symbolic thought, one should look for the evolutionary pressures that led to the acquisition of this ability. Overcoming the limitations of iconic representational systems and developing representational systems that could support advanced forms of thought-structures and communicative abilities offer the more plausible explanations for the selective pressures that led to the development of symbolic thought. At the same time, symbol acquisition and use relate to advanced cognitive capacities, such as those required, for example, to support recursive structures in thought and language, and these cognitive capacities had to be selected as well so that symbolic thought could develop. In view of this, one should extent their search to include the cognitive development that allowed the gradual emergence of increasingly complicated symbolic systems.

The success in planning and executing long distance travel across oceans, the success in tool-construction and social learning, etc., shows that *Homo erectus* had the ability to think abstractly, that is in a spatio-temporal context-free way about the conditions that they could meet along their journey and at their destination. This means that H.E starting from 1.7MA had discovered symbols, at least in some initial form (Everett, 2017, 218). In addition, planning ahead presupposes that one can foresee the results of their actions and this presupposes that they had the capacity, albeit limited, to understand recurring patterns in their environment (a theme to which I shall shortly return). Moreover, this amount of planning certainly involves exchange of information and discussion of the plan and this presupposes some form of language (Everett, 2017), even if some have objected that even though the vocalizing capacity of H.E may have been considerably enhanced it remained too simple to be called a language (Mithen, 1996, 162). However, later in his book, Mithen (1996, 238–239) calls the communication among the members of H.E groups a language, when he notes that although *Homo erectus* possessed linguistic capabilities, its language lacked the vast lexicon and set of grammatical rules seen in languages today. These two features of modern languages seem to be related to the second spurt of brain enlargement between 500 and 200 Ka. It has been argued that grammar emerged 70 Ka (Bolhuis et al., 2014). In the next chapter, I address the problem of when language first appeared in Hominins relating language to consciousness and self-consciousness. All these point to the conclusion that H.E had symbolic thought capacities. This capacity in the beginning consisted in the ability to form simple symbolic structures and gradually developed to allow hierarchically structured thoughts that are certainly required for the advanced forms of planning that H.E behavior suggests that H.E were engaged in. Let us return to the question asked in the beginning, namely, why symbols?

All animals, including Hominins, use indices, and Hominins also use icons to refer to objects and events, and to communicate with the other hominis through gestures. At a certain point in their evolution Hominins start using symbols. What made such a move possible, evolutionary speaking? That is, what environmental (including social organization) factors favored the use of symbols so that symbol-users had an evolutionary advantage and transmitted their genes? We have already discussed some of the reasons why representational systems more effective that icons were necessary and these included the need to express abstract, mostly, social concepts, and the need for more effective social learning. One should add to these factors, the need to relieve social tensions within the group and, of course, the need for effective communication. To these social factors, one should add the environmental pressures due to the encounter of new habitats with varying environmental conditions and food sources. Addressing successfully these challenges requires careful planning and, thus, communication that only relatively advanced communicative abilities could support. These abilities, in turn demand extensive symbol use. I return to elaborate on how social factors influenced the development of symbolic cognition and language later on.

A second argument that managing social relations led to the development of symbols because the need was created to express abstract social relations and commitments was made by Deacon (1997). Deacon argues that the complexity in social relations that resulted from eating meat and having to scavenge or hunt to find it led to the need for some of the male members of the group to be away for some considerable time from the rest of the group. In order to secure the sexual and other social relations in the absence of these males, social commitments, including monogamy and some form of marriage, had to be made. In other words, the need for some kind of social contract that would impose rights, restrictions and punishment arose. Marriage contracts, for example, Deacon argues, establish both vertical lineal symbolic relationships and horizontal affinal symbolic relationships. Marriage is a form of regulation of sexual and, thus, reproductive relationships by symbolic means. 'It is preeminently a symbolic relationship, and owing to the lack of symbolic abilities, it is totally absent in the rest of the animal kingdom. What I am suggesting here is that a related form of regulation of reproductive relationships by symbolic means was essential for early hominins to take advantage of a hunting-provisioning subsistence strategy' (Deacon, 1997, 406). To solve these social problems, symbolization was the only viable answer because only it could support a social contract.

It is plausible that the need to establish and maintain social relationships that would benefit a group was a sufficient environmental pressure to lead to the appearance of a symbolic capacity. Symbols could serve this need because they can express abstract relationships, such as marriage or monogamy. Icons, by being firmly attached to what can be experienced perceptually due to the need to resemble physically what they represent, could not go beyond surface properties to express abstract relations. We see here a manifestation of the principle of socio-cognitive co-evolution. Since the symbolic capacity eventually allowed the emergence of proto-languages, first, and fully developed modern languages later on, the members of a community could express these abstract symbols linguistically. This suggests that language is highly likely closely related to social complexity 'in broad evolutionary terms, the more complex the social relations of a species, the more complex its language and communication systems' (Lewis-Williams, 2002, 89).

Closing, there are two further factors that put selection pressures for symbolic thought to develop. These are the need for social learning, especially as it relates to teaching tool-construction techniques, and the need for the group to come up with an effective management of intragroup social tensions and stress that resulted from the increase of the size of the group, and the demand for more cooperative activities. I will discuss these two factors when I examine the protolanguage of H.E, because symbolic capacities developed to support effective communication that could address these two problems.

Finally, new foraging and hunting techniques necessitated by new environments, and the amount of planning required to develop them, also calls for more effective communication that only some form of language (or protolanguage) could support. Since the development of language requires symbol use, these environmental factors apply selection pressures for the development of symbolic thought.

Hence, to complete the picture, the usage of symbols allows for increased social complexity because the capacity to use linguistically in communication the abstract terms expressing social relations in discussions and argumentation opens the possibility for more complex social relations be forged.

The second problem that we should address is what cognitive mechanisms and capacities enabled the acquisition of the symbolic capacity and allowed the transition from using indices and icons to using symbols. To do this properly, one must revisit the differences between indices and symbols, and icons and symbols.

References

Bolhuis, J., tattersall, I., Chomsky, N., & Berwick, R. C. (2014). How could language have evolved. *Plos Biology, 12*(8), e1001934.

Christiansen, M., & Chater, N. (2016). *Creating language: Integrating evolution, acquisition, and processing*. MIT Press.

Crane, T. (2003). *The mechanical mind* (2nd ed.). Routledge.

Crane, T. (2009). Is perception a propositional attitude? *The Philosophical Quarterly, 59*, 452–469.

Deacon, T. W. (1997). *The symbolic species: The co-evolution of language and the brain*. W. W. Norton & Company.

Everett, D. L. (2017). *How language began*. Liverlight.

Gärdenfors, P. (1995). Cued and detached representations in animal cognition. *Behavioral Processes, 35*(1–3), 263–273.

Heck, R. G., Jr. (2007). Are there different kinds of content? In B. P. Mclaughlin & J. Cohen (Eds.), *Contemporary debates in the philosophy of mind* (pp. 117–139). Blackwell.

Kulvicki, J. V. (2006). *On images: Their structure and content*. Clarendon Press.

Lewis-Williams, J. D. (2002). *The mind in the cave*. Thames & Hudson.

Mithen, S. (1996). *The prehistory of the mind: A search for the origins of art, religion and science*. Thames and Hudson Ltd.

Sainsbury, R. M. (2005). *Reference without referents*. Oxford University Press.

Sterelny, K. (2020). Afterword: Tough questions; hard problems; incremental progress. *Topics in Cognitive Science, 12*(3), 766–783.

Tomasello, M. (1999). *The cultural origins of human cognition*. Harvard University Press.

Tomasello, M. (2014). *A natural history of human thinking*. Harvard University Press.

Tomasello, M. (2016). The ontogeny of cultural learning. *Current Opinion in Psychology, 8*, 1–4.

Torrey, E. F. (2017). *Evolving brains emerging Gods: Early humans and the origins of religion*. Columbia University Press.

Chapter 7
Cognitive Capacities and Symbol Usage

In the previous chapter, we examined the evolutionary reasons that favored the emergence of symbolic thought. In this chapter, I address the problem of which cognitive capacities were necessary for symbolic thought to emerge from a background of iconic representations. Firstly, I discuss the relations and differences between symbols and indices. Secondly, I discuss the relations between icons and symbols, and, then, I turn to discuss the cognitive factors required for the emergence and usage of symbols. Specifically, I discuss the cognitive factors that enabled the move from iconic representations to symbolic representations. I am searching for a plausible sort of "cognitive revolution", that is a set of processes in the mind, that allowed the move from iconic representations to symbols, and, in this sense, went beyond the cognitive capacities required for the employment of iconic representations. For example, iconic representations are not recursive, whereas advanced symbolic systems are recursive. It follows that agents that use recursive representational systems must have a cognitive capacity that the predecessors who could not use recursive representational systems did not have. Finally, I address the problem of whether H.E possessed some sort of language or protolanguage.

To address this problem, one must first examine the differences between indices and symbols, and between icons and symbols, and use the results to explain the cognitive development required so that symbolic representations could emerge out of indices and iconic representations. To put it differently, what are the cognitive differences between agents that use symbols and agents that use only icons?

A. Raftopoulos, *The Origins of Representations: Cognitive and Brain Development*, Studies in Applied Philosophy, Epistemology and Rational Ethics 73,
https://doi.org/10.1007/978-3-031-91213-9_7

7.1 Indices and Symbols

Signs are symbolic if, among other things, they refer to some entity through a convention. This convention establishes a link between the symbol and its referent so that the activation of the former invokes the latter. The basic idea is that a symbolic association is formed when one learns the convention and, thereby, reliably pairs a syntactic entity, be it a thought, a sound or a typed string, with something else. Thus, the relation between the signifier and the signified is a mere correlation between objects. However, as the reader recalls, this is also what an indexical association between an index and its referent consists in. Indices are objects that have a form that bears a physical, factual, or causal relationship with what they indicate. Similarly, when the association between a symbol and its referent is learned (notice that indexical relations may be learned or inherited), the association becomes involuntary and a factual or causal relation is formed between the symbol and what it signifies, in the sense that the occurrence of the symbol causes the occurrence of the referent. In this sense, symbols function like indices.

However, symbolic, and indexical signs are not equivalent. First, symbols are representational signs, whereas indices are not. Second, agents that process symbols have much more representational and expressive capabilities that agents that simply use indices. What are their differences between indices and symbols, and how do these differences explain the representational sophistication of symbol users? The main difference is that symbols form a system in which its members are interrelated. Eventually symbols become concepts and concepts do not stand alone (barring Fodor, of course) but are engaged in various inferential and associative relationships. Indices do not have this property. This explains a significant difference between symbols and indices. For an indexical relationship to hold, there must be a correlation in time and place between the sign and its object. If the correlation breaks down, the association is eventually forgotten, and the indexical power of that sign to refer is lost. Symbolic reference, in contradistinction, remains stable and nearly independent of any such correlations. The reason is that the referential relation is sustained through the interconnections with the other symbols in the system. A symbol can refer to an object even if the physical association between the symbol and the object is very rare, or even an impossibility, as with unicorns. No indexical association could have been developed with so little correlation, or no correlation at all.

As a consequence of this difference, indexical associations are largely independent of one another. If one indexical association is lost, this has no, or almost no, repercussions for the other indexical associations. This is not the case with symbols. Symbols are incorporated into quite specific individual relationships to other symbols in the system. Change in the use of one may require a restructuring of the whole system, as it happens in cases of strong conceptual change. This is why symbols need to be in context with other words (a predicate term, for example, must have an argument and both together have a truth-value) in order to have any truth value and determinate reference.

In the words of Deacon (1997, 82–83).

> Symbolic reference derives from combinatorial possibilities and impossibilities, and we therefore depend on combinations both to discover it (during learning) and to make use of it (during communication). Symbols cannot be acquired one at a time, the way other learned associations can, except after a reference symbol system is established. A logically complete system of relationships among the set of symbol tokens must be learned before the symbolic association between anyone symbol-token and an object can even be determined. The learning step occurs prior to recognizing the symbolic function, and this function only emerges from a system; it is not vested in any individual sign-object pairing.

Symbolic systems, to repeat a claim made above, differ from indices, and have emerged from a ground of indexical relations between objects when these indexical relations are organized so as to form a set of interconnected, through logical/conceptual connections, symbols. Unlike indices, symbols do not merely accumulate in unstructured sets of indexical relations, but they form a nexus of interrelated objects that is a hierarchic network of nodes and connections that defines a vast and constantly changing semantic space. This vindicates Planer's (2021, 243) remark that 'one "has" symbolic cognition when one possesses a rich, combinatorial symbol system for thought and thinking.' If symbols are organized in a hierarchical and recursive manner, what are the cognitive mechanisms and, thus, cognitive capacities that could sustain such symbolic systems? We shall discuss this issue after I have examined the differences between icons and symbols.

7.2 Icons and Symbols

We have seen that initially icons represented through their physical resemblance to some of the physical properties of the represented entities. This imposes the restriction that icons can represent only entities that have observable properties; abstract entities could not be represented iconically at that stage. One could object that one could represent "freedom" by means of picture or drawing of birds flying in the open sky. This is correct but it is possible to do so because we know beforehand what freedom is (we have a concept or symbol for it), and we can see the metaphor behind the icon. All these require the usage of symbols, and, thus, the representation is not purely iconic. One could insist that even the cave-drawings in the Paleolithic most probably had some ceremonial or functional significance for the painters and their viewers, which also involves the usage of symbols; these icons were not purely iconic either. The first part of the argument is probably correct, but the conclusion does not follow. Even the uninitiated, or even the members of another tribe that did not share the same culture, upon seen an icon of a bison would recognize the animal depicted in the icon, even though they would probably miss the cultural, social, or functional significance of the symbol. The referential association to the animal was ensured by the icon, something that does not hold with the iconic representation of freedom above. The reason is that the drawing of a bison resembles the important physical features of the animal, and this is what allows the recognition to take place. Birds flying, on the other hand, resemble some of the properties of freedom only in a metaphorical sense, because "freedom" as an entity does not have any observable properties,

The restrictions of icons in terms of representational power that we discussed in previous sections, especially the facts that they have to resemble physically their referents, that they cannot combine logically to form complex symbols, and that they cannot be combined across contexts to represent new items because of their flat structure, precludes them from eventually expressing more complex situations and abstract properties, are also part of the explanation of the need for groups of agents to move from icons to symbols, and, in writing, from pictograms to ideograms. Sumerians used a sign for womanhood that resembles the female pubic area (an icon), and an icon for mountain that resembles mountains. These signs are pictograms since they depict what they represent. The complex sign consisting of the icon of womanhood on the left of the icon for mountain means a woman slave (since Sumerians got their slaves by attacking mountainous tribes). However, these two icons put together do not resemble a slave woman and their combination cannot explain the logical relations between 'woman' and 'slave'. For this reason, the combined sign is not a pictogram but an ideogram, a symbol. The same holds for abstract terms describing social relations, such as "monogamy" or "marriage," whose possible role in the emergence of symbols was discussed in the previous chapter.

Deacon (1997, 32–33) observes that symbols exhibit a combinatorial form in which distinguishable elements recur in different combinations (which, in philosophical parlance, is the *Generality Constraint*) exactly because their form is not tied through resemblance to some physical form and, therefore, they can express the same thing in different contexts. This property, in turn, explains the basic properties of symbolic systems emphasized by Fodor and Pylyshyn (1988), namely, their combinatorial semantic and syntactic structure. Deacon goes on to note that although there is a high degree of variety in the possible combinations of atomic symbols, the majority of combinatorial possibilities would be systematically excluded because they do not follow the rules of the syntax that underlies the symbolic system; this is, of course the result of the fact that symbols are composed according to some syntactic or logical rules, a trait that icons do not have.

All these do not mean that symbolic representations and symbolic representational systems are orthogonal to iconic representations. First, many symbolic systems also have an iconic component, and t many of the symbolic representations in memory and cognitive functions have an iconic component as well (Burge, 2022; Raftopoulos, 2019). It is well known, for example, that sometimes the activation of symbols carries with it the activation of images that form part of the meaning of the symbol. This is true, but it would be misleading to suggest that sensory images exhaust the meaning of symbols because this meaning is also determined by the nexus of relations among concepts.

Second, symbols emerged on the background of icons. Peirce has remarked that symbols were preceded by icons and are a later step in the semiotic progressions. In other words, symbols were based on icons. Symbols, as the reader recalls, are arbitrary conventions that refer through this convention, but before symbols there was the intentionality (that is, the ability to be mentally directed towards something) of icons that referred through the physical resemblance of their forms to what they referred. Since the first contact of an organism with the world is through its senses, the first representations that appeared are naturally based on these perceptual representations, and these are iconic representations.

Third, symbols are dependent on the support of sensory-based images. Even abstract symbols frequently activate images; when we read stories images are generated, sometimes we need imagery to understand the meaning of an expression, etc. All these phenomena point to a strong ancestral relation between symbols and icons. Neuroscience confirms this by showing that damage to posterior perceptual areas cause severe semantic disturbances. Work on the way abstract concepts appear suggests the crucial role of visual perceptual experiences in the foundation of these concepts. Barsalou (1999) suggested that conceptual schemes and perceptual representations are tightly interwoven.

> During perceptual experience, association areas in the brain capture bottom-up patterns of activation in sensory-motor areas. Later, in a top-down manner, association areas partially reactivate sensory-motor areas to implement perceptual symbols. The storage and reactivation of perceptual symbols operates at the level of perceptual components – not at the level of holistic perceptual experiences. Through the use of selective attention, schematic representations of perceptual components are extracted from experience and stored in memory (e.g., individual memories of green, purr, hot). Barsalou (1999, 577)

With experience, memories of the same component are organized around a common frame and 'implement a simulator that produces limitless simulations of the component (e.g., simulations of purr).' These simulators 'implement a basic conceptual system that represents types, supports categorization, and produces categorical inferences' (Barsalou, 1999, ibid.). It follows that each instance of a (conceptual) type is accompanied by a perceptual component, activated in a top-down manner.

Fourth, even though language presupposes symbols and their combinations, language is much more than a mere computational system, no matter how sophisticated, because the main concern of language is communication and, thus, understanding. If this is the case, icons play a pivotal role in the emergence of language and the symbols language employs because it is the icons that provide a first form of understanding and communication. Recall the discussion of the Sumerian cuneiform discussed above. The linguistic form of the sign was an icon of the object to which the sign referred. When Aramaic became the language written (Aramaic is an alphabetic language), a symbol was associated with an image and came to mean the referent. However, this symbol got its meaning through its association with the iconic sign.

Fifth, language is more than speech in that it includes gestures and mimesis, and gesture and mime are iconic elements. Gesturing and mimesis are characteristic traits of animal communication that does not rely on speech and language, and, thus, it is an older form of communication. As such, it is probably the base-line form which language gradually emerged as the gesture/mime arsenal expanded to satisfy the needs of the Hominin communication that grew increasingly more demanding. If this view of the emergence of language is correct (see Planer and Sterelny (2021) for a thorough defence), it suggests that symbols, which underlie any language, emerged from the gradual expansion of gestures and mimes that are iconic representations. In this sense, symbols are grounded on icons.

Let me expand on this point. If communication among hominis evolved as a system of gestures, gestures being manual acts, the expansion of gestural capacities went hand in hand with the elaboration of manual skills and, therefore, with the appearance of more advanced toolmaking techniques. Toolmaking and communication are, thus, naturally related and this explains why the need for social learning required for the transmission of tool making techniques was closely tied to the advancement of communicative capacities, a process that culminated, if Planer and Sterelny (2021) are right, to the mental internalization of the hierarchically structured manual dexterity in toolmaking that resulted in the emergence of hierarchically organized thought and, eventually, language. It should be noted that the idea that communication and eventually language was expanded through the expansion and elaboration of gestural signs is also shared by Tomasello (2008) and Corballis (2000, 2011). In so far as gestural signs are mostly iconic, the elaboration of gestures resulted in conventionalized icons that became symbolic signs, a typical example of which are most of road signs. Even in this case, the meaning of the symbol is aided by the initial iconic content because it is easier to recognize iconic signs owing to the fact that they involve the perceptual systems. It is easy to see (see Planer and Sterelny (2021), 50–51, for a discussion of this point) how the iconic ingredient of gestures could have enormously helped learning and remembering the associations between arbitrary symbols and their referents during the first stages of symbol acquisition in hominin groups when no adaptations for symbol use were still present, and there were no established forms of social learning of symbolic usage.

In Philosophy, a similar remark was made by Putnam (1983), who argued in his causal theory of reference that the meaning of natural kind terms/symbols is partly fixed through a perceptual encounter of some member of the linguistic community with a specimen of that kind, an encounter which, being perceptual, created an icon of the represented thing. Suppose that language is a set of computations over symbols that refer through some, arbitrary mapping, to entities in the world. Therefore, the mind is a machine that processes symbols based on their physical characteristics, exactly like a computer that runs a program, but, unlike the computer, the human mind assigns meaning to these symbols through some correspondence to entities in the world. Putnam (1980) showed that any such, model-theoretic, semantics is inconsistent since the truth of as sentence (satisfaction of a sentence in a model) underdetermines reference. A sentence may remain true in a model even though the references of its terms change. Putnam's argument relies on the Lowenheim-Skolem theorem, which tells us that there is no unique interpretation for a formal set of sentences that can be singled out—i.e., our sentences are satisfied by non-intended models. Since the symbols acquire their meaning by referring to entities in a model, the underdetermination of reference undermines the whole enterprise of assigning unique meanings to symbols.

The problem, therefore, is how to explain the meaningfulness of linguistic representations, if the model-theoretic semantics approach fails. Putnam (1980, 442) and Dennett (1987, 144–147, 154) identify the source of the problem, which is none other than the fact that we conceive a language as something whose use is completely specified and yet still lacks an interpretation. Once one starts with this assumption, then one 'can only have crazy solutions' (Putnam, 1980, 442). The only way out is to assume that some kind of interpretation of the symbols, or at least of the basic or atomic symbols, exists from the beginning, an interpretation that grounds the representational system. According to Putnam (1983, 71), the extension of certain kinds of terms is not fixed by a set of criteria, but is, in part, fixed by the world. In other words, symbolic descriptions in propositional form ascribing properties to some terms do not suffice to fix the referents of the signs. The world has a saying on the fixing. This interpretation can only be the product of some causal chains that ground the symbol directly in the world through our perceptual links with the world, that is, through icons.

7.3 What Is Cognitively Required for Symbolic Thought?

We have examined the differences between indices and symbols and icons and symbols, and we discussed the representational benefits that symbolic structures afford compared to indices and iconic representations. We can turn now to examine the cognitive underpinnings of this advance representational capacity. If, as I have argued, H.E did have symbolic thought, what cognitive capacities should they have had in order to be able to have symbolic thought and what evidence do we have that they did indeed have these capacities? To be more specific, what cognitive capacities are required for symbolic thought and are added to those required for representations and iconic representations that we discussed in the previous chapters? A natural suggestion is that since symbolic capacities go hand in hand with language, one could turn to examining the linguistic capacities of H.E to draw conclusions about their cognitive capacities. This has the advantage that language is more likely to leave behind behavioral traces and, thus, it allows an empirical examination of the linguistic capabilities of H.E.

Admittedly, H.E most likely did not have a syntactically rich structured language, but this does not mean that they did not a language and speech, or that they did not understand and use simple syntactic structures. As we have seen, Deutscher (2005, 214–224) argues that the ordering of the words in languages to form more complex linguistic forms was initially determined by some principles that echo reality and are ingrained in Homos' cognitive and perceptual mechanisms that allow them to perceive and interpret the world. These principles are, first, that things that belong together in reality ought to appear side by side in language. Second, the order in which events occur should be mirrored in the order in which events occur in language. Third, the words that are included in a narrative are those that cannot be inferred from the context. Fourth, the actor of an act comes first as s/he is the focus of the narrative. The second principle concerns the order of worldly events and can be easily extended to apply to all sorts of combining signs to form more complex signs. Deutscher remarks that these principles are ingrained in the cognitive and perceptual apparatus of the hominins that first developed complex signs. No matter what these cognitive capacities consist of, it is highly likely that early hominins were cognitively primed to notice and represent the sequence of events in the world. Perceiving the casual order is shared by almost all animals and, thus, it is part of the perceptual apparatus of early hominins as well. In addition, since apes use the order of signs to interpret complex signals in the wild (Cissewski & Luncz, 2021; Planer & Sterelny, 2021), and, of course, the encultured famous Kanzi (Truswell, 2017), this means that the requisite cognitive ability is part of the baseline from which hominis evolved. Kanzi, for example, regularly interprets requests using the hierarchical structure of sentences, although her abilities are limited to simple, linearly structured sentences, and do not extend to syntax involving embedded clauses or logical connectives. The cognitive capacities underlying this capability consist not only in the ability to use the ordering of signs for semantic purposes but, also, in the ability to parse complex signs into parts whose combination yields the complex sign, because only then could an agent use the meanings of the atomic constituents along with the semantic value of the ordering to fix the meaning of the complex signals. Hominins inherited the cognitive capacities that allow the use and understanding of simple syntactic hierarchical structures from the common ancestor of hominins and the great apes 12 Ma.

Planer and Sterelny (2021) convincingly argue that he hierarchical structure inbuilt in the actions involved in the tool-construction technique of H.E shaped the mind in such a way as to allow the gradual emergence of the cognitive capacities required by hierarchically structured representations, and, eventually, language. Thus, syntax, construed as an organizing capacity resulting in hierarchical structures, depends on cognitive skills acquired through the evolution of technical skills The idea is that the toolmaking activity and the complex technical skills involved in constructing tools and, particularly, hand-axes that appeared during H.E era, gradually shaped the brain in such a way as to acquire the cognitive skills that can support hierarchically structured patterns of thought, that is, syntax, viewed as the capacity to organize and recombine representations into structured sequences. The tool-activity when internalized results in putting complex-motor sequences that are hierarchically structured under executive control, guided by the mental templates associated with the tool to be constructed, which is a mental image, and not by the actual, external tool itself, which means that the whole process is emancipated from the here-and-now. At the same time, successful learning of the activities associated with tool-construction requires that the agent can follow, attend to, and analyze the actions of those who already know how to make the tools, which means that the agent is capable of parsing and recomposing the actions of others in a manner that reflects the hierarchical structure of the motor sequences that have to be learned. Needless to say that all these require a well-developed WM and executive control that not only were not themselves parts of the base-line cognitive capacities (but were developed from the base-line, "primitive" WM and executive control capacities that characterize the great apes), but, also, were more advanced than those required for iconic representations. To be able to entertain iconic representations, one need not have hierarchically structured thought because icons, owing to their flat structure, do not require recursive capabilities, or the capability to compose complex representations based on a set of recursively hierarchical rules. Recursive capabilities, however, are required for recursive symbolic thought, and, thus, advanced forms of WM and executive control are needed to support recursive symbolic forms. One could object that icons have a compositional structure true. This is true, but, as the reader should recalls from our discussion in Chap 4, the compositionality involved is a parts-whole compositionality which is not recursive but follows the mathematical rules of Mereotopology.

It should be noted that the capability for symbolic thought is not an all-or-nothing affair; symbolic thought gradually developed from less advanced forms of symbolic thought like those displayed by Kanzi, for instance, or those required for G1-type languages that do not have G3 (hierarchical and recursive) or G2 (hierarchical) structurer. In these cases, mastering of symbols is required, but the structure of complex symbols is not hierarchical or recursive, but, rather, reflects the environment and the nature of actions themselves (Deutscher, 2005). H.E were capable of symbolic thought and had inherited the ability to decompose complex scenes from the common ancestor of hominins and great apes, and had also inherited the capacity to analyze simple hierarchically structured symbols, but it took time before H.E gradually developed hierarchically structured symbolic thought—in Kanzi's case it was instruction and not evolution that led her to recognize and use simple, hierarchically structured complex symbols. Before this, H.E at the beginning probably had a simple protolanguage that was a mixed of iconic gestures and simple symbols. As the representational capabilities gradually evolved, concomitantly with the increase in WM capacity and attention control, H.E's language became increasingly more symbolic and less iconic, and relied on hierarchically, recursively structured symbols.

Planer and Sterelny (2021, 149) conclude that '[t]echnological evolution in our line drove the evolution of a system for the hierarchical control of complex, extended, and precise action plans, and in social learning for the parsing of the complex actions of other agents.' This view is also shared by Ambrose (2010) who argues that the evolution of modern human behavior was accompanied by neurological changes that enhanced capacities for innovation in technology, language, social organization, and strategic planning associated with WM. Ambrose focuses on two stages of the evolution of cognitive faculties for planning. The first involves executing complex sequences of actions involving manufacture of multicomponent artifacts, such as the hand-axes, especially those constructed during the later phases of H.E era. The second involves enhanced planning through information sharing that requires the establishment of extended regional social interaction networks based on trust and cooperation. Ambrose thinks that both stages were probably accompanied by important innovations in grammatical speech, apparently as a result of the capability for syntactically, hierarchically structured thought.

Welshon (2024) thinks along similar lines and argues that hierarchically structured tool production and construction techniques developed over several hundred thousand years led to the development of jointly attentive tutorials that played a crucial role in the initial occurrence and subsequent spread of reflective self-consciousness, which, as we have seen, are necessary for representations and symbols to become possible. However, Welshon thinks that this took place around 100 Ka, whereas

Planer and Sterelny (2021) and I argue that this happened gradually much earlier during the H.E era.

Finally, Corballis (2014) argues that what distinguishes us in the animal kingdom is recursive capacities in thought, that is, the ability to embed our thoughts within other thoughts. Recursion enables agents to conceive of their own minds and the minds of others. It also gives them the ability to interpolate past experiences, or imagined future ones, into present consciousness, which enhances planning capabilities. Corballis argues that recursive structures led to the emergence of language and speech, which ultimately enabled us to share our thoughts, to plan with others, and foster social cohesion. Language itself adapted to recursive thinking, first through manual gestures, then later, through sounds.

We have already encountered the idea that manual activities shape the brain when we discussed Lombard's (2021) exploration of the relationship between attention development in modern archers and attention as a cognitive requirement for early bow hunting. Lombard argues that practicing to use bimanual technologies shapes aspects of our cognition, including attention. We have also seen that the physical and geometrical properties of our environment have shaped the hardwiring of our perceptual systems.

We examined psychological evidence for the effects of actions on cognition and perception in Chap. 5, Sect. 5.1, in assessing the role of action-files and their interplay with the object files in perception. We saw that in actions involving tools, the brain has seen to it not only that there be bimodal visual-haptic cells but also that these cells can expand their receptive fields to encompass the tool as well as the hand. This way the brain solves the problem of the displacement of perception from action. When a tool is held by the hand to manipulate an object, visual focus is on the end of the tool, which is the source of feedback about the contact with the manipulated object, and away from the movements of the hand itself that does the manipulatory movements. This is a case of perception/action decoupling since the eye is separated from the hand (Klatzky et al., 2010). By expanding their receptive fields, the cells integrate hand movements and the tools used. This enables the registration of the hierarchical structure inherent in tool use and hand movements in the brain, shaping, thus, the brain so as to facilitate this registration and representation.

Concerning the effects of actions on language, Coventry and Diessel (2024) argued that actions such as navigating in the environment and locating objects in it have shaped our spatial communication systems and that, in this sense, action is at the center of spatial communication and in the development of spatial demonstratives. It is not difficult to see why spatial demonstratives are essential in communicating the presence of objects that are of interest for the success of some communal endeavor; if one wishes to alert others to the presence of prey at some location in their visual field, one must employ some sort of spatial demonstrative. This may explain why spatial demonstratives develop very early in ontogeny and are the starting point of spatial communication. The objects that an agent can act upon usually are "within reach", that is, they are situated in the agent's peripersonal space (as opposed to the "extrapersonal" space). These two spaces are supported by distinct neural circuits, as several studies with patients with neural damages suggest since they show a double dissociation, that is, patients with damage to their right hemisphere experience near-space visual neglect when objects are presented within their reach, but not when they are out of reach. Conversely, other patients exhibit far-space neglect, with the opposite pattern of performance. The core neural circuits subserving peripersonal space both in humans and primates include the Front Eye Fields (FEF), the Ventral Premotor Cortex (PMV), the Lateral, Medial, and Ventral Intraparietal Sulcus (LIP, MIP, and VIP) (Basile et al., 2024). LIP and VIP are involved in visual attention and saccadic eye movements, and VIP and MIP are involved in visual control of reaching and pointing, that is, they all are areas crucially involved in guided action. This shows the neural underpinnings of the intricate relationship between action and spatial cognition.

In discussing the differences between the three kinds of signs, we saw that an important difference between indices and symbols is that the latter form a nexus of interconnected items so that a complete system of relationships among the symbols has to be learned before the symbolic association between symbols and objects can be securely established. Symbols refer to all sort of entities and often these entities belong to different domains (some are tools, other are animals, other members of the group, etc.) This entails that connections are eventually established between symbols that refer to entities that belong to different domains of reality. This is not easy to accomplish, and, with regard to the evolution of our species, it took considerable time (millions of years) before the fully fledged inter-connectivity across domains (natural history, social, technological, etc.) could be achieved.

Chimpanzees and other big apes have a considerable symbolic network that manages their inter- and intra-group relations and allows flexibility in their social relations, but this flexibility stops there. There is little, if any, connection with other domains of activity, such as toolmaking or foraging. Even though in the latter domain these apes have considerable knowledge of the resources in their environment, they display no flexibility that would attest to the fact that there is a network of interrelated symbols in the two domains, allowing, thus, the use of symbols in one domain in creative ways in another domain. One could say that the symbolic framework in the social domain is like a cognitive module closed to the other cognitive areas.

A similar phenomenon can be observed with respect to Hominins. Even though the ever-growing size of the social groups constituted an evolutionary pressure that probably led to the development of symbolic thinking and language, and H.E most likely had a rich set of interconnected symbols that allowed them considerable flexibility in managing their social interrelations, this flexibility did not transfer to their technologies that remained, with very few modifications, largely unchanged for millions of years. From the Oldowan artifacts 2 million years ago, and the Acheulean artifacts 1.75 Ma (constructed by H.E), it took bout 1.250.000 years for the Levallois technique to appear 500 Ka, also created by late H.E. This means that the flexibility of thought in the social domain did not characterize the other domains of thought, since the same inflexibility in the technological domain also applies to their foraging and hunting activities. I hasten to note that this does not mean that H.E did not have advanced hunting techniques. We have seen that H.E used mental models to plan ahead their hunting expeditions and they also had all the cognitive capacities required for effective communication during planning. It only means that these techniques did not show any considerable development during the H.E era, which is quite unlike with what happened in the social domain. It seems, thus, that the modularity and lack, thereof, of interconnectivity between symbolic systems concerning different domains, characterized H.E as well. Ideas and ways of thinking in one domain could not transfer to other domains, except perhaps by accident.

The same restriction holds true for later Hominins (after 500 Ka), such as Archaic Homo sapiens in Africa or Homo Heidelbergensis in Europe. There was a rich symbolic niche concerning the social domain, but there seems to be no or very little transfer of knowledge across different domains. One has to await until the appearance of Early Homo Sapiens (150–100 Ka) for some kind of cognitive fluidity to appear. The full-fledged cognitive fluidity that characterizes modern humans appears with Homo Sapiens about 60.000 Ka (Mithen, 1996).

The problem of cognitive fluidity and how it emerges from a general-purpose knowledge capacity and from a set of cognitive modules has been extensively discussed in Psychology (Carey, 2009; Carey & Spelke, 1994; Gardner, 1993; Greenfield, 1991; Karmiloff-Smith, 1992; Sperber, 1994). It is generally accepted that cognitive fluidity became possible with the development and increase in size of the white brain matter, that is, with the increase in the number of glial cells and connecting fibers that allow intra-brain communication. Human brains have 31% more white matter than chimpanzees, but only 2% more of gray matter (the neurons in the brain), which seems to indicate that the cognitive differences between the two species are likely due, mostly, to the difference in the size of white matter. It is interesting to note that the order of myelination of the pathways in the brain during ontogeny follows the order in which the pathways were developed phylogenetically. As far as ontogeny is concerned, there is evidence (Benes, 1989) that, in accordance with Yakovlev and Lecours (1967) early findings, primary motor and sensory cortices show robust myelination early in the first decade, while associative cortical areas show increased myelination in the second decade. It is interesting that the cingulate and basofrontal cortex remain poorly myelinated throughout life.

How are these brain developments related to the cognitive capacities allowing the modern symbol-usage? Williams-Lewis (2002, 76–7) discusses stone tools making techniques in the Upper Palaeolithic and notes that in addition to having a much clearer mental picture of what their tools look like, this mental image was related by Homo sapiens to the social group they belonged. Thus, 'human creativity and symbolism was linked to social diversity and change.' By the time of the Upper palaeolithic, the mind of modern humans has been developed and was characterized by four main types of behavior (Williams-Lewis, 2002, 97) that allowed them, among many other things to produce icons. These are important because they underlie, in addition to symbolic thought, the ability to draw icons and may shed light on the question concerning whether H.E could draw icons that we discussed in Chap. 5, Sect. 4.2. The first is abstract thinking, that is, the ability to use abstract concepts not limited in time or space. The second is the ability to plan in depth. That is, the ability to formulate strategies based upon past experiences and to act based on them in the context of a social group. The third is behavioral, economic, and technological innovation. Finally, fourth, is the ability to display symbolic behavior, that is the ability to represent objects, people, and abstract concepts with arbitrary symbols and to reify such symbols in cultural practise.

Although Williams-Lewis traces these capacities with the advent of Homo sapiens, our discussion thus far suggests that H.E undoubtedly displayed the first two types of behavior, and, thus, had the cognitive capabilities required by such behaviors. The third behavior cannot be found in H.E because there is no evidence of significant technological or economic innovations during most of their life-time. Their economy, for example, remained a form of immediate-return mutualism, where agents act for a cooperation profit (Tomasello, 2014). We have to wait for Homo Heidelbergensis (500 Ka) and Homo Neanderthal for a new form of economic cooperation, named reciprocal or delayed return reciprocation, which changed socioeconomic life profoundly and required much more sophisticated communication capabilities and conceptual arsenal. However, the social complexity of H.E "communities" and their economic life demanded significant advances in communication capabilities and, thus, in symbol use. So, even if technological and economic innovations are very rare, the fact remains that that the social and economic activities of H.E were possible only on account of a system of symbols used for referring to entities and in communication. As for the fourth type of behavior, we shall see in a while that H.E eventually became able to associate arbitrary symbols with persons, objects etc., when, for example, thinking about the various roles that can be played by agents in hypothetical hunting, foraging, travelling, etc., scenarios. As Planer and Sterelny (2021, 91–92) argue, increasing vocalization was related to increasing use of arbitrary signs. Vocal control seems to have been in place by 800 Ka and, thus, between 1.8 Ma and 800 Ka vocal control was gradually increasing. Since most vocal sounds do not bear some similarity relation with what they refer to, are arbitrary symbols rather than iconic signs.

Thus, even though H.E were capable of symbolic thought, the symbols used were restricted in their respective domain of application and could not be elaborated and extended so that they could be used either to apply to other domains, or to represent increasingly abstract items across various domains. In view of the fact that cognitive fluidity is of paramount importance, because it opens channels for interdomain communication, H.E should have exhibited only some limited sort of cognitive fluidity, because, as it was noted above, it seems that H.E could not, at least in a systematic way, transfer knowledge from one domain to another. This means that they had enough cognitive fluidity to use concepts across different contexts only within the social domain; this no small feat, however, because it allows understanding and communication with other groups.

Advanced WM capacity and executive attention control are crucial for symbolic thought not only because they enable recursive, hierarchical thinking, but for other reasons too. Using a symbol to represent some entity requires remembering the context in which such the initial pairing of symbol/entity originally took place because the same scene in different contexts may elicit different representations, and, thus, an entity may be represented differently across contexts. To able to use the same symbol to represent an entity across different contexts presupposes that one can store both object/entity-knowledge, and previous encounters with the entity in long-term memory, activate these representations in some relevant context in WM, and focus on them by means of sustained attention.

Several studies with different experimental procedures suggest that the capacity of WM dramatically increases between 6 months of age, when infants can respond well on procedures with only a single item to be remembered, and at most two months later, when infants can respond well on procedures with several items to be remembered. Similar developmental trends are found with respect to attention. This suggests, in turn, that the development of attention and WM in hominins was likely gradual and spanned the whole period of the hominins' split from chimpanzees 7–10 Ma. When we examined the ontogenetic and phylogenetic development of attention and WM, some conclusions were drawn concerning the phylogenetic development of these capacities, and, thus, of the behavioral capabilities these cognitive capacities underwrite. These conclusions emphasized that H.E certainly had sufficiently developed attention and WM, and, thus, the cognitive capabilities attention and WM render possible. Furthermore, the PFC of H.E must have been developed sufficiently to be able to subserve such cognitive capabilities.

If, as I said, to be able to use a symbol, one should retrieve information about the context of a past, probably, perceptual encounter, this requires the ability to travel back in time mentally to locate a past event and to be able to provide information about the *what*, *when*, and *where* of that event (Tulving, 2002). This type of memory is called episodic memory. Episodic memory depends on individual awareness of a continuous self over time and the ability to mentally re-experience the recalled event (Wheeler et al., 1997). Episodic memories that are self-relevant form the autobiographical memory (Nelson & Fivush, 2004). This sort of memory requires the ability to project oneself backward and forward in time. The age at which true episodic/autobiographical memory emerges in development is debateable. Some theorists claim that episodic memory emerges between the age of 4 and 6 (Perner & Ruffman, 1995; Perner, 2001; Tulving, 2005). Tulving proposed that episodic memory is preceded by development of semantic memory. Others hold that episodic memory is possible earlier, by age 3 (Scarf et al., 2013), age 2 (Howe & Courage, 1997) or even birth (Conway, 2009).

Turning to phylogeny, one could assume that it should have taken some time for hominins to develop sufficiently rich semantic memory and then episodic and autobiographical memory. Torrey (2017, 3–4) argues that awareness of self (which, as the reader recalls, is necessary to form representations) appeared with *Homo erectus* 1.8 Ma. The ability to think of others as having thoughts, a first-order theory of mind, appeared 200 Ka with Archaic Homo sapiens. 100 Ka, when Early Homo Sapiens came to the scene, introspection, that is, the ability to reflect on our own thoughts, first appeared. Finally, episodic memory and autobiographical memory appeared with modern Homo sapiens 40 Ka. That is, the ability to integrate past and present thoughts and think about the future, which requires autobiographical memory, appeared 40 Ka with modern Homo Sapiens. If true, in view of the analysis above about what is required for symbols to emerge, Torrey's views entail that before Modern Homo sapiens, Archaic Humans and H.E could not have developed symbolic thought such that would allow them to recreate past events. In addition, H.E even lacked an introspective ability that would have given them access to the contents of their own perceptual states and thoughts, which imposes severe restrictions to planning and other activities. In other words, H.E could not have had joint intentionality. Moreover, lacking autobiographical memory and, thus, the ability to integrate past and future, H.E could not understand the recurrence of natural phenomena, because this integration underlies the ability to understand the recurrence of natural phenomena. Beings that lacked autobiographical memory could not understand recurring natural phenomena, and, thus, they were unable to make plans that were based on such an understanding. Let me immediately say that this thesis is wrong. Chimpanzees and other apes, not to mention many other animals, have episodic and autobiographical memory (Lewis et al., 2019; Martin-Ordas et al., 2010; Mulcahy & Call, 2006), and, thus, some form of these types of memory is among the base-line cognitive capabilities that Hominins inherited and gradually developed and expanded as a response to environmental and social pressures.

Before I turn to examine in some detail Torrey's views, let me say a few more things about autobiographical memory. It is likely that a factor that should be taken into consideration in discussions of the development of autobiographical memory is the role of inner speech in the construction of the autobiographical self. One interesting suggestion is that cognitive fluidity was facilitated by the development of inner speech, that is, our conscious experience of internally produced, natural language sentences because inner speech may make information from one part (module) of the brain available to another. Carruthers (2013) claims that inner speech, or mental rehearsal of speech, as he calls it, is necessary to explain the cognitive differences between humans and non-human primates, despite the fact that, as Carruthers thinks, humans and non-human primates have the same WM capacities. Rehearsing a piece of information paves a way to communicate this information to some cognitive areas of the brain that do not, as yet, have access to the information within the module in which this piece of information originated (Carruthers, 2006; Fields, 2002). As a matter of course, the same role in creating cognitive fluidity is being played by natural language itself, because expressing orally some thought originating in some module makes it accessible to other modules in the mind with which the originating module could not directly communicate (Mithen, 1996).

Fields (2002) argues that inner speech plays a critical role in the construction of the autobiographical self, that is, a sense of self as a person who persists through time, has a unique life history, and a unique collection of personality traits. One should also add to this, a sense of a person who has a set of memories that are of that person. Inner speech, that is, our conscious experience (or semi-conscious experience to include cases in which one does not have access to all parts) of internally produced, natural language sentences. If this view is correct, it is possible that the phylogenetic development of autobiographical memory goes hand in hand with the development of language, on account of the fact that inner speech requires some form of a language of thought (perhaps a primitive language of thought) and, thus, of natural languages. Inner speech may have followed directly the development of languages through some internalisation process. Alternatively, it could be an indirect result if inner speech requires some additional cognitive mechanisms that had to be built first (Planer, 2021, 243). Nevertheless, since linguistic capacities rest on the possession of a system of symbols that displays all the properties of symbolic systems that we have discussed thus far, inner speech too depends on such capacities.

Güneş et al. (2024) examined the relation between gesturing and autobiographical memory. This study is important because, first, as we have seen, gestures are the first (iconic) representations used for communication among great apes, and the expansion of gestures led gradually to the emergence of symbols, and language. It is also important because, second, autobiographical memory is important for the development of representational capacities and especially, for symbolic representations. Güneş et al. (2024) asked 35 children and 46 adults to report autobiographical memories, and rated the recalled memories on phenomenological qualities. The episodic and non-episodic details of autobiographical memories and representational gestures produced during memory narration were coded. The study showed that the use of representational gestures was positively correlated with the episodic details of adult memories, but the same correlation was not present in child memories. Thus, gesture use may be related to the retrieval of autobiographical memories in adults but not in children. It is, thus, plausible to suggest that the expansion of gesturing capabilities during either ontogeny or phylogeny is associated with more developed autobiographical memory and, thus, with more advanced representations.

The notion of autobiographical self is interlinked with that of self-consciousness because most of the elements that make up self-consciousness are already included in what constitutes an autobiographical self, most notably the notion of self as an acting agent that persists in space and time and has experiences some of which can be remembered as being that person's own experiences. Edelman (1993) distinguishes two types of consciousness, primary and higher-order consciousness. The first refers to a state of being in which one is aware of things in the world and of having mental images in the present. It is not accompanied by any sense of a person with a past and future (this means that no autobiographical memory is involved). It is limited to a small memorial interval around the present and does not afford the ability to model the past or the future as part of a correlated scene. These organisms have a long-term memory, but they cannot be aware of that memory, and they cannot use it to plan for an extended future. Finally, even though these organisms possess mental images, they cannot view these images from the vantage point of a socially constructed self.

Creatures with higher-order consciousness, on the other hand, are thinking creatures who recognize their own acts or affections. These creatures are conscious of being conscious and have a model of themselves in space and time as creatures with a will and unified experiences. They see themselves as parts of a social group and construct, thus, a notion of social self. Edelman notes that to acquire higher-order consciousness and break, thus, the tyranny of the remembered present, new forms of symbolic memory and new systems serving social communication need to develop. Without a symbolic memory system, one cannot store and use, for example, symbolic relations acquired through interactions with other individuals in the same social group. Recall that I examined above the idea that an environmental social pressure to the development of symbols was the necessity to think and communicate concepts expressing the complex social relations of large social groups. According to Edelman, this development led gradually to the acquisition of higher-order consciousness.

As we have seen, and we shall revisit this problem next, empirical evidence strongly suggests that H.E did perceive themselves as members of a social group and had a notion of social self. They were able to plan long-distance trips crossing oceans, they had elaborated stone knapping techniques and techniques for quarrying the appropriate material, which often required exceedingly long trips. They most probably also had apprenticeship learning systems for transmitting tool-related knowledge (Hiscock, 2014). They also planned for ambush-hunting. All these activities require integrating past experience, present environmental situation, and some sort of forecast or anticipation of what might happen in the future as the activity unfolds, that is, the ability to think about future situations. This requires planning ahead for eventualities, which entails that H.E could use models of situations that integrated many sources of information and engage in what-if kinds of hypothetical thought. All these presuppose that a robust and relatively rich symbolic system be in place.

Before I discuss further these forms of consciousness, I wish to digress a bit and point out that the distinction between these two types of consciousness brings to the fore another difference between icons and symbols. Creatures with primary consciousness can have and use within a limited time-frame mental images, that is, iconic representations and indices of course. However, they cannot have symbolic systems owing to the fact that symbols are detached from their spatio-temporal context in the sense explained above.

Let us return to the two types of consciousness. It is obvious that higher-order consciousness developed on the background of primary consciousness after our split for chimpanzees. As every student of developmental psychology knows, and as most cognitive archaeologists think, cognitive capacities are very rarely all or nothing, and also, develop gradually. This means that between the primary consciousness and the full-fledged higher-level consciousness of modern Homo sapiens there is a spectrum of intermediate states that gradually diverge from primary consciousness and become more complex. That is, there is a grading scale from primary consciousness to full human consciousness with many inter-between states of increasing order of consciousness. Thus, H.E certainly have some form of higher-order consciousness compared to their ancestors. Homo Neanderthals had higher-level consciousness compared to H.E, which was less that the consciousness of modern Homo Sapiens. This means that these populations were endowed with different representational capabilities, in addition to any anatomical, genetic, or epigenetic evolutionary changes. A factor that played an important role in this development was the social niche and the complexity of the societies. Recall that, as we saw earlier, it is estimated that the average size of groups of hominins with modern human cranial capacities was 120 individuals (for archaic Homo sapiens 200 Ka), 150 individuals (for modern Homo Sapiens 60–40 Ka). In comparison, Neandertal average group comprised 10–13 individuals, and H.E groups had a size of 110 individuals.

Using symbols to represent reality raises the problem of how arbitrary symbols could come to represent some entity. To examine this problem, it would help to go back to the difference between icons and symbols. Let us consider the first form of an iconic representation, namely, the content of an early visual perceptual state. This state presents the viewer with some of the objects in the visual scene and their properties and, also, the spatial arrangement of these objects. As we saw, the perceptual icons of early vision are highly contextual and cannot be taken out of the context of their formation and used in other contexts. So, how do viewers remember the contents of their perceptual contacts with a visual scene and are able to recreate them in the absence of the scene, and very often across different contexts?

We have already seen that iconic templates in working memory are required for the formation of iconic representations that are relatively independent of the context of the particular perceptual encounter, because they abstract away from many of the specific physical properties of objects by expressing statistically constructed mean values of these physical properties. Some of the items held in working memory are eventually consolidated in long-term memory. As we go higher in the brain, visual details are gradually lost, and more abstract representations appear. This allows categorization of various objects in categories irrespective of some of the differences in their physical form. The template that allows categorization is a network of interconnected neurons that have certain activation values. These values form a n-dimensional matrix (where "n" is the number of neurons) that corresponds to a vector in the activation subspace. The tip of this vector can be seen as the symbol that corresponds to this representation, and which is assigned a linguistic term. This term is an arbitrary attribution of a name to a template that can be used in thought independently of the template. It is like upon seeing a cat, a community decides to call the animal as it appears in their visual image a "CAT" and to signify that animal in discussion uses the term "CAT", which bears no resemblance to cats and, thus, is detached from ay perceptual context and can be used across different contexts conferring to the symbol all the benefits of symbolic representations.

Notice that even though the symbol "CAT" is an arbitrary assignment, the representational perceptual iconic template which it names is not, as it resembles to a certain extent the animal. This is an example of the point made in the previous section that symbols are grounded on, and thus, presuppose, icons. This is why even if the representation of a category in thought is a symbol, this does not mean that rich perceptual information is not stored and cannot be retrieved in long term memory, although this information is not stored in the higher cognitive areas of the brain that can store only abstract information, but in the neuronal networks in early visual areas that are activated in a top-down manner when the relevant representation has been activated in the higher network and becomes, thus, part of working memory. In sum, one can retrieve from long-term memory the details of a visual scene and recreate a representation of this scene in working memory. It follows that working memory is pivotal for the capacity to build iconic and symbolic representations and for their interactions.

Everett (2017, 103) extends this account in terms of gradual removal of resemblance to what is represented and draws the following picture:

> The evolutionary path to the development of symbols. The mistaken association of one things for another (the root for a snake) can lead later on the intentional use of the object of the false impression to represent the things it was falsely associated with. One could draw a root to represent or mean a snake. As early hieroglyphic writing systems show, this use of representation based on resemblance can further evolve so that all resemblance is gone, thus leading from intentional use based on resemblance to an intentionally arbitrary symbol. As this happens, a drawing changes from an icon to a symbol.

Everett's mention of the evolution early hieroglyphic systems, to which one could add the cuneiform system of Sumerians, is telling. A look at a table of the evolution of cuneiform signs from archaic Sumerian (that used graphs for representing numerals, names of objects, and names of persons) to New Babylonian cuneiform writing systems shows that eventually the physical resemblance between the sign and its referantum all but disappears and an arbitrary symbol appears, arbitrary in the sense that it bears no physical similarity to the depicted object. Recall that in Archaic Sumerian, signs represented things by being stylized icons of the things they represented.

Concluding this section, let us examine Torrey's arguments that purport to show, among many other things that before modern Homo sapiens the other earlier hominins could not have had symbolic though and, thus, any form of language. Torrey makes most of the fact that burial practices and also the practice of caring for others first appeared with Archaic Homo sapiens (including Homo Neanderthals) some 200.00 years ago. Since these practices require empathy and, thus, a theory of mind, Torrey concludes that it is only 200 Ka that the ability of thinking that other have thoughts too, a first-order theory of mind appeared.

However, it is hard to understand the success of *Homo erectus* in populating the world and adapting to radically different ecological niches in the absence of an advanced theory of mind, one that allows at least the joint intentionality that Tomasello thinks necessary for cooperation and planning. It is also difficult to explain the success not only of emigration, but also of the whole industry of tool production and teaching the required skills for it, and of ambush hunting, if H.E could not plan successfully ahead. This, however, would have been impossible if they could not have a rudimentary at least grasping of the relevant natural phenomena and their patterned occurrences. Even if Torrey is right and introspection was available after 100.000 with Homo Sapiens (and it is certainly true that a fully developed introspective capacity that gave rise to a fully developed self-consciousness is attested with Homo Neanderthal and modern Homo Sapiens around 67–45 Ka respectively as evidenced by their rock paintings), self-consciousness and introspective capacity is not an all or nothing capacity but is comes in degrees of sophistication. Most capacities are like this. The Theory of Mind in young children, for instance, develops gradually between 2 and 5 years (Kavanaugh, 2006). Why should these capacities have developed differently in the phylogenetic scale? We saw that there is growing evidence that Neanderthals painted the rock of their caves both icons (Pike et al., 2012), and spatially structured, non-figurative marks, which most probably were some sort of symbols (Marquet et al., 2023). If these abilities require episodic and autobiographical memory, as Torrey claims, this means that Neanderthals had them too, contrary to Torrey's belief that they had only a rudimentary theory of mind.

In addition, stretching back in time, to be able to plan for long travels across oceans *Homo erectus* would certainly have to discuss their plans, exchange views, and reach some agreement. This presupposes that not only were they self-conscious, but that they also had some form of theory of mind because they understood both that others had minds and thoughts like them and, second, that it was likely that their interlocutors would be persuaded by the same arguments that they themselves would have accepted. Thus, the evidence we have about the groups of *Homo erectus* points to the fact that they must have possessed some sort of self-consciousness and also an early form of a theory of mind and an autobiographical memory.

Finally, planning successfully for crossing the oceans and choosing other ecological niches to live in certainly requires understanding of the recurrence of natural phenomena. In fact, the H.E social structures were certainly based on some understanding of social laws or regularities, which means that they were aware of regular recurrences in their societies. Given that understanding of natural laws was based on understanding of social structure and its regularities (as any student of Ancient Greek Philosophy can tell you), it is only a small step to applying these patterns to the natural domain.

If H.E had these cognitive capacities, why did they not have burial practices, why did they not care for the injured or sick members of their groups, and why did they not use their knowledge of natural occurring phenomena to develop more efficient and sophisticated tools as Torrey claims? Torrey takes the lack of evidence to that effects to entail lack of the respective cognitive capacities that they require. There are problems with Torrey's analysis. First, there is another explanation of these phenomena, assuming that they did in effect occur, that does not entail the absence of these cognitive capacities. This is that these capacities existed within one domain, most probably the social domain, but owing to the lack of cognitive fluidity did not transfer to other domains (as Mithen, 1996 has argued). A rich social life certainly requires a developed theory of mind, which as such should have appeared much earlier than Torrey thought, that is earlier than 200 Ka. Mithen (1996) argues that from 500 Ka Early Humans had a rich social and natural history intelligence, even though these two could not interconnect yet. This means that by at least 500.000 Ka, our ancestors have the capacity of symbolic thought, even if in isolated domain areas. Moreover, our discussion of *Homo erectus* exploits pushes the date of symbol usage and language usage back to 2.1–1.7 Ma or shortly after that. Instead of assuming that Homo Sapiens developed these cognitive capacities after they emigrated from Africa to Europe, the evidence suggests that Homo Sapiens brought its cognitive capacities from Africa, where they were gradually developed (Renfrew, 2008, 87). These cognitive capacities determined a set of practices, such as toolmaking, the use of fire, cooking, the making of cloths, some form of language, boat-making, childhood learning, social practices like monogamy or marriage, etc. Homo sapiens emerged from Africa 60.000 years ago with this cognitive arsenal in their disposal, an arsenal some elements of which they probably shared in its basic form with *Homo erectus*. There were many differences, of course. *Homo erectus*, for example, probably did not wear cloths since genetic studies on lice differentiation suggest that cloths were used for the first time about 170.000 years ago in the preceding ice age (Toups et al., 2011).

It is certainly true that it is with Homo Sapiens that modern speech developed in which language was structured by syntax. These enriched grammatical structures enabled the formation of new modes of thought and some think that it is about 40.000 years ago that Homo Sapiens fully mastered, first, the capacity for hypothetical and imagined actions that could be expressed through conditional speech (Renfrew, 2008, 81), and which characterize modern humans. These, however, were the gradual development of cognitive capacities that Homo Sapiens brought with them from Africa. The view that such capacities, such as the capacity for hypothetical thought, were present in some form with earlier Hominins is supported by the fact that planning to travel long distances in new unknown territories, as *Homo erectus* did, required careful planning and this presupposes the capacity for hypothetical thought in order to account for possible eventualities.

A second problem with Torrey's analysis is that fossil evidence suggests that H.E likely did care about the injured or sick members of their groups. A 1.7 million-year-old female *Homo erectus* skeleton (known as KNM-ER 1808) found in 1973 had a debilitating disease, probably hypervitaminosis A. She had lived for some time with this disease, indicating that she was looked after by members of her group (Hammond et al., 2021). An elderly male belonged to a population of H.E that spread from Africa to the Caucasus Mountains in western Asia. Most of his teeth fell out long before he died, and his jaw deteriorated as a result. Members of his social group must have taken care of him (Lordkipanidze et al., 2006). This is some of the earliest known evidence for this kind of group care and compassion in the human fossil record. If caring about the sick and injured relies on empathy and, thus, on a developed theory of mind, *Homo erectus* seem to have had it. Finally, the 1.5 Ma Turkana boy was diagnosed with juvenile spinal disc herniation that caused some scoliosis. These usually cause recurrent lower back pain and pain running down the leg, and likely restricted Turkana boy in walking, bending, and other daily activities. The boy had survived into adolescence, which means that the other members of the group looked after him (Hauesler et al., 2013).

7.4 Did *Homo erectus* Have a Language?

In the preceding chapters, I have referred on several occasions to the fact that H.E most likely had some sort of protolanguage. The main evidence for this conclusion comes from behavioral traces left behind by H.E and pertain to their tool techniques and stone quarrying, ambush hunting, and their migrations from Africa throughout the world. All these activities require planning and, thus, communication which would have much more benefited from spoken communication through symbols, rather than from relying on mere gesturing and random vocalizing. If this is true, given that, as I have argued in the preceding chapters, H.E had the requisite cognitive arsenal for symbol using, it is highly plausible that evolution should see to it that protolanguage was selected for. In this section, I expand more on the issue of protolanguage in H.E. As I said in the introduction, I adopt a gradualist approach to explain the appearance of the various cognitive capacities in hominins. It follows that I hold that protolanguage emerged gradually from the communication forms of hominin's ancestors whose baseline cognitive capacities constitute the springboard from which our cognitive capacities emerged. Let us, thus, start from our closest relatives, namely, chimpanzees and early hominins.

Modern chimpanzees often vocalize at the same time as they produce gestures (Heesen et al., 2019). It is therefore likely that the early hominins did the same. Clark and Henneberg (2017) have analysed the craniofacial morphology of *Ardipithecus ramidus* from ~ 4.4 Ma. They suggest that the larynx lay deeper in the vocal tract than in modern chimpanzees, and that it has increased mobility. They interpreted this to mean that *Ardipethecus* would have been better able to modulate vocalizations. Later, when the hominins starting hunting rather than simply scavenging for food, there would have been a selection pressure towards better communication, whether by gesture or vocalization. There are chimpanzees in the Tai National Park that hunt colobus monkeys, and to do so they take separate roles during the hunt. These chimpanzees show an increase in cooperation and communication compared with the chimpanzees in neighbouring areas of the Tai park that do not hunt (Luncz & Boesch, 2015). As H.E, Homo Neanderthalensis, and Homo Sapiens depended increasingly for food on animals both big and small (Ben-Dor et al., 2021; Kübler et al., 2015; Marín-Arroyo & Sanz-Royo, 2022; Wedage et al., 2019), there would have been a similar need to communicate. Differences between communication in H.E, on the one hand, and Neanderthals and our direct ancestors, on the other hand, probably relate to the production of grammatical strings of symbols rather than sounds alone.

If, as Edelman (1993) thinks, modern higher-order language and high order consciousness are interlinked, since both require symbolic systems with their properties, it follows that language also developed gradually. As I have claimed, the development of language, whose emergence was made possible by the development of symbolic thought, certainly shaped in its turn the development of the symbolic abilities of hominins, especially early (100 Ka) and modern Homo Sapiens (60 Ka). To be able to express linguistically the complex social relations among the members of a group requires a richer symbolic system and a more complex grammar. Bolhuis et al. (2014) argued that grammar appears 70 Ka, but it is likely that H.E gradually developed a grammatically structured protolanguage, even if that grammar was simple. The language of H.E probably had a very limited grammar, if at all, (Everett, 2017; Lewis-Williams, 2002; Mithen, 1996; Torrey 2017), but even this minimal grammar restricted the possible ways symbols could be combined to express new situations and concepts, although, as we shall see next, restrictions on linguistic expressions can be placed by the world itself in the absence of grammatical rules, creating, thus, a primitive grammar. Thus, grammar was the first to develop gradually acquiring a richer structure in the form of increasingly more advanced grammatical rules, eventually acquiring recursive structure. Spoken language could be one factor that accounted for the rapid expansion of the brain of hominins from ~ 400 Ka. However, as I have argued above, this does not disprove the view that protolanguages may have existed long before this time (Dunbar, 2014).

Returning to early hominin communication, several researchers suggest that early hominin communication involved some form of protolanguage, which appeared 1.8–1.5 Ma, and evolved from the vocalizations and gestures of the great apes (Everett, 2017; Planer, 2017). This is a form of language that is simpler than full language, with a proto-lexicon. Protolanguage may have utilized vocal, gestural, and mimed components. Tallerman (2012) argues that a compositional or lexical protolanguage consists of single protowords, initially uttered separately and slowly, and subsequently joined in short, fairly random sequences, in which the order of words does not follow any grammatical rules. This does not mean that there is no-ordering, obviously ordering becomes important as the situations to be linguistically expressed become more complex, but that whatever ordering exists is not determined by a set of formal rules. Any ordering comes from restrictions imposed from outside (Deutscher, 2005) and the ordering of the words was initially determined by some principles that echo reality and are ingrained in Homos' cognitive mechanisms that allow them to perceive the world.

Thus, protolanguages started with little (if any) hierarchical structure, no formal syntactic combinatorial principles, and only a loose pragmatic relationship between protowords. These protolanguages exhibit several properties. One of the properties is that, at least for very early protolanguages, the ordering of elements is relatively random, especially if these languages were a combination of words and gestures; 'Me Tarzan you Jane', where "me" and "you" are expressed by gestures does not convey a different meaning from "You jane, me Tarzan". Different word orders do not necessarily relate to information structure. The most ancient form of protolanguage putatively contained various purely semantically based principles that map into linear adjacency (recall Deutscher's views mentioned above) without using anything syntactic, and lacked a mechanism for constructing complex lexical units out of words. This form of protolanguage also lacks a distinction between lexical elements primarily verbs, nouns, adjectives and functional elements such as grammatical items, including determiners, auxiliaries, and sub-words. Gradually, for the reasons explained above, these protolanguages were succeeded by more elaborate, syntactically structured forms that eventually led to the language of modern humans.

Cognitive development resulting from the interplay between social pressures and cognitive structures allowed the formation of much more complex symbolic networks whose constituents entered in a dramatically increased number of interrelations. Hierarchical structure and recursive operations on symbols were eventually made possible. Recursiveness means that one could construct sentences containing clauses within clauses. This leads naturally to the conclusion that grammar gave the capability of the language speakers to construct sentences that contained more information, and, most importantly, were much more precise. It is not difficult to see how this could have benefited, say, instruction and social learning in general (Kolodny & Edelman, 2018; Passingham, 2021). The more precise the instructions, the less reliance on learning by trial and error is required and, thus, instruction takes less, time, is more effective, and, importantly, as the need of learning through trial and error is reduced the less material is wasted in failed trials. Populations having languages with grammar, thus, have a distinctive evolutionary selective advantage over those that did not.

Suppose that H.E had some form of proto-language that has very little syntactic structure in the form of grammatical rules. One might wonder how a proto-language with a minimal grammar was possible and how it could be used to express in linguistic terms, say, events in the world. These events are structured and if the first languages did not have any significant structure, how could they had captured the structure of the world? There is a more basic question, however; are such languages even possible? The answer to this basic question is a straightforward yes. There are indeed languages that do not have a recursive structure and they do not allow for hierarchical and recursive constructions, and do not even have any syntactic rules (Deutscher, 2005; Everett, 2017) these are the 'Mc Tarzan, you Jane" languages, the G1 languages in linguistic theory, in which sentences are merely conjoined together to yield complex stories without the use of any complex syntactic rules. '[S]ubordinating clauses, complex noun phrases, word-compounding and others, are not crucial for language and are later additions made for cultural reasons.' (Everett, 2017, 74) In addition, there are expressions in modern languages that lack syntactic structure and, yet, are perfectly intelligible because it is the context and common sense that determine the meaning (recall that semantic compositionality fails in natural languages); A typical example is the expression "you do the crime, you do the time."

We have seen that even in these primitive languages or protolanguages in order to be able to express complex situations the ordering of symbols is important, and that this ordering ned not be determined by some sort of syntax in the form of grammatical rules. In addition to the fact that order can be impose by the world itself in the way explained above, there is another way that the ordering of linguistic symbols can be imposed in the absence of syntactic rules. As we saw, according to Peirce, there is a historical evolution from iconic signs to symbolic signs. This means that at the interface of the transition some of the properties of iconic representations may have been retained in the symbolic representational system. Recall Everett's (2017, 103) discussion of how an iconic representation, a root, became a sign for snakes and then that sign became a symbol for snakes. The root resembles to some extend the shape of snakes, and, thus, the symbol itself retains this similarity. Since icons were initially resembling their representatum in some physical respects, they were in some sense reflecting reality. It is plausible that such a reflection of reality passed onto languages, and it also plausible to assume that the languages at their first stages relied more on reflections of reality than on syntactic rules that took many years to develop, in the way suggested by Deutscher (2005). If this is correct, at the initial stages of languages, or even in some modern languages that lack intricate syntactic structure, it is reality that determines the ordering of words and the ordering of sentences and not some syntax.

In view of the fact that the sentences produced this way purport to represent reality and since their meaning is the state of affairs they represent, it follows that meaning determines the form of the representation, that is, its syntax. At the interface of iconic and symbolic representations, one sees that symbolic structures display a basic property of icons, namely that meaning constrains the form and not the other way around, as is the case in mature symbolic systems. As we have seen, Deutscher (2005, 214–224) argues that the ordering of the words was initially determined by some principles that echo reality and are ingrained in Homos' cognitive mechanisms. If Deutscher is right, it follows that in early languages, the way words are put together to form relatively complex sentences, that is, their semantic compositionality, does not depend on some syntax or a set of rules but is determined by the fact that sentences should mirror reality, as it is perceived by the users of the language through their perceptual system. In other words, semantic compositionality resembles, as it were, the structure of the world. Grammar comes later and, through its syntactic structure, facilitates semantic compositionality and allows more complex linguistic structures to be constructed. The preceding analysis shows the close link between icons and symbols and how the former enables the appearance of the latter.

Let me add here that H.E were anatomically adept at having language, as they had a well-developed Broca area (Begun & Walker, 1993; Wu et al., 2011), although differences concerning the vocal tract due to the different structure of the hyoid bone meant that the sounds produced were different (less vowels, lower intensity, etc.). For example, *Homo erectus* skull anatomy suggests that they lacked the muscle control essential for the fine respiration in human speech. They, therefore, could produce a wide range of sounds in the context of social interaction, but their vocabulary should have been limited, and they probably had little (if at all) grammatical rules. Note in passing that Homo Heidelbergensis had probably the anatomical substratum that could support vocal control approximating that of modern humans (Levinson & Dediu, 2018).

A word of caution is needed, however. The existence of the Broca area, by itself, is not proof that *Homo erectus* spoke a language. Only in conjunction with their behavior and travels can we safely conclude that they have possessed some language and, thus, symbolic thought. The reason is that Broca area is not associated exclusively with language but is also involved in the ability for sequential motor-related activities (Everett, 2017) and in the processing of hierarchically structured representations in domains ranging from music to arithmetic (Fadiga et al., 2009) that certainly predate language. Moreover, non-human primates have a Broca and a Wernicke area (in fact, they have homologue[1] areas to that of Broca and Wernicke in humans), but only humans possess a linguistic ability (Rilling, 2014). Hence, homology by itself does not entail similarity of function. Rizzolatti et al., (1996) show, however, that the premotor cortex F5 area in macaques, which is the homologue of Broca' area in humans, is involved in action types such as tearing, eating, and communicating, which suggests that F5 is involved in regulating structure in actions, as Broca's area does in humans.

Planer (2017) argues that H.E and perhaps Homo Habilis have developed a form of proto-language. He bases his argument on empirical evidence concerning tool-usage in the Oldowan and Acheulean periods within the time span of Early Pleistocene Hominins (2.5 Ma and 800 Ka). (Although Planer includes Homo Habilis, I restrict my discussion to H.E, who entered the scene about 1.8 Ma.) Planer and Sterelny (2021, 92) observe that it is likely that the H.E proto-language initially consisted more in sequences of gestures and less in arbitrary signs, that is, symbols. It is also likely that later H.E had an increased number of symbols in their proto-language and that there was a gradual increase in the numbers of symbols used. Gestures work only if they resemble to some extent their targets, and, as we saw above, resemblance poses severe restrictions on the expressive capabilities of the communication system, making planning, social learning etc., more difficult. A communication system based on signs that do not have to resemble what they intend to express is much more informative and, thus, enhances communication capabilities and information sharing benefiting all activities requiring collaboration. For species that have the requisite cognitive capacities to support communication systems relying more on symbols than icons, selective pressures would see to it that such a communicative system will gradually develop.

Planer's (2017) arguments to the effect that H.E had a proto-language are based on a set of assumptions concerning some cognitive abilities that empirical evidence and argumentation suggest that the Early Pleistocene Hominins most likely possessed and which are necessary for the possession of a proto-language. These are (Planer, 2017, 212–213), first, an informational theory of mind, which entails that an agent can represent other agents as possessing certain information-carrying, or intentional, states, as well as states that specify an environmental condition that the other agent is motivated to effect. Another cognitive prerequisite, second, is the ability to construct multifaceted mental representations or models of a situation, that is, the ability to integrate in the representation distinct kinds of information, be it spatial, temporal, causal, etc. As we saw in the preceding chapters, it is highly likely that H.E had these cognitive capacities.

The third cognitive prerequisite concerns the ability to imitate interpersonal acts, that is, the ability to imitate acts that others have performed either towards oneself or towards other agents. This ability depends, in turn, on the ability to construct abstract representations of social interactions (Tomasello, 2008). These representations are abstract in the sense that they represent interactions in terms of abstract roles, each of which can be filled by some agent or other. Put differently, these representations represent the structure of an interaction, which means that abstract ways the specific identities of the participants. Tomasello (2009) calls them "bird's eye view" representations and, obviously, they are crucial in any kind of cooperative activity in which the participants are often required to switch roles. The existence of such abstract representations in important when collaboration among the members of a groups is required, because collaboration is based on an understanding of distinct roles in an activity, as well as understanding of the possibility of role reversals. This understanding requires the capacity to represent social interactions abstractly in terms of abstract roles.

It is interesting to note that the ability to form this sort of abstract representations has a non-conceptual predecessor in the perceptual system and, perhaps, it is grounded in it. Petitot (1995) talks of the positional (local) content structure of the scene. The positional content does not require concepts to be had that express the parts of the content, and it may also convey information about non-visual properties, such as causal relations (e.g., X "transfers" something to Y). Suppose that one witnesses a scene in which X gives Z to Y. The semantics of the scene consists of (i) the semantic lexical content of X, Z, Y and 'give' as a specific action and (ii) the purely positional local content. The latter is the image scheme of the "transfer" type. X, Y, and Z occupy specific locations in the space occupied by the scene (just as they are the arguments in the three-place predicate "to give"). In the image scheme, X, Y, Z are thus reduced to featureless objects that occupy specific relative locations, and, in that sense, they can be viewed as pure abstract places, in the way Descartes reduced matter to space by denuding it of all its properties except that it occupies space. More specifically, X, Y, and Z, which in a linguistic description of the scene are the semantic roles, "are reduced to pure abstract places, locations that must be filled by 'true' participants." These places are related by means of an action of a "transfer" type (ibid., pp. 251–252). The reader is advised to remember the discussion on Mereotopology, where, as we saw, the principles guiding the construction of wholes out of parts refer only to the spatial properties of the objects or object parts involved.

Petitot's places do not refer to the actual locations occupied by the objects in a scene. The whole description can take place without any reference to the actual locations occupied by the objects. What Petitot seems to allude to using the spatial metaphor is the notion of an object devoid of all features (including actual location) except that it persists in time retaining its numerical identity. In that sense, Petitot construes the first representation of an object that is formed in perception the same way that the theories of object-files do. It is this individuation of an object that Petitot seeks to describe by saying that the objects' only property is that they occupy their own space. This is what the notion of objects as "pure abstract places" means.

Petitot draws from Langacker's (1987) cognitive grammar, in which things, relations, and processes are schematic structures, or parts of domains. The domains have a geometric structure. They are built on a set of fundamental "entities" (space, time, color, shapes, etc.), which provide the elementary representational space upon which the rest of our concepts will be built. In this framework, the concepts that are used in the linguistic descriptions of a scene are locational configurations, or spatial structures. When one perceives a visual scene, one's visual system "scans" the scene, picking up qualitative discontinuities (abrupt discontinuities in luminance, for instance). From the local discontinuities in the retinal image, contours are detected, boundaries are set, and so forth, until the various image schemes that categorize perceptual events are progressively built. What is important is that the global positional information contained in a scheme be retrieved from the morphology of the scene by the mechanisms of early vision in a bottom-up manner. Petitot gives a detailed exposition of the routines and algorithms of early vision that might perform this task.

It follows that no conceptually rich mechanisms are required for the representation of abstract roles in social interactions and, thus, *Homo erectus*, were perfectly capable of constructing these representations on the basis of their perceptual capacities. The abstract representations thus produced were, needless to say, at a minimal level of complexity. However, they formed the scaffolding on which gradually more elaborate representations of social interactions could be produced, as the conceptual arsenal of *Homo erectus* kept increasing. This gradual process should be considered along with the process of increasing self-consciousness and self-consciousness that also accompany symbol construction and linguistic development

Returning to Planer (2017), his argument for H.E having a protolanguage, in view of the fact that they most likely possessed the aforementioned cognitive abilities, namely, informational theory of mind, mental model construction, and imitation, is that these abilities allowed *Homo erectus* to develop learning systems for transmitting various kinds of knowledge concerning knapping and quarrying techniques, strategies for big game ambush hunting, etc. Then

> if such learning systems did indeed exist, then we should also think that social circumstances were favorable to the creation and maintenance of a protolanguage; "communicative tools" would have piggybacked on the transmission of tool-related skill and knowledge. Lithic activities—which we know were central to hominin lifeways during this period of prehistory, and so must have occurred at high frequency—would have featured numerous communicative interactions between experts and apprentices, but also among experts and among apprentices. (Planer, 2017, 219)

There are two more reasons to think that H.E probably had to have some sort of proto-language. The first is that increasing brain size and cultural complexity along with with technological refinement of the H.E, as evidenced by the Acheulean culture attested first at Kenya 1.7–18 Ma, and, in addition, the hypothesis that articulate Neanderthals and modern Homo Sapiens who had linguistic capabilities may have inherited speech capabilities from a common ancestor, could indicate that H.E used some proto-language and constructed the basic framework on which fully fledged languages would eventually be built around (Hillert, 2015). Of course, between H.E and Neanderthals and modern Homo Sapiens is Homo Heidelbergensis and this species may have been the common ancestor that bequeathed Homo Sapiens and Neanderthals their linguistic capabilities. However, the above argument that places the origin of protolanguages in H.E does not work in isolation but in tandem with all the other considerations discussed thus far concerning the possibility that H.E used a proto-language.

According to Hillert (2015), one should draw a distinction between the biology of modern language, that is, the *language-ready brain* and the availability of a *linguistic code*. A particular hominin species might have had a language-ready brain, but the language that was used at that time may have been different from modern language. Modern languages, as we have seen, are recursive and hierarchical, which means that they have combinatory, compositional as well as complex hierarchical structures, which are cultural products that may have co-evolved along with genetic changes over a long period of time. Modern language may require not only cortical wiring for language-specific operations, which the particular hominin species might have had, but also a certain cortical mass to express thoughts and complex concepts, since it is plausible that the complex properties of modern languages are, from an evolutionary viewpoint, interrelated to the principles of complex concept formations. The hominin species did not have this critical mass, and, thus, could not have had a modern language, but nothing precludes the possibility that they had a much simpler, perhaps even a simple-grammatical, proto-language.

The second reason we have to believe that H.E used a proto-language is based on arguments about the origins of language that have focused on the reconstructed history of genes associated with language production in humans. The recovery of ancient genetic sequences from Neanderthals and other archaic human specimens (e.g., Denisovans) have shed light onto the genetic history of language production. The human *FOXP2* gene exists in a derived form in humans today and appears to play a critical role in language development. The identification of the human form of *FOXP2* in both Neandertal and Denisovan genomes suggests this gene likely goes back at least to the Middle Pleistocene, with H.E a possible source lineage. Of course, one may wish to argue that the *FOXP2* gene appeared by some genetic miracle in the Archaic Homo Sapiens endowing them with linguistic capabilities. However, this view, once popular, based on Chomsky's (2012, 2015) revolutionary work on language, according to which grammar and, thus, language (because for Chomsky there can be no languages without grammar), emerged by a sudden mutation 50100 Ka. However, this theory is not accepted today (Deacon, 1997; Deutscher, 2005; Everett, 2017; Hillert, 2015).

Let us return to the selective pressures that led to the emergence of language in our lineage. We have already discussed in previous sections two reasons for which symbolic cognition and eventually language may have been selected, namely, the need to instruct tool-making techniques to the next generation, and the need to communicate more effectively to come up with better plans for various collaborative activities. These, however, were not the only social pressures that led to the appearance of protolanguages and eventually of languages. Social relations among the members of fair-sized groups, such as those of H.E., in order to be maintained at a level allowing the group to function successfully require mechanisms of stress relief, since stress inevitably occurs for various reasons in such groups mostly as a result of the intensification of cooperation. By relieving stress, these mechanisms allow for better communication, and, thus, planning within the group. Boyd (2016) and Sterelny (2021) have extensively studied this problem and both emphasize the crucial role of language, which, by enhancing communicative abilities, works towards alleviating stress, and, consequently, allowing society to function smoothly.

Furthermore, there is a growing consensus that symbolic thought, and language along with it, became necessary to allow our ancestors to deal with the ever-growing number of their groups and the social demands this expansion entailed, predominantly with respect to communication, planning for activities requiring collaboration, social learning required for the transmitting of useful knowledge, and managing social tensions within the group. As I said in the introduction, in this book I adopt without examining them two main theses, namely, that of gene-culture coevolution, and that of cognition-culture coevolution. Demography, culture (including learning) and evolution in general are intertwined.

> Evolution is driven by the propagation of genes, which is determined by the key processes of survival (mortality), fertility, and dispersal (migration) of individuals. Demographic trends influence the age- and sex-structure of a society, altering the balance between kin and non-kin within individuals' social networks, with consequences for cooperation, parental investment, and information exchange. Demography further influences the development and variability of human culture, through its recognized role in trait transmission and cumulative culture. (Page & French, 2020, 1)

To put things in a demographic perspective, using estimates of brain size and extrapolating to the average group size (it is generally assumed that a larger brain size allows for more complex social interactions and thus, may lead to the formation of larger groups. It is estimated that the average size of groups of hominins with modern human cranial capacities was 120–131 (for archaic Homo sapiens 200 Ka), 150 individuals (for modern Homo Sapiens 60.0–40 Ka). In comparison, Neandertal average group comprised 100–130 individuals (Duveau et al., 2019) or 140 individuals (Mithen, 1996). There is evidence dated 1.5 Ma for an all-male *Homo erectus* groups of 20 individuals (Hatala et al., 2016) in some special mission, which points to a group size of around 100–110 individuals. Mithen (1996, 150) estimates that *Homo erectus* groups had a mean size of 111.

It is known that among apes the way to preserve the social order or to enhance one's social status in the group is grooming. This is how apes built social relations and maintain them (and, of course, an efficient way to get rid of parasites and lice). It is also the way to pass along social information and send social messages. Whom one chooses to groom, how long does this grooming last, whom does one let watch this activity, all these are important activities. Grooming, thus, is the primary means of social bonding. As groups grow larger grooming becomes difficult. Language was developed as a substitute for grooming as a means of social interaction eventually replacing it altogether. As Mithen (1996, 124) remarks,

> [L]anguage has evolved to provide exchange of information, initially as a supplement to grooming, and then as a replacement for it. It is also a much more efficient way to transferring information. An ape can groom two of his mates at once, but an articulate human can chat ways to whoever is listening.

Grooming takes time and it is estimated that the larger the size of the group the longer the time spent in grooming not because the parasites are more but because there are more members to communicate with. Dunbar (1991) has estimated that 30% of the time budget of an ape is the upper maximum of time that the ape has to spend in grooming activity. More than that and the ape will go hungry and lack energy. Given the average group size of Homo Habilis, Dunbar estimated that they should spend about 23% of their time in grooming activity, just below the acceptable threshold. 250 Ka the group size of Archaic Homo Sapiens demanded that grooming should take 40% of the time budget, which is unacceptable. Aiello and Dunbar (1993) argue that some form of social language should have developed to alleviate the situation, which means that by 250 Ka the latest symbols were in use even if they were restricted within the social domain. In view of these, Dunbar (1991, 1993) proposed the social brain hypothesis, according to which social demands lead to larger brains and more sophisticated representational systems. It goes without saying that the social evolutionary pressures were accompanied by pressures due to environmental conditions, dietary changes (for instance, the consumption of meat and the changes it induced in social structure is, according to Deacon (1997), the main reason behind the development of symbolic capabilities.)

In sum, our discussion in this section points towards the conclusion that social and linguistic complexity coevolved by forming a loop of inextricable interactions between the two domains. This coevolution, in turn, changed profoundly the ways of cooperation of our ancestors, by allowing for more effective communication and more effective handling of social complexities and their problems.

Note

1. Homology means that there are areas with the same features across different species.

References

Aiello, L., & Dunbar, R. I. M. (1993). Neocortex size, group size and the evolution of language. *Current Anthropology, 36*, 199–221.

Ambrose, S. H. (2010). Coevolution of composite-tool technology, constructive memory, and language: Implications for the evolution of modern human behavior. *Current Anthropology, 51*(s1), S135–S147.

Barsalou, L. W. (1999). Perceptual symbol systems. *Behavioral and Brain Sciences, 22*(4), 577–609.

Basile, G. A., Tatti, E., Bertino, S., Milardi, D., et al. (2024). Neuroanatomical correlates of peripersonal space: Bridging the gap between perception, action, emotion and social cognition. *Brain Structure and Function, 229*, 1047–1072.

Begun, D., & Walker, D. (1993). The endocast. In D. Walker & R. Leakey (Eds.), *The Nariokotome Homo erectus skeleton* (pp. 326–358). Springer Verlag.

Ben-Dor, M., Sirtoli, R., & Barkai, R. (2021). The evolution of the human trophic level during the Pleistocene. *American Journal of Physical Anthropology, 175*(72), 27–56.

Benes, F. M. (1989). Myelination of cortical-hippocampal relays during late adolescence. *Schizophrenia Bulletin, 15*(4), 585–593.

Bolhuis, J., Tattersall, I., Chomsky, N., & Berwick, R. C. (2014). How could language have evolved? *Plos Biology, 12*(8), e1001934.

Boyd, R. (2016). *A different kind of animal: How culture made humans exceptionally adaptable and cooperative*. Princeton University Press.

Burge, T. (2022). *Perception: The first form of mind*. Oxford University Press.

Carey, S. & Spelke, E. (1994). Domain-specific knowledge and conceptual change. In L. A., Hirschfeld & S. A. Gelman (Eds.), *Mapping the mind: Domain specificity in cognition and culture* (pp. 169–200). Cambridge University Press.

Carey, S. (2009). *The origin of concepts*. Oxford University Press.

Carruthers, P. (2013). Evolution of working memory. *Proceedings of the National Academy of Science, U. S. A., 110*, 10371–10378.

Carruthers, P. (2006). *The architecture of the mind*. Oxford University Press.

Chomsky, N. (2012). *The science of language. Interviews with James McGilvray*. Cambridge University Press.

Chomsky, N. (2015). Some core contested concepts. *Journal of Psycholinguistic Research, 44*, 91–104. https://doi.org/10.1007/s10936-014-9331-5

Cissewski, J., & Luncz, L. V. (2021). Symbolic signal use in wild Chimpanzee gestural communication? A theoretical framework. *Frontiers, Psychology, 12*. https://doi.org/10.3389/fpsyg.2021.718414

Clark, G., & Henneberg, M. (2017). Ardipithecus ramidus and the evolution of language and singing: an early origin for hominin vocal capability. *Homo, 68*(2), 101–121.

Conway, M. A. (2009). Episodic memories. *Neuropsychologia, 57*(11), 2305–2313.

Corballis, M. C. (2014). *The recursive mind: The origins of human language, thought, and civilization* (Updated ed.). Princeton University Press.

Corballis, M. C. (2009). The evolution of language. *Annals of New York Academy of Sciences, 1156*(March), 19–43.

Coventry, K. R., & Diessel, H. (2024). Spatial communication systems and action. *Trends in Cognitive Science*. https://doi.org/10.1016/j.tics.2024.10.002

Deacon, T. W. (1997). *The symbolic species: The co-evolution of language and the brain*. W. W. Norton & Company.

Dennett, D. (1987). *The intentional stance*. MIT Press.

Deutscher, G. (2005). *The unfolding of language: An evolutionary tour of mankind's greatest invention*. Picador.

Dunbar, R. I. M. (1991). Functional significance of social grooming in primates. *Folia Primatologica, 57*, 121–131.

Dunbar, R. I. M. (1993). Size of human groups during the Paleolithic and the evolutionary significance of increased group size. *Behavioral and Brain Sciences, 16*, 681–735.

Dunbar, R. (2014). *Human evolution*. Penguin.

Duveau, J., Berillon, G., & Verna, C. H. (2019). The composition of a Neandertal social group revealed by the hominin footprints at Le Rozel (Normandy, France). *PNAS, 116*(39), 19409–19414.

Edelman, G. M. (1993). *Bright air, brilliant fire: On the matter of the mind*. Basic Books.

Everett, D. L. (2017). *How language Began*. Liverlight.

Fadiga, L., Craighero, L., & D'Ausillo, A. (2009). Broca's area in language, action, and music. *Annals of the New York Academy of Sciences, 1169*(1), 448–458.

Fields, C. (2002). Why do we talk to ourselves? *Journal of Experimental Theoretical Artificial Intelligence, 14*(4), 255–272.

Fodor, J., & Pylyshyn, Z. (1988). Connectionism and cognitive architecture: A critical analysis. *Cognition, 28*, 3–71.

Gardner, H. (1993). *Multiple intelligences: The theory in practice*. Basic Books.

Greenfield, P. M. (1991). Language, tools and brain: The ontogeny and phylogeny of hierarchically organized sequential behavior. *Behavioral and the Brain Sciences, 14*, 531–595.

Güneş, A. N., Göksun, T., & Tekcan, A. İ. (2024). *Details in hand: How does gesturing relate to autobiographical memory?* Advance online publication. https://doi.org/10.1080/23273798.2024.2385029

Hammond, A. S., Mavuso, S. S., Biernat, M., Braun, D. R., Jinnah, Z., Kuo, S., Melaku, S., Wemanya, S. N., Ndiema, E. K., Patterson, D. B., & Uno, K. T. (2021). New hominin remains and revised context from the earliest *Homo erectus* locality in East Turkana, Kenya. *Nature Communications, 12*, 1939. https://doi.org/10.1038/s41467-021-22208-x

Hatala, K., Roach, N., Ostrofsky, K., et al. (2016). Footprints reveal direct evidence of group behavior and locomotion in *Homo erectus. Scientific Reports, 6*, 28766. https://doi.org/10.1038/srep28766

Hauesler, M., Schiess, R., & Boeni, T. (2013). Evidence for juvenile disc herniation in Homo erectus boy skeleton. *Spine, 38*(3), E123–E128.

Heesen, R., Hobalter, C., Ferrer, R., Semple, S. (2019). Linguistic laws in chimpanzee gestural communication. *Proceedings of the Royal Society B, Biological Sciences, 286*, 20182900. http://dx.doi.org/10.1098/rspb.2018.2900

Hillert, D. G. (2015). On the evolving biology of language. *Frontiers in Psychology, 6*, 1796. https://doi.org/10.3389/fpsyg.2015.01796

Hiscock, P. (2014). Learning in lithic landscapes: A reconsideration of the hominid "toolmaking" niche. *Biological Theory, 9*, 27–41.

Howe, M. L., & Courage, M. L. (1997). The emergence and early development of autobiographical memory. *Psychological Review, 104*, 499–523.

Karmiloff-Smith, A. (1992). *Beyond modularity: A developmental perspective on cognitive science.* Cambridge University Press.

Kavanaugh, R. D. (2006). Pretend play and theory of mind. In L. Balter L, & C. S. Tamis-LeMonda (Eds.), *Child psychology: A handbook of contemporary issues* (2nd ed., pp. 153–166). Psychology Press.

Klatzky, R. L., Wu, B., & Stetten, G. (2010). The disembodied eye: consequences of displacing perception from action. *Vision Research, 10*, 2618–2626.

Kolodny, O., & Edelman, S. (2018). The evolution of the capacity for language: The ecological context and adaptive value of a process of cognitive hijacking. *Philosophical Transactions of the Royal Society London B Biological Science, 373*, 20170052. https://doi.org/10.1098/rstb.2017.0052

Kübler, S., Owenga, P., Reynolds, S. C., Rucina, S. M., & King, G. C. P. (2015). Animal movements in the Kenya Rift and evidence for the earliest ambush hunting by hominins. *Scientific Reports, 5*, Article Number 14011.

Langacker, R. (1987). *Foundations of cognitive grammar* (Vol. 1). Stanford University Press.

Levinson, S., & Dediu, D. (2018). Neanderthal language revisited. Not only us. *Current Opinion in Behavioral Sciences, 21*, 49–56.

Lewis, A., Berntsen, D., & Call, J. (2019). Long term memory of past events in great apes. *Current Directions Psychological Science, 28*(2), 117–123.

Lewis-Williams, J. D. (2002). *The mind in the cave.* Thames & Hudson.

Lombard, M. (2024). *Paying attention: The neurocognition of archery, middle Stone Age bow hunting, and the shaping of the sapient mind.* Advance online publication. https://doi.org/10.1007/s11097-024-09980-z

Lordkipanidze, D., Vekua, A., Ferring, R., Rightmire, Ph., Zollikofer, Ch., Ponce de Leon, M. S., Agusti, J., Kiladze, G., Mouskhelishvili, A., Nioradze, M., & Tappen, A. (2006). A fourth hominin skull from dmanisi, Georgia. *The Anatomical Record, PART A 288A*, 1146–1157.

Luncz, L., & Boesh, Ch. (2015). The extent of cultural variation between adjacent chimpanzee (Pan troglodytes verus) communities: a microecological approach. *American Journal of Physiological Anthropology, 156*(1), 67–75.

Marín-Arroyo, A. B., & Sanz-Royo, A. (2022). What Neanderthals and AMH ate: Reassessment of the subsistence across the Middle-Upper Palaeolithic transition in the Vasco-Cantabrian region of SW Europe. *Journal of Quaternary Science, 37*, 320–334.

Marquet, J.-C., Freiesleben, T. H., Thomsen K. J., Calligaro, M., Macaire, J.-J. (2023). The earliest unambiguous Neanderthal engravings on cave walls: La Roche-Cotard, Loire Valley, France. *PLoS One, 18*(6), e 0286568.

Martin-Ordas, G., Haun, D., Colmenares, F., & Call, J. (2010). Keeping track of time: Evidence for episodic-like memory in great apes. *Animal Cognition, 13*, 331–340.

Mithen, S. (1996). *The prehistory of the mind: A search for the origins of art, religion and science.* Thames and Hudson Ltd.

Mulcahy, N. J., & Call, J. (2006). Apes save tools for future use. *Science, 312*, 1038–1040.

Nelson, K., & Fivush, R. (2004). The emergence of autobiographical memory: A social cultural developmental theory. *Psychological Review, 111*, 486–511.

Page, A., & French, J. C. (2020). Reconstructing prehistoric demography: What role for extant hunter-gatherers? *Evolutionary Anthropology*. https://doi.org/10.1002/evan.21869

Passingham, R. E. (2021). *Understanding the prefrontal cortex: Selective advantage, connectivity, and neural operations*. Oxford Psychology Series (Oxford, 2021; online edn, Oxford Academic, 19 Aug. 2021). https://doi.org/10.1093/oso/9780198844570.001.0001

Perner, J. (2001). Episodic memory: Essential distinctions and developmental implications. In C. Moore & K. Lemmon (Eds.), *The self in time: Developmental perspectives* (pp. 181–202). Lawrence Erlbaum Associates.

Perner, J., & Ruffman, T. (1995). Episodic memory and autonoetic consciousness: Developmental evidence and a theory of child amnesia. *Journal of Experimental Child Psychology, 59*, 516–548.

Petitot, J. (1995). Morphodynamics and attractor syntax: Constituency in visual perception and cognitive grammar. In R. Port & T. Van Gelder (Eds.), *Mind as motion: Explorations in the dynamics of cognition.* MIT Press.

Pike, A. W., Hoffmann, D. L., García-Diez, M., Pettitt, P. B., Alcolea, J., De Balbín, R., González-Sainz, C., de las Heras, C., Lasheras, J. A., Montes, R., & Zilhão, J. (2012). U-series dating of Paleolithic art in 11 caves in Spain. *Science, 15, 336*(6087), 1409–13.

Planer, R. J. (2017). Talking about tools: Did early Pleistocene Hominins have a protolanguage? *Biological Theory, 12*, 211–221.

Planer, R. J. (2021). What is symbolic cognition. *Topoi, 40*, 233–244.

Planer, R. J., & Sterelny, K. (2021). *From signal to symbol: The evolution of language.* MIT Press.

Putnam, H. (1983). Reference and truth. In *Realism and reason: Philosophical papers* (Vol. 3). Cambridge University Press.

Putnam, H. (1980). Models and reality. *Journal of Symbolic Logic, 45*, 464–482.

Raftopoulos, A. (2019). *Cognitive penetrability and the epistemic role of perception.* Palgrave, McMillan.

Renfrew, C. (2008). *Prehistory: The making of the human mind*. Phoenix.

Rilling, J. K. (2014). Comparative primate neurobiology and the evolution of brain language systems. *Current Opinions in Neurobiology, 28*, 10–14.

Rizzolatti, G., Fadiga, L., Gallese, V., & Fogassi, L. (1996). Premotor cortex and the recognition of motor actions. *Cognitive Brain Research, 3*(2), 131–141.

Scarf, D., Gross, J., Colombo, M., & Hayne, H. (2013). To have and to hold: Episodic memory in 3- and 4-year-old children. *Developmental Psychobiology, 55*, 125–132.

Sperber, D. (1994). The modularity of thought and the epidemiology of representations. In L. A., Hirschfeld & S. A. Gelman (Eds.), *Mapping the mind: Domain specificity in cognition and culture* (pp. 39–67). Cambridge University Press.

Sterelny, K. (2021). *The Pleistocene social contract: Culture and cooperation in the hominin lineage.* Oxford University Press.

Tallerman, M. (2012). Chap. 51. Protolanguage. In M. Tallerman, & K. Gibson (Eds.), *The Oxford Handbook of Language Evolution* (pp. 479–491). Oxford University Press.

Tomasello, M. (2008). *Origins of human communication.* MIT Press.

Tomasello, M. (2009). *Why we cooperate.* MIT Press.

Tomasello, M. (2014). *A natural history of human thinking.* Harvard University Press.

Torrey, E. F. (2017). *Evolving brains emerging Gods: Early humans and the origins of religion*. Columbia University Press.

Toups, M. A., Kitchen, A., Light, J. E. & reed, D. L. (2011). Origin of clothing lice indicates early clothing use by anatomically modern humans in Africa, *Molecular Biology and Evolution, 28*(1), 29–32.

Truswell, R. (2017). Dendrophobia in Bonobo comprehension of spoken English. *Mind and Language, 32*(4), 395–415.

Tulving, E. (2002). Episodic memory: From mind to brain. *Annual Review of Psychology, 53*, 1–25.

Tulving, E. (2005). Episodic memory and autonoesis: Uniquely human? In H. S. Terrace & J. Metcalfe (Eds.), *The missing link in cognition: Origins of self-reflective consciousness* (pp. 3–56). Oxford University Press.

Wedage, O., Amano, N., Langley, M. C., et al. (2019). Specialized rainforest hunting by Homo sapiens ~45,000 years ago. *Nature Communications, 10*, 739.

Welshon, R. (2024). Tool production, joint attention, and the emergence of reflective self-awareness. In T. Wynn, K. A. Overmann, & F. Coolidge (Eds.), *Oxford handbook of cognitive archaeology* (pp. 1–18). Oxford University Press.

Wheeler, M. A., Stuss, D. T., & Tulving, E. (1997). Toward a theory of episodic memory: The frontal lobes and autonoetic consciousness. *Psychological Bulletin, 121*, 331–354.

Wu, X., Holloway, R. L., Schepartz, L. A., & Xing, S. (2011). A new brain endocast of *Homo erectus* from Hulu Cave, Nanjing, China. *American Journal of Biologicall Anthropology, 145*(3), 452–460.

Yakovlev, P. I., & Lecours, A. R. (1967). The myelogenetic cycles of regional maturation of the brain. In A. Minkowski (Ed.), *Regional developments of the brain in early life* (pp. 3–70). Blackwell.

Chapter 8
Concluding Discussion

I have argued that the cognitive capacities that are required for having iconic, first, and then, building upon the icons, symbolic representations were present in *Homo erectus*. This means that it is highly likely that these ancestors had some form of language that started without syntactic rules but was structured so as to mirror the structure of the world. This language undoubtedly became increasingly complex and grammatical rules started to appear to allow for the construction of more complex expressions required to keep up with the increasing complexity of their social niches. This, in turn, created evolutionary pressures that led eventually to increasing cranial size and restructuring of the brain. For the majority of researchers, the evolution of human brain and the development of language were both a response to the acquisition of symbolic thinking in a process in which all three components were intrinsically entangled.

The average absolute (that is, non-relative to body mass) brain size of *Homo erectus* was 960 cm^3, 60% larger than that of Homo Habilis. The average absolute brain size of Neanderthals 450–300 Ka was almost the same as that of Homo Sapiens, but it had an expanded parietal area. The average absolute brain size of early Homo Sapiens was 1.350 cm^3 similar to that of modern Homo Sapiens. The brain size relative to body mass of Neanderthals is smaller than that of anatomically modern humans. Relative brain size of *Homo* did not change from 1.8 Ma to 600 Ka. After about 600 Ka it increased until about 35 Ka, when it began to decrease.

The restructuring of the brain was also significant, with the general trend being the increase of the parietal, prefrontal, and frontal brain regions with a relative decrease in the perceptual occipital regions. Thus, even though the former regions expanded significantly during the development of our species, the occipital cortex remained the same as with the other apes. Needless to say, that clusters of areas in the parietal, pre-frontal, and frontal cortex are necessary for many of the cognitive and social capacities that, as we have seen, are required for iconic and, especially, symbolic, representations. The emergence of new areas and new functions also meant that there should be an increase in the number of synaptic connections needed to transmit

A. Raftopoulos, *The Origins of Representations: Cognitive and Brain Development*, Studies in Applied Philosophy, Epistemology and Rational Ethics 73,
https://doi.org/10.1007/978-3-031-91213-9_8

signals among these areas. Indeed, this was the case. White matter, that is, glial cells and connecting fibers in humans is 31% more than in chimpanzees, while the gray matter (that is, neurons) in humans is only 2% larger than that of chimpanzees (Schoenemann et al., 2005). This strengthens the thesis that it is the interconnectivity of our brain areas that gives us the requisite cognitive fluidity that makes us unique in the animal kingdom with regard to our cognitive capacities.

During the evolution of our species increasing symbolic capacities evolved, which allowed the use of increasingly sophisticated symbolic representational systems. However, the link with the iconic representations in which symbols were grounded was never lost. At a last analysis, icons keep endowed symbols with meaning.

I have discussed the cognitive capacities and their development along the hominin line that allowed the emergence of representations and eventually the emergence of symbolic representations. To that effect, I discussed the development, both ontogenetic and phylogenetic, of cognitive capacities such as working memory, sustained attention, autobiographical memory, etc., that likely underly the development of hominins' representational capacities. I also said something about the development of the brain structures, emphasizing the role of the PFC that rsubserves many of these cognitive capacities. I adopted a gradualist approach according to which the development of these capacities, and, thus, the development of representational capabilities did not appear suddenly as the result of some fortuitus mutation, but, instead, through the interplay between social and cognitive factors. This interplay was based on the baseline cognitive capacities that hominins inherited from the common ancestor that we shared with chimpanzees. Sterelny's main thesis reflects this view nicely. Sterelny (2012, 2021) sees the various stages of human evolutions not as hinging on a single key innovation, but as emerging from a positive feedback loop caused by smaller divergences from other great apes. Starting from a baseline of cognitive capacities, divergences such as bipedalism, better causal and social reasoning, reproductive cooperation, and changes in diet and foraging style gradually led to the development of modern humans' cognitive capacities.

However, this does not mean that there are no intrinsic cognitive differences between the various sub categories of hominins. Differences between Homo Sapiens and Homo Neanderthals have been extensively studied, and various socio-ecological-economic factors have been proposed to explain these differences that include more advanced planning, reasoning, and tool-making activities, as well as a more advanced social cognition on the part of Homo Sapiens. Cranial differences have also been proposed to explain these differences. Even though it is undeniable that the physical environment and the social reality affect neuroanatomy, the fact remains that the neuroanatomical differences are relatively small and it is far from certain that they, alone, could explain the significant differences in life—styles and, thus, in cognitive capacities, between Homo Sapiens and Homo Neanderthals. However, the interplay between these small differences and the social, environmental, and economic factors could explain the significant cognitive differences. I think that the same holds from the whole hominin lineage.

References

Schoenemann, P. T., Sheenan, M. J., & Glotzer, L. D. (2005). Prefrontal white matter volume is disproportionately larger in humans than in other primates. *Nature Neuroscience, 8*, 242–252.
Sterelny, K. (2012). *The evolved apprentice: How evolution made humans unique*. MIT Press.
Sterelny, K. (2021). *The Pleistocene social contract: Culture and cooperation in the hominin lineage*. Oxford University Press.

Uncited References

Brunner, E., Sracino, B., Ricci, E., Tafuri, M., Passarello, P., & Manzi, C. (2004). Midsagittal cranial shape variation in the genus Homo by geometric morphometrics. *Coll Anthropology, 28*, 99–112.
Burnston, D. (2017). Cognitive penetration and the cognition-perception interface. *Synthese, 194*, 3645–3668.
Dretske, F. (1995). *Naturalizing the mind*. MIT Press.
Dunbar, R. I. M. (1993). Size of human groups during the Paleolithic and the evolutionary significance of increased group size. *Behavioral and Brain Sciences, 16*, 681–735.
Fodor, J. A. (1975). *The language of thought*. Crowell.
Fodor, J. A. (1983). *The modularity of mind*. MIT Press.
Fodor, J. A. (1987). *Psychosemantics. The problem of meaning in the philosophy of mind*. MIT Press.
Glasser, M. F., Goyal, M. S., Preuss, T. M., Raichle, M. E., & Van Essen, D. C. (2014). Trends and properties of human cerebral cortex: Correlations with cortical myelin content. *NeuroImage, 93*, 165–175.
Haugeland, J. (1981). Analog and analog. *Philosophical Topics, 12*(1), 213–225.
Haeusler, M., Schiess, R., & Boeni, T. (2013). Evidence for juvenile disc herniation in a *Homo erectus* boy skeleton. *Spine, 38*(3), E123–E128.
Husserl, E. (1898–1925/2006). *Phantasy, image consciousness, and memory (collected works (1898–1925)*. J. Brough, translation. Springer.
Kelso, J. A. S. (2007). *Dynamic patterns: The self-organization of brain and behavior*. MIT Press.
Lee, K., Bull, R., & Ringo Ho, M. H. (2013). Developmental changes in executive functioning. *Child Development, 84*(6)
Morrison, R., & Reiss, D. (2018). Precocious development of self-awareness in dolphins. *PLoS ONE, 13*(1), E0189813.
Moncel, M.-H., Chiotti, L., & Gaillard, C. (2012). Non-utilitarian lithic objects from the European Paleolithic. *Archaeology, Ethnology and Anthropology of Eurasia, 40*(1), 24–40.
Postle, B. R. (2015). The cognitive neuroscience of visual short-term memory. *Current Opinions in Behavioral Science, 1*, 40–46.
Pylyshyn, Z. (2003). *Seeing and visualizing: It's not what you think*. MIT press.
Rescorla, M. (2014). The causal relevance of content to computation. *Philosophy and Phenomenological Research, 88*(1), 173–208.
Sowell, E. R., Thompson, P. M., & Toga, A. W. (2004). Mapping changes in the human cortex throughout the span of life. *The Neuroscientist, 10*, 372–392.
Spelke, E. S. (1988). Object perception. In A. I. Goldman (Ed.), *Readings in philosophy and cognitive science* (pp. 447–461). MIT Press.
Tallerman, M. (2011a). Did our ancestors speak a holistic protolanguage? *Lingua, 117*, 579–604.

Tallerman, M. (2011b). Protolanguage. In K. R. Gibson, & M. Tallerman (Eds.), *The Oxford handbook of language evolution*. Oxford University Press.

Wang, X. J. (2001). Synaptic reverberation underlying mnemonic persistent activity. *Trends in Neuroscience, 24*, 455–463.

The manufacturer's authorised representative in the EU is Springer Nature Customer Service Centre GmbH, Europaplatz 3, 69115 Heidelberg, Germany. If you have any concerns regarding our products, please contact ProductSafety@springernature.com

Printed and bound by CPI Group (UK) Ltd, Croydon, CR0 4YY
01/10/2026
02225865-0008